State of the Art in Molecular Catalysis in Europe

State of the Art in Molecular Catalysis in Europe

Guest Editor

Carl Redshaw

 Basel • Beijing • Wuhan • Barcelona • Belgrade • Novi Sad • Cluj • Manchester

Guest Editor
Carl Redshaw
Department of Chemistry
University of Hull
Hull
United Kingdom

Editorial Office
MDPI AG
Grosspeteranlage 5
4052 Basel, Switzerland

This is a reprint of the Special Issue, published open access by the journal *Catalysts* (ISSN 2073-4344), freely accessible at: www.mdpi.com/journal/catalysts/special_issues/Europe_Catalysis.

For citation purposes, cite each article independently as indicated on the article page online and using the guide below:

Lastname, A.A.; Lastname, B.B. Article Title. *Journal Name* **Year**, *Volume Number*, Page Range.

ISBN 978-3-7258-3368-9 (Hbk)
ISBN 978-3-7258-3367-2 (PDF)
https://doi.org/10.3390/books978-3-7258-3367-2

© 2025 by the authors. Articles in this book are Open Access and distributed under the Creative Commons Attribution (CC BY) license. The book as a whole is distributed by MDPI under the terms and conditions of the Creative Commons Attribution-NonCommercial-NoDerivs (CC BY-NC-ND) license (https://creativecommons.org/licenses/by-nc-nd/4.0/).

Contents

About the Editor vii

Preface ix

Carl Redshaw
State of the Art in Molecular Catalysis in Europe
Reprinted from: Catalysts 2024, 14, 459, https://doi.org/10.3390/catal14070459 1

Khairil A. Jantan, Gregor Ekart, Sean McCarthy, Andrew J. P. White, D. Christopher Braddock and Angela Serpe et al.
Palladium Complexes Derived from Waste as Catalysts for C-H Functionalisation and C-N Bond Formation
Reprinted from: Catalysts 2024, 14, 295, https://doi.org/10.3390/catal14050295 4

Antigoni G. Margellou, Stylianos A. Torofias, Georgios Iakovou and Konstantinos S. Triantafyllidis
Valorization of Chlorella Microalgae Residual Biomass via Catalytic Acid Hydrolysis/Dehydration and Hydrogenolysis/Hydrogenation
Reprinted from: Catalysts 2024, 14, 286, https://doi.org/10.3390/catal14050286 21

Yizhou Wang, Zheng Wang, Qiuyue Zhang, Yanping Ma, Gregory A. Solan and Yang Sun et al.
Non-Symmetrically Fused Bis(arylimino)pyridines with para-Phenyl Substitution: Exploring Their Use as N',N,N-Supports in Iron Ethylene Polymerization Catalysis
Reprinted from: Catalysts 2024, 14, 213, https://doi.org/10.3390/catal14030213 38

William Clegg, Mark R. J. Elsegood and Carl Redshaw
Extended Hydrogen-Bonded Molybdenum Arrays Derived from Carboxylic Acids and Dianilines: ROP Capability of the Complexes and Parent Acids and Dianilines
Reprinted from: Catalysts 2024, 14, 214, https://doi.org/10.3390/catal14030214 57

Evgeny V. Rebrov and Peng-Zhao Gao
Molecular Catalysts for OER/ORR in Zn–Air Batteries
Reprinted from: Catalysts 2023, 13, 1289, https://doi.org/10.3390/catal13091289 75

Mark R. J. Elsegood, William Clegg and Carl Redshaw
Vanadium Complexes Derived from O,N,O-tridentate 6-bis(o-hydroxyalkyl/aryl)pyridines: Structural Studies and Use in the Ring-Opening Polymerization of -Caprolactone and Ethylene Polymerization
Reprinted from: Catalysts 2023, 13, 988, https://doi.org/10.3390/catal13060988 100

Alexandra Jakab-Nácsa, Viktória Hajdu, László Vanyorek, László Farkas and Béla Viskolcz
Overview of Catalysts with MIRA21 Model in Heterogeneous Catalytic Hydrogenation of 2,4-Dinitrotoluene
Reprinted from: Catalysts 2023, 13, 387, https://doi.org/10.3390/catal13020387 116

Matthew J. Andrews, Sebastian Brunen, Ruaraidh D. McIntosh and Stephen M. Mansell
Preformed Pd(II) Catalysts Based on Monoanionic [N,O] Ligands for Suzuki-Miyaura Cross-Coupling at Low Temperature
Reprinted from: Catalysts 2023, 13, 303, https://doi.org/10.3390/catal13020303 133

Mattia Forchetta, Francesca Valentini, Valeria Conte, Pierluca Galloni and Federica Sabuzi
Photocatalyzed Oxygenation Reactions with Organic Dyes: State of the Art and Future Perspectives
Reprinted from: *Catalysts* **2023**, *13*, 220, https://doi.org/10.3390/catal13020220 . . . . **147**

Muhammad Zada, Desalegn Demise Sage, Qiuyue Zhang, Yanping Ma, Gregory A. Solan and Yang Sun et al.
Thermally Stable and Highly Efficient *N,N,N*-Cobalt Olefin Polymerization Catalysts Affixed with *N*-2,4-Bis(Dibenzosuberyl)-6-Fluorophenyl Groups
Reprinted from: *Catalysts* **2022**, *12*, 1569, https://doi.org/10.3390/catal12121569 . . . . **168**

Alexandra Barnes, Richard J. Lewis, David J. Morgan, Thomas E. Davies and Graham J. Hutchings
Improving Catalytic Activity towards the Direct Synthesis of H_2O_2 through Cu Incorporation into AuPd Catalysts
Reprinted from: *Catalysts* **2022**, *12*, 1396, https://doi.org/10.3390/catal12111396 . . . . **192**

About the Editor

Carl Redshaw

Carl Redshaw (now Emeritus) was the Chair of Inorganic Materials at the University of Hull prior to the closure of the Chemistry Department. He has been a Visiting Professor at the Northwest University (Xi'an, China), Sichuan Normal University (Chengdu, China), the Shanghai Institute of Organic Chemistry (SIOC), Chinese Academy of Sciences (Shanghai, China), the Institute of Chemistry (ICCAS), Chinese Academy of Sciences (Beijing, China), and the National Institute of Technology (Akashi, Japan). Earlier in his career, Carl was a Lecturer, Senior Lecturer, and Reader at the University of East Anglia, Norwich (U.K.). In his research, Carl utilizes coordination chemistry to tackle issues ranging from catalysis to cancer. He has published 530 publications in the primary literature.

Preface

Europe remains an important region in the fields of both heterogeneous and homogeneous catalysis, with work including but not limited to catalysts for chemical synthesis, the biorefinery process, environmental solutions, and possible future sustainable energy strategies. This Special Issue includes both reviews and original research articles on aspects of heterogeneous and homogeneous catalysis, with an emphasis on fundamental and applied research that is being conducted across Europe. Topics include the following: catalysts and value-added products derived from waste, catalysts for biodegradable polymer formation, olefin polymerization, catalytic hydrogenation, catalysts for crosscoupling reactions, improved catalysts for hydrogen peroxide synthesis, molecular catalysts in Zn–air batteries, and organo-photocatalysis in oxygenation reactions. We hope that the reader will enjoy these wide-ranging and exciting topics.

Carl Redshaw
Guest Editor

Editorial

State of the Art in Molecular Catalysis in Europe

Carl Redshaw

Department of Chemistry, University of Hull, Hull HU6 7RX, UK; c.redshaw@hull.ac.uk

Citation: Redshaw, C. State of the Art in Molecular Catalysis in Europe. *Catalysts* **2024**, *14*, 459. https://doi.org/10.3390/catal14070459

Received: 5 July 2024
Accepted: 12 July 2024
Published: 16 July 2024

Copyright: © 2024 by the author. Licensee MDPI, Basel, Switzerland. This article is an open access article distributed under the terms and conditions of the Creative Commons Attribution (CC BY) license (https:// creativecommons.org/licenses/by/ 4.0/).

In this editorial, I would like to provide an overview of the eleven contributions to the Special Issue entitled "State of the Art in Molecular Catalysis in Europe", which is part of the Organic and Polymer Chemistry Section of *Catalysts*. The Special Issue requested contributions involving aspects of heterogeneous and homogeneous catalysis with an emphasis on fundamental and applied research conducted across Europe. As anticipated, the contributions to this Issue cover a wide range of topics with two reviews associated with catalytic aspects of Zn-air batteries and organic dyes. The nine scientific research papers focus on topics ranging from palladium catalysts for C-H functionalisation/C-N bond formation as well as for Suzuki–Miyaura cross-coupling at low temperature, to catalytic acid hydrolysis/dehydration, and to catalysts for polymerization processes (α-olefins and ROP of cyclic esters). There are also contributions discussing catalysts of relevance to the hydrogenation of 2,4-dinitrotoluene and how the use of incorporated Cu into AuPd catalysts can improve H_2O_2 synthesis.

Contribution 1 is a review by Rebrov and Gao entitled "Molecular Catalysts for OER/ORR in Zn-Air Batteries". The review describes types of catalysts employed for OER/ORR reactions including those based on mixed metal oxides, perovskites, alternative carbons, monometallic noble metals, inorganic–organic composites, and spinels. The limitations and advantages of current systems are discussed, and future directions are proposed. Contribution 2 is a review entitled "Photocatalyzed oxygenation reactions with organic dyes: State of the art and future perspectives" by Conte, Sabuzi et al., which describes recent advances in the use of organo-photocatalysis in the area of the selective oxygenation of organic substrates. The photocatalysts discussed include flavinium salts, cyano-arenes, Eosin Y, Rose Bengal, acridinium salts, and quinone-based dyes, and their deployment in transformations such as the oxygenation of amines, phosphines, silanes, alkanes, alkenes, alkynes, aromatic compounds, and thioethers. Their benefits over metal counterparts are discussed and future perspectives are highlighted.

Contribution 3 is a research article that reports that the palladium species recovered from waste catalyst, namely, $[N^nBu_4]_2[Pd_2I_6]$, can be readily converted into PdI_2(dppf) (dppf = 1,1′-bis(diphenylphosphino)ferrocene). Both palladium catalysts were found to be active in processes such as Buchwald–Hartwig amination reactions, with the bio-derived solvent cyclopentyl methyl ether offering a more sustainable approach. It was also reported that the catalyst $[N^nBu_4]_2[Pd_2I_6]$, in the presence of an oxidant, performed well in the oxidative functionalization of benzo[*h*]quinoline to 10-alkoxybenzo[*h*]quinoline and 8-methylquinoline to 8-(methoxymethyl)quinoline, and could be reused multiple times. In contribution 4, a new palladium catalyst bearing a five-membered chelating [*N*,*O*] ligand, derived from the condensation of 2,6-diisopropylaniline and maple lactone, is reported. The catalyst was active in the Suzuki–Miyaura cross-coupling reaction and operated under mild conditions for a variety of aryl bromides, as well as boronic acids and pinacol esters. Under slightly more robust conditions, the catalyst was also capable of the cross-coupling of aryl chlorides and phenylboronic acid. In contribution 5, the conversion of *Chlorella vulgaris* biomass into useful materials such as organic acids, sugars, and furanic compounds via hydrolysis–dehydration–rehydration reactions catalyzed by dilute sulfuric acid is discussed. The type of product formed was found to be dependent on the conditions employed. The use of metallic ruthenium catalysts supported on activated carbons (5%Ru/C) for the catalytic hydrogenation/hydrogenolysis of this residual carbohydrate biomass is also discussed. Contributions 6 and 7 describe catalysts based on the bis(imino)pyridine ligand set for ethylene polymerization. In particular, non-symmetrical [N,N-diaryl-11-phenyl-1,2,3,7,8,9,10-heptahydrocyclohepta[b]quinoline-4,6-diimine]iron(II) chloride complexes, upon activation with either MAO or MMAO as a co-catalyst, exhibited exceptional activities with values as high as 35.92×10^6 g (PE) mol^{-1} (Fe) h^{-1}. β-H elimination and chain transfer to aluminum were identified by ^{1}H/^{13}C NMR spectroscopy. In contribution 7, a family of cobalt bis(imino)pyridine catalysts containing at least one N-2,4-bis(dibenzosuberyl)-6-fluorophenyl group is reported. Again, upon activation with either MAO or MMAO, very high activities of the order of 1.15×10^7 g PE mol^{-1} (Co) h^{-1} were achievable at 70 °C. Contributions 8 and 9 to this Special Issue also involve the use of catalysts for polymer production. High valent vanadium oxo complexes bearing 6-bis(o-hydroxyaryl)pyridine derived ligation were found to be capable of the ring opening polymerization (ROP) of ε-caprolactone (ε-CL), δ-valerolactone (δ-VL), and rac-lactide (r-LA), with the best results being achieved when the catalysts were deployed in the melt form. These species were also capable, in the presence of DMAC (co-catalyst)/ETA (reactivator), of ethylene polymerization, albeit with moderate activities ($\leq$8600 Kg·mol·V^{-1}bar^{-1}h^{-1}). In contribution 9, the use of H-bonded molybdenum arrays, which are derived from dianilines, amine-functionalized acids, as catalysts for the ROP of ε-CL and δ-VL is reported. Such systems, when used as melts under N$_2$ or air, afforded relatively high-molecular-weight polymers (10,420–56,510 Da) with a variety of end groups. The parent dianilines were also capable of the ROP of δ-VL, whilst both PCL and PVL were formed when using the parent acids alone. In contribution 10, the Miskolc Ranking 21 (MIRA21) model was applied to compare 58 catalysts in 2,4-dinitrotoluene catalytic hydrogenation to 2,4-toluenediamine. The results placed eight catalysts in the high (D1) class, and 80% of the catalysts afforded excellent conversions with 45% revealing selectivity above 90% n/n%. A comparison of the various systems revealed that catalysts with oxide and/or magnetic supports performed better than carbon-based supports when utilized under laboratory conditions. Work in contribution 11 found that catalytic activity for H$_2$O$_2$ synthesis can be improved by the incorporation of low concentrations of the metals Ni, Cu, or Zn into supported AuPd nanoparticles. The best results were achieved using Cu versus the use of a Pt promotor, and this was attributed, based on XPS and CO-DRIFTS experiments, to changes in the surface composition, including the formation of mixed Pd^{2+}/Pd^0 domains and the electronics of the system. Upon reuse, some deactivation was noted, and this was attributed to reduction to Pd^0 species via H$_2$ rather than Cl loss.

Finally, we would like to express our sincere thanks to all those who submitted articles of such high quality to this Special Issue on the "State-of-the-art in Molecular Catalysis in Europe". We hope that you, the readership of *Catalysts*, can also enjoy this Special Issue.

Conflicts of Interest: The authors declare no conflicts of interest.

List of Contributions:

1. Rebrov, E.V.; Gao, P.-Z. Molecular Catalysts for OER/ORR in Zn–Air Batteries. *Catalysts* **2023**, *13*, 1289. https://doi.org/10.3390/catal13091289.
2. Forchetta, M.; Valentini, F.; Conte, V.; Galloni, P.; Sabuzi, F. Photocatalyzed Oxygenation Reactions with Organic Dyes: State of the Art and Future Perspectives. *Catalysts* **2023**, *13*, 220. https://doi.org/10.3390/catal13020220.
3. Jantan, K.A.; Ekart, G.; McCarthy, S.; White, A.J.P.; Braddock, D.C.; Serpe, A.; Wilton-Ely, J.D.E.T. Palladium Complexes Derived from Waste as Catalysts for C-H Functionalisation and C-N Bond Formation. *Catalysts* **2024**, *14*, 295. https://doi.org/10.3390/catal14050295.
4. Andrews, M.J.; Brunen, S.; McIntosh, R.D.; Mansell, S.M. Preformed Pd(II) Catalysts Based on Monoanionic [N,O] Ligands for Suzuki-Miyaura Cross-Coupling at Low Temperature. *Catalysts* **2023**, *13*, 303. https://doi.org/10.3390/catal13020303.

5. Margellou, A.G.; Torofias, S.A.; Iakovou, G.; Triantafyllidis, K.S. Valorization of Chlorella Microalgae Residual Biomass via Catalytic Acid Hydrolysis/Dehydration and Hydrogenolysis/Hydrogenation. *Catalysts* **2024**, *14*, 286. https://doi.org/10.3390/catal14050286.
6. Wang, Y.; Wang, Z.; Zhang, Q.; Ma, Y.; Solan, G.A.; Sun, Y.; Sun, W.-H. Non-Symmetrically Fused Bis(arylimino)pyridines with para-Phenyl Substitution: Exploring their use as N′,N,N″-Supports in Iron Ethylene Polymerization Catalysis. *Catalysts* **2024**, *14*, 213. https://doi.org/10.3390/catal14030213.
7. Zada, M.; Sage, D.D.; Zhang, Q.; Ma, Y.; Solan, G.A.; Sun, Y.; Sun, W.-H. Thermally Stable and Highly Efficient N,N,N-Cobalt Olefin Polymerization Catalysts Affixed with N-2,4-Bis(Dibenzosuberyl)-6-Fluorophenyl Groups. *Catalysts* **2022**, *12*, 1569. https://doi.org/10.3390/catal12121569.
8. Elsegood, M.R.J.; Clegg, W.; Redshaw, C. Vanadium Complexes Derived from O,N,O-tridentate 6-bis(o-hydroxyalkyl/aryl)pyridines: Structural Studies and Use in the Ring-Opening Polymerization of ε-Caprolactone and Ethylene Polymerization. *Catalysts* **2023**, *13*, 988. https://doi.org/10.3390/catal13060988.
9. Clegg, W.; Elsegood, M.R.J.; Redshaw, C. Extended Hydrogen-Bonded Molybdenum Arrays Derived from Carboxylic Acids and Dianilines: ROP Capability of the Complexes and Parent Acids and Dianilines. *Catalysts* **2024**, *14*, 214. https://doi.org/10.3390/catal14030214.
10. Jakab-Nácsa, A.; Hajdu, V.H.; Vanyorek, L.; Farkas, L.; Viskolcz, B. Overview of Catalysts with MIRA21 Model in Heterogeneous Catalytic Hydrogenation of 2,4-Dinitrotoluene. *Catalysts* **2023**, *13*, 387. https://doi.org/10.3390/catal13020387.
11. Barnes, A.; Lewis, R.J.; Morgan, D.J.; Davies, T.E.; Hutchings, G.J. Improving Catalytic Activity towards the Direct Synthesis of H_2O_2 through Cu Incorporation into AuPd Catalysts. *Catalysts* **2022**, *12*, 1396. https://doi.org/10.3390/catal12111396.

Disclaimer/Publisher's Note: The statements, opinions and data contained in all publications are solely those of the individual author(s) and contributor(s) and not of MDPI and/or the editor(s). MDPI and/or the editor(s) disclaim responsibility for any injury to people or property resulting from any ideas, methods, instructions or products referred to in the content.

Article

Palladium Complexes Derived from Waste as Catalysts for C-H Functionalisation and C-N Bond Formation

Khairil A. Jantan [1,2], Gregor Ekart [1], Sean McCarthy [1], Andrew J. P. White [1], D. Christopher Braddock [1], Angela Serpe [3] and James D. E. T. Wilton-Ely [1,*]

[1] Department of Chemistry, Imperial College, Molecular Sciences Research Hub, White City Campus, London W12 0BZ, UK; khairil0323@uitm.edu.my (K.A.J.); gregor.ekart15@imperial.ac.uk (G.E.); c.braddock@imperial.ac.uk (D.C.B.)
[2] Faculty of Applied Sciences, Universiti Teknologi MARA (UiTM), Shah Alam 40450, Malaysia
[3] Department of Civil and Environmental Engineering and Architecture (DICAAR), INSTM Unit, University of Cagliari, Via Marengo 2, 09123 Cagliari, Italy; serpe@unica.it
* Correspondence: j.wilton-ely@imperial.ac.uk

Citation: Jantan, K.A.; Ekart, G.; McCarthy, S.; White, A.J.P.; Braddock, D.C.; Serpe, A.; Wilton-Ely, J.D.E.T. Palladium Complexes Derived from Waste as Catalysts for C-H Functionalisation and C-N Bond Formation. *Catalysts* 2024, *14*, 295. https://doi.org/10.3390/catal14050295

Academic Editor: Jacques Muzart

Received: 25 March 2024
Revised: 22 April 2024
Accepted: 25 April 2024
Published: 29 April 2024

Copyright: © 2024 by the authors. Licensee MDPI, Basel, Switzerland. This article is an open access article distributed under the terms and conditions of the Creative Commons Attribution (CC BY) license (https:// creativecommons.org/licenses/by/ 4.0/).

Abstract: Three-way catalysts (TWCs) are widely used in vehicles to convert the exhaust emissions from internal combustion engines into less toxic pollutants. After around 8–10 years of use, the declining catalytic activity of TWCs causes them to need replacing, leading to the generation of substantial amounts of spent TWC material containing precious metals, including palladium. It has previously been reported that $[N^nBu_4]_2[Pd_2I_6]$ is obtained in high yield and purity from model TWC material using a simple, inexpensive and mild reaction based on tetrabutylammonium iodide in the presence of iodine. In this contribution, it is shown that, through a simple ligand exchange reaction, this dimeric recovery complex can be converted into $PdI_2(dppf)$ (dppf = 1,1'-bis(diphenylphosphino)ferrocene), which is a direct analogue of a commonly used catalyst, $PdCl_2(dppf)$. $[N^nBu_4]_2[Pd_2I_6]$ displayed high catalytic activity in the oxidative functionalisation of benzo[*h*]quinoline to 10-alkoxybenzo[*h*]quinoline and 8-methylquinoline to 8-(methoxymethyl)quinoline in the presence of an oxidant, $PhI(OAc)_2$. Near-quantitative conversions to the desired product were obtained using a catalyst recovered from waste under milder conditions (50 °C, 1–2 mol% Pd loading) and shorter reaction times (2 h) than those typically used in the literature. The $[N^nBu_4]_2[Pd_2I_6]$ catalyst could also be recovered and re-used multiple times after the reaction, providing additional sustainability benefits. Both $[N^nBu_4]_2[Pd_2I_6]$ and $PdI_2(dppf)$ were also found to be active in Buchwald–Hartwig amination reactions, and their performance was optimised through a Design of Experiments (DoE) study. The optimised conditions for this waste-derived palladium catalyst (1–2 mol% Pd loading, 3–6 mol% of dppf) in a bioderived solvent, cyclopentyl methyl ether (CPME), offer a more sustainable approach to C-N bond formation than comparable amination protocols.

Keywords: palladium; recovered metals; C-H functionalisation; C-N amination; sustainability

1. Introduction

Palladium-mediated reactions are among the most frequently employed transformations in synthetic chemistry and are used widely for C-C, C-N and C-O bond formation [1–6]. However, the extremely low natural abundance of palladium in the Earth's crust, its limited distribution and geopolitical factors threaten future supply [7]. This has led to a substantial palladium deficit due to the high demand and limited supply [8]. The route of this metal from ore to its application in numerous fields features many acknowledged environmental repercussions on water and air quality, biodiversity, and population displacement [9–11]. Furthermore, the low palladium content in unrefined ore (<10 g/tonne) requires energy-intensive treatments (3880 kg CO_2e per kg of Pd [9]) in the pyrometallurgical extraction process [12], which involves smelting in blast and electric furnaces at high temperatures

(>2000 °C) [13] and the use of hazardous chemicals. As a less energy-intensive and more selective alternative, hydrometallurgical processes typically involve acid (*aqua regia* or HNO_3) or alkaline (cyanide) leaching, followed by separation phases exploiting solid-to-liquid or solvent extraction processes [14]. However, the well-established processes use toxic and/or hazardous reagents, producing toxic gaseous emissions (i.e., NO_x, Cl_2 and NOCl for aqua regia and HCN for cyanide leaching) and producing significant amounts of contaminated wastewater [14].

The exceptional performance of palladium as an oxidation catalyst has made it a critical component in automotive three-way catalytic converters (TWCs) [15]; however, the relatively short lifetime (around 10 years or 100,000 miles) of TWCs has resulted in the accumulation of palladium-rich waste, with up to 4 g of Pd in each catalytic converter, along with smaller amounts of Pt and Rh [16]. The concentrations of palladium in TWCs (2000 g/tonne) are far higher than those found in natural ore (10 g/tonne), indicating the huge potential of these waste streams as more sustainable sources of palladium. Attempts to transform this scrap into a valuable but underutilised source of potentially recoverable raw materials is often termed 'urban mining' [17]. Unfortunately, current recovery methods rely on the environmentally harmful processes described above for ore treatment [18]. As a result, research has focused on developing milder, more environmentally friendly palladium recovery methods. Serpe, Deplano and colleagues reported the first selective and acid-free recovery method of palladium from a model spent catalytic converter, achieving 90–99% palladium recovery. This method employed a combined dithiooxamide-iodide lixiviant, $Me_2dazdt\cdot 2I_2$ (Me_2dazdt = *N,N'*-dimethylperhydrodiazepine-2,3-dithione), in an organic solvent under mild conditions to form $[Pd(Me_2dazdt)_2][I_3]_2$ as the recovery product [19]. While this approach to palladium is highly effective, it yields a molecular recovery product that cannot viably be used to prepare new TWCs. In addition, longer times are needed to achieve palladium dissolution than is required by the conventional acidic and alkaline leaching mixtures, and the use of Lawesson's reagent to make Me_2dazdt leads to poor atom economy, which has led to the use of commercially available dithiooxamides as a less effective but viable alternative [20].

The solution to this was to valorise the recovery complexes in other important applications, such as catalysts for organic synthesis. Our previous work demonstrated that the $[Pd(Me_2dazdt)_2][I_3]_2$ recovery product could successfully mediate the regioselective C-H bond functionalisation of benzo[*h*]quinoline and 8-methylquinoline (Figure 1) [21], competing favourably with the results reported by Sanford and co-workers using $Pd(OAc)_2$ sourced from conventional mining [22].

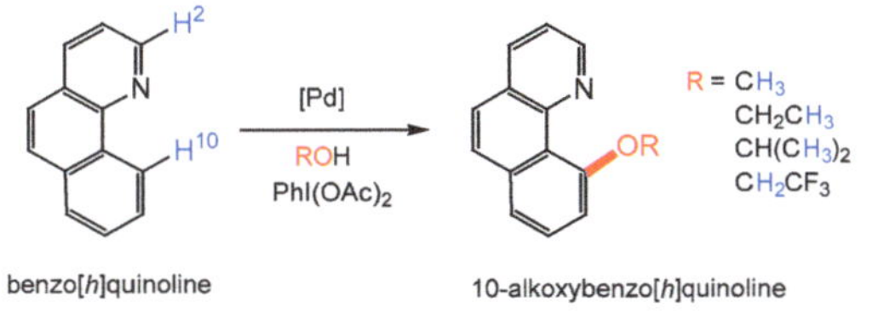

Figure 1. Oxidative C-H functionalization of benzo[*h*]quinoline in the presence of different alcohols, a sacrificial oxidant and a palladium complex.

Commercial (pre)catalysts for palladium-catalysed reactions are typically produced using damaging mining practices that employ dangerous leaching additives and/or require high energy pyrometallurgical treatments to extract the very low palladium content from the mined ore. Further transformations, often involving hazardous and costly agents, are also needed to refine and produce marketable materials before transport around the world to its end use. All of these steps require significant energy and use dangerous and environmentally harmful chemical treatments. In contrast, the generation of $[N^nBu_4]_2[Pd_2I_6]$ (**1**) requires far fewer steps (deconstruction and milling of the locally sourced TWC material followed by treatment with the selective, low-cost and safer polyiodide salt), which require much less energy and are based on a feedstock that can contain up to 200 times more palladium than in mined ore. A recent cost analysis for catalysts based on recovered gold [23] illustrated that even unoptimized, small-scale production of the catalyst leads to a significantly lower cost than commercial catalysts derived from environmentally damaging mining.

In this contribution, we present the use of the dimer $[N^nBu_4]_2[Pd_2I_6]$ (**1**) as an even more effective and sustainable catalyst for oxidative functionalisation reactions, such as those shown in Figure 1. Compound **1** is obtained selectively from mixed-metal model TWC material in over 70% yield using a low-energy route based on the reaction with iodine in the presence of organic iodides (such as $[N^nBu_4]I$), as shown in Figure 2 [24]. Selective precipitation of **1** from the organic solvent (typically acetone or MIBK) allows separation from the other metals present in TWCs. The inexpensive reagents and low-energy conditions are a substantial improvement over the use of $Me_2dazdt·2I_2$; however, the process of returning the Pd content of **1** to its metallic form is still energy-intensive and thus costly. Therefore, a more effective way of valorising this recovery product is to use it directly (or with simple modification) in catalysis [16,25,26]. This is a strategy that led to the first direct application in homogeneous catalysis of gold recovery products sourced from e-waste [23]. Using simple ligand exchange reactions, $PdI_2(dppf)$ (**2**), the iodide analogue of the well-known catalyst, $PdCl_2(dppf)$, was synthesised from compound **1** to provide a phosphine-supported derivative. The results obtained for oxidative C–H functionalisation and Buchwald–Hartwig amination reactions using **1** and **2** are compared both in terms of catalytic activity and sustainability.

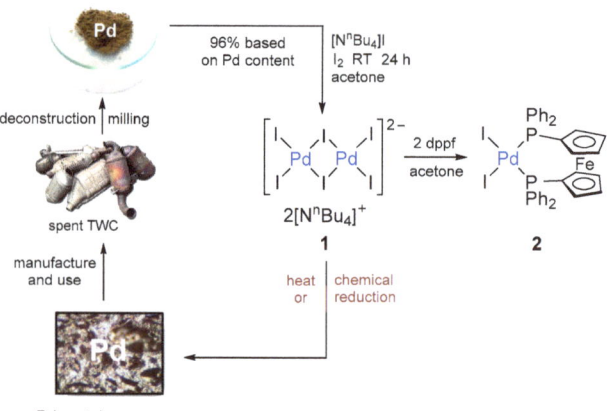

Figure 2. Palladium recovery process from TWCs to form **1** and the ligand exchange reaction used to prepare **2**.

2. Results and Discussion

2.1. Synthesis of Palladium Complexes

Complex **1** is the main product obtained from the selective Pd leaching and recovery process carried out on model samples of spent TWC using tetrabutylammonium iodide in the presence of iodine in an acetone solution [24] (Figure 2). The iodine/iodide system offers a 'greener' approach to Pd recovery compared to traditional pyrometallurgical or hydrometallurgical processes due to its versatility, mild conditions, minimal environmental impact, use of inexpensive reagents and easily recyclable solvents. These attributes also allow it to be employed in low-income regions where there is currently no local recycling infrastructure for TWCs. Since the use of $[N^nBu_4]I/I_2$ to recover Pd from model TWC material is already well established [24], compound **1** was obtained directly by leaching palladium metal on a small scale using the same reagents. The black crystalline product

(86% yield) was usually obtained by Et$_2$O diffusion into the acetone leachate over a period of 3 days. A more rapid crystallisation process was also used to obtain a less crystalline but analytically identical product through the addition of ethanol and concentration of the solvent volume until precipitation had been achieved. Compound **1** obtained from both approaches leads to characteristic features in the UV-vis spectrum at 2960–2869 cm^{-1}, 1457 and 1379 cm^{-1}, with the latter two absorptions attributed to the tetrabutylammonium units. The presence of the dimeric palladium complex [Pd$_2$I$_6$]$^{2-}$ was indicated by an absorption at 340 nm. The overall formulation was confirmed by an abundant molecular ion in the electrospray (negative ion) mass spectrum at m/z 487 (ESI, Section S1.2) and good agreement of the elemental analysis with calculated values.

Palladium phosphine complexes are the most widely used ligand-supported (pre)catalysts [4–6,16,27], with PdCl$_2$(dppf) being used in many settings, including for Buchwald–Hartwig amination reactions [28]. Thus, the recovery complex [N^{n}Bu$_4$]$_2$[Pd$_2$I$_6$] (**1**) was treated with dppf in acetone to produce the iodide analogue, PdI$_2$(dppf) (**2**), in 97% yield. A singlet was observed in the ^{31}P{^{1}H} NMR spectrum at 24.3 ppm, while the presence of the dppf ligands was clearly observed in the ^{1}H NMR spectrum through ferrocenyl resonances at 4.14 and 4.35 ppm. In addition, typical aromatic features were observed in the solid-state FTIR spectrum (1092 and 1480 cm^{-1}) and the overall composition was confirmed by mass spectrometry (ESI, Section S1.3) and elemental analysis. Crystals suitable for single-crystal X-ray diffraction measurements were obtained through the slow diffusion of ethanol into a chloroform solution of PdI$_2$(dppf) (Figure 3). This revealed a distorted square planar arrangement at the Pd(II) centre due to the steric requirement of the bulky dppf ligand. The bond lengths and angles are similar to those reported for the same complex in a lower-quality structure in 2001 [29]. Further discussion and crystal data are provided in the ESI (Section S2).

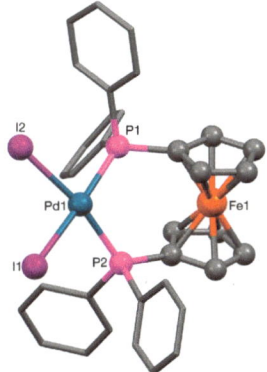

Figure 3. Molecular structure of **2**. Selected bond lengths (Å) and angles (°): Pd(1)–I(1) 2.6495(5), Pd(1)–I(2) 2.6454(6), Pd(1)–P(1) 2.3281(13), Pd(1)–P(2) 2.3245(15), P(2)–Pd(1)–I(1) 83.83(3), P(1)–Pd(1)–I(2) 88.60(4), P(1)–Pd(1)–P(2) 99.77(5).

2.2. Oxidative C–H Functionalisation

Palladium(II) compounds are known to catalyse a variety of oxidative C–H functionalisation processes, particularly the conversion of benzo[*h*]quinoline to 10-alkoxybenzo[*h*] quinoline [2,3,21,22]. The catalytic potential of recovery product **1** and its derivative **2** was assessed for this class of transformation using the conversion of benzo[*h*]quinoline to 10-methoxybenzo[*h*]quinoline as a benchmark reaction (Figure 1). The term 'conversion' will be used here rather than yield, as the amount of desired product formed was determined using ^{1}H NMR spectroscopy, rather than as an isolated yield of each product. NMR spectroscopy has been shown [21] to allow sufficient accuracy (2–5% error) to determine the impact of changes in the reaction conditions on the formation of the product. Isolated yields have been reported previously and show only a small decrease compared to the conversion determined spectroscopically [21]. The conditions reported in the literature for this transformation were used (1.1 mol% catalyst loading, 2 equivalents PhI(OAc)$_2$, 100 °C, 22 h). Under these conditions, complexes **1** and **2** provided near-quantitative conversions to the desired product, as determined by ^{1}H NMR spectroscopy (97 and 96%, respectively), of 10-methoxybenzo[*h*]quinoline, competing favourably with the 96% yield reported in the literature (and confirmed here) using Pd(OAc)$_2$, sourced from conventional mining (ESI, Section S4). These results also matched those we previously reported for [Pd(Me$_2$dazdt)$_2$][I$_3$]$_2$ [21]. The proposed mechanism [22,30] for these reactions involves a Pd(II)-Pd(IV) manifold, though Pd(III) intermediates have also been postulated [31].

Heating methanol at 100 °C in a sealed vessel for 22 h leads to both safety and sustainability concerns and so attempts were made to perform the reaction under milder conditions. Following the initial screening reactions, the product conversion over time was investigated for the methoxylation of benzo[*h*]quinoline at different temperatures, first using the literature catalyst Pd(OAc)$_2$ at a loading of 1.1 mol% (Figure 4). At high temperatures (100 °C), no induction period was observed, with the fastest rate of reaction occurring at the start of the reaction. However, heating flammable solvents above their boiling point in a confined space generates potential danger related to pressure build-up. At the lower temperature of 50 °C, the Pd(OAc)$_2$ reaction proceeds more slowly but then achieves 86% conversion to 10-methoxybenzo[*h*]quinoline after 5 h before reaching 95% after 22 h, a conversion reached in only 2 h at 100 °C (ESI, Table S4-1). Perhaps due to a tendency to use established protocols involving longer reaction times, it appears that previous investigations did not reveal that high yields were already formed after much shorter reaction times.

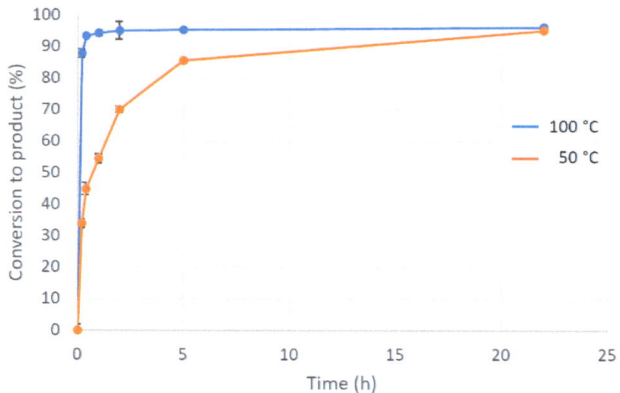

Figure 4. Formation of 10-methoxybenzo[*h*]quinoline from benzo[*h*]quinoline over time using 1.1 mol% Pd(OAc)$_2$ in the presence of PhI(OAc)$_2$ in MeOH at 100 °C (blue line) and 50 °C (orange line). Conversion was determined from an average of three independent experiments by comparing the integration of H-2 and H-10 resonances in benzo[*h*]quinoline with the diagnostic methoxy resonance in the 10-methoxybenzo[*h*]quinoline product.

The appearance of a black residue at the bottom of the vials used in the reaction at 100 °C after 22 h was noted, and this material was analysed using transmission electron microscopy (TEM). This 'palladium black' was revealed to also contain palladium nanoparticles (Figure 5A) with an average diameter of 2.6 nm and a relatively broad size distribution of ±1.1 nm. Though palladium nanoparticles (PdNPs) are well known to catalyse many Pd(0)-mediated reactions, such as Suzuki–Miyaura couplings [32], they have also been observed in C-H functionalisation reactions [33,34]. The formation of the PdNPs was investigated in separate control experiments (ESI, Section S3.2.1), in which Pd(OAc)$_2$ was heated for 22 h at 100 °C in a methanol solution with or without PhI(OAc)$_2$ to yield palladium nanoparticles with an average diameter of 1.2 nm (±0.3 nm) and 1.5 nm (±0.5 nm), respectively (Figure 5B,C). The presence of the sacrificial oxidant does not appear to prevent the reduction of Pd(OAc)$_2$. These results suggest that the Pd(OAc)$_2$ catalyst undergoes (at least partial) in situ reduction in the alcohol solution to produce PdNPs under the reaction conditions for C-H functionalisation.

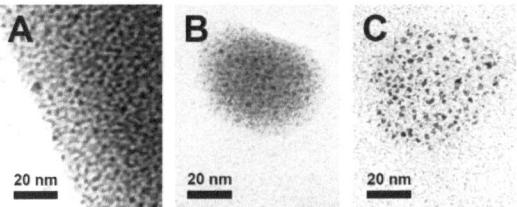

Figure 5. TEM images of Pd nanoparticles obtained (**A**) after methoxylation of benzo[*h*]quinoline using Pd(OAc)$_2$ as a catalyst, (**B**) after Pd(OAc)$_2$ was heated with the sacrificial oxidant PhI(OAc)$_2$ in methanol and (**C**) after heating Pd(OAc)$_2$ in methanol. All reactions were conducted at 100 °C for 22 h.

Further investigation was carried out to investigate whether the formation of palladium nanoparticles aided or hindered the methoxylation of benzo[*h*]quinoline. Palladium acetate was heated in methanol in air for 2 h at 100 °C followed by the addition of benzo[*h*]quinoline and PhI(OAc)$_2$ to the solution. The reaction mixture was stirred and heated at 100 °C for another 22 h. A 67% conversion to 10-methoxybenzo[*h*]quinoline was obtained after 22 h of reaction, which is substantially lower than the 95% obtained without the initial heating step in methanol. One suggestion is that a fraction of Pd(OAc)$_2$ survives the reduction to Pd nanoparticles and is then able to catalyse the C-H functionalisation reaction. Alternatively, as proposed by Fairlamb and co-workers [33,34], the PdNPs formed could be acting as a 'reservoir' of active Pd that performs the transformation but the larger PdNPs formed are inactive. The palladium black that can be observed by the naked eye consists of much larger, micrometre-sized particles, which are themselves unlikely to be catalytically active. Taken together, these results suggest that the use of Pd(OAc)$_2$ at high temperatures in an alcohol solvent, as reported in the literature [22], actually contributes to the deactivation of the catalyst and that the good conversions obtained after 22 h were actually due to the residual catalytic species that managed to operate under the conditions employed.

Following this investigation of the literature catalyst, Pd(OAc)$_2$, the catalytic activity of [N^{n}Bu$_4$]$_2$[Pd$_2$I$_6$] (**1**) was explored in the functionalisation of benzo[*h*]quinoline using the conditions that had provided the highest conversions (c.f. Figure 4) over shorter reaction times (1–2 mol% [Pd] loading, 100 °C, 2 h). Near-quantitative conversions to the desired product were recorded for the selective alkoxybenzo[*h*]quinoline for two alcohols, ROH (R = Me, CH$_2$CF$_3$), using 1 mol% Pd loading. However, a two-fold increase in catalyst loading was required to provide better conversion to 10-ethoxybenzo[*h*]quinoline (81%) and 10-isopropoxybenzo[*h*]quinoline (75%) (Figure 6 and ESI, Table S4-2). The lower activity with ethanol and isopropanol could be due to the differing speciation of the active catalyst species in solvents of different polarity.

Similarly to the reactions catalysed by Pd(OAc)$_2$, a black residue was observed at the bottom of the reaction vials after 2 h of the reaction using [N^{n}Bu$_4$]$_2$[Pd$_2$I$_6$] (**1**) at 100 °C for all substrates except for the trifluoroethanol reaction mixture. The black precipitate was centrifuged, washed with methanol and dried under a vacuum. TEM images (Figure 7) revealed the formation of small palladium nanoparticles (PdNPs). Average diameters of 2.0 nm (methanol), 1.8 nm (ethanol) and 2.3 nm (mixture of isopropanol/acetic acid) were recorded, confirming that the formation of PdNPs was not exclusive to the use of Pd(OAc)$_2$. The lack of formation of nanoparticles in the trifluoroethanol reaction mixture could be due to the electron-withdrawing CF$_3$ unit stabilising the palladium(II) complex effectively, hindering the formation of PdNPs at high temperatures.

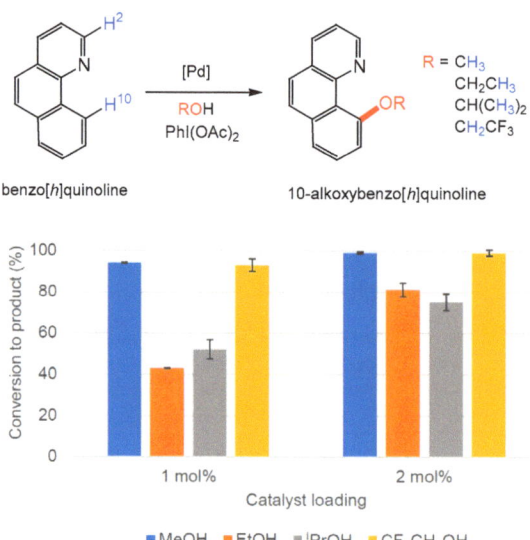

Figure 6. Results for the formation of alkoxybenzo[*h*]quinoline using **1** as a catalyst in the alcohols shown at 100 °C. The conversion to 10-alkoxybenzo[*h*]quinoline was determined by ^{1}H NMR spectroscopy using the average of three independent experiments by comparing the integration of H-2 and H-10 resonances in benzo[*h*]quinoline with the diagnostic resonances of methoxy, ethoxy, isopropoxy and trifluoroethoxy groups in the products. 10-isopropoxybenzo[*h*]quinoline is formed from a mixture of isopropanol and acetic acid, according to the original literature protocol [22].

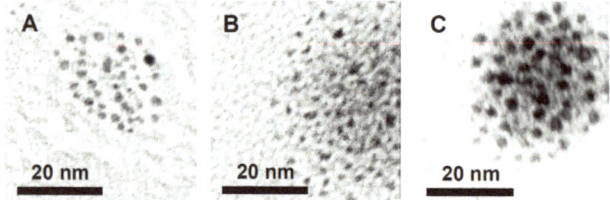

Figure 7. TEM images of Pd nanoparticles formed after heating benzo[*h*]quinoline and PhI(OAc)$_2$ with **1** at 100 °C for 2 h in the presence of (**A**) MeOH, (**B**) EtOH, and (**C**) ^{i}PrOH (and acetic acid).

The same transformations were repeated with [N^{n}Bu$_4$]$_2$[Pd$_2$I$_6$] (**1**) as the catalyst at 50 °C while modifying the reaction conditions (loading and reaction times) to attain satisfactory conversions within an acceptable reaction time (Figure 8 and ESI Table S4-3). Significantly, in all cases at this lower temperature, there was no visual evidence for the formation of PdNPs. At 1 mol% [Pd] catalyst loading, low to moderate conversions (as determined by ^{1}H NMR spectroscopy) were observed (Figure 8A), which improved over longer reaction times for reactions with MeOH and CF$_3$CH$_2$OH. Doubling the catalyst loading to 2 mol% [Pd] led to essentially quantitative conversions to the desired product in 2 h for all products, except for 10-isopropoxybenzo[*h*]quinoline, which showed a steady increase in the conversion to 82% after 24 h (Figure 8B). A 3 mol% loading of **1** and a reaction time of 4 h were found to be necessary to achieve a high conversion (92%) to the isopropoxy-functionalised product (ESI Table S4-3). The standard operating conditions were thus defined as being 2 mol% [Pd] catalyst loading at 50 °C for 2 h. An isolated yield of 97% of 10-methoxybenzo[*h*]quinoline was obtained using these conditions, reflecting the ease of isolation of these products through flash column chromatography.

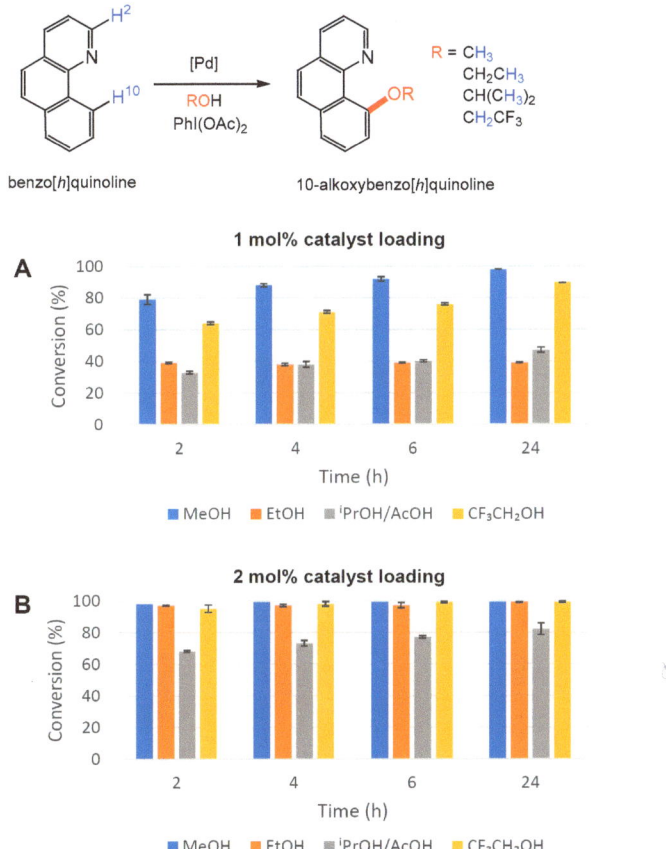

Figure 8. Alkoxy functionalisation of benzo[*h*]quinoline in different alcohol solvents at 50 °C employing **1** as a catalyst at (**A**) 1 mol% and (**B**) 2 mol% [Pd] loading. The conversion to 10-alkoxybenzo[*h*]quinoline was determined by ^{1}H NMR spectroscopy using the average of three independent experiments by comparing the integration of H-2 and H-10 resonances in benzo[*h*]quinoline with the diagnostic resonances of methoxy, ethoxy, isopropoxy and trifluoroethoxy groups in the products.

The same conditions were then employed to investigate the use of PdI$_2$(dppf) (**2**) as a catalyst for the same reaction. Using 1 mol% loading (Figure 9A), 98% conversion to 10-methoxybenzo[*h*]quinoline was obtained after only 2 h. Increasing the catalyst loading to 2 mol% (Figure 9B) improved the conversions of the more challenging reactions to produce 10-trifluoroxybenzo[*h*]quinoline (94%) and 10-isopropoxybenzo[*h*]quinoline (95%) after 4 h and 24 h, respectively (see also ESI Table S4-4).

Encouraged by the successful result for the alkoxylation of benzo[*h*]quinoline, the synthesis of 8-(methoxymethyl)quinoline from 8-methylquinoline was briefly investigated. The data obtained suggest that compound **1** appears to deliver superior catalytic performance compared to the other catalysts under investigation (ESI, Table S4-6). An isolated yield of 94% of 8-(methoxymethyl)quinoline was obtained using 1 mol% of **1** in methanol at 50 °C for 2 h following a flash column (ESI, Section S3.2.2).

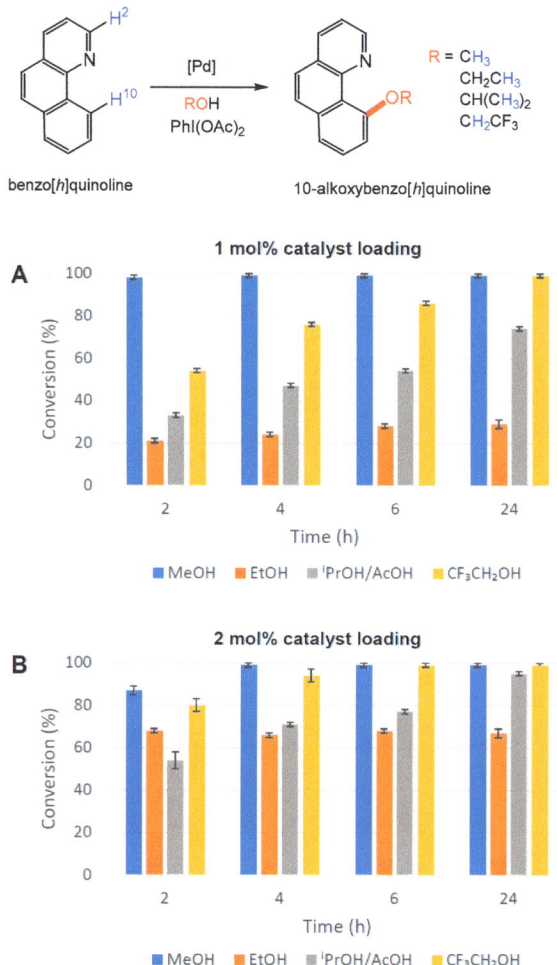

Figure 9. Alkoxy functionalisation of benzo[*h*]quinoline in different alcohol solvents at 50 °C employing PdI$_2$(dppf) (**2**) as a catalyst at (**A**) 1 mol% and (**B**) 2 mol% [Pd] loading. The conversion to 10-alkoxybenzo[*h*]quinoline was determined by ^{1}H NMR spectroscopy using the average of three independent experiments by comparing the integration of H-2 and H-10 resonances in benzo[*h*]quinoline with the diagnostic resonances of methoxy, ethoxy, isopropoxy and trifluoroethoxy groups in the products.

While the use of a recovery product obtained under mild conditions from waste as the catalyst enhances the 'green' credentials of the approach, recycling of the catalyst would further enhance the sustainability of the process. This was investigated briefly in the reaction to form 10-ethoxybenzo[*h*]quinoline, shown in Table 1. Conditions were optimised to produce 10-ethoxybenzo[*h*]quinoline over a shorter reaction time (2 h). This required an increase in the reaction temperature to 75 °C and the use of 2 mol% catalyst loading. A near-quantitative conversion to the desired product was obtained in the first run, with a palladium black precipitate forming within the first 30 min of the reaction. This precipitate was isolated, washed and converted back to [N^{n}Bu$_4$]$_2$[Pd$_2$I$_6$] (**1**) in 75% yield using the same procedure for its synthesis from Pd powder using [N^{n}Bu$_4$]I/I$_2$. The recovered compound **1** was used for the second run, which led to the same quantitative conversion to the 10-

ethoxybenzo[*h*]quinoline product. Following the reaction, after treatment with [N^{n}Bu$_4$]I/I$_2$, compound **1** was recovered for a second time with 72% yield. This approach is being investigated further to assess the breadth of its application in different catalytic reactions.

Table 1. Use and re-use of [N^{n}Bu$_4$]$_2$[Pd$_2$I$_6$] (**1**) in the synthesis of 10-ethoxybenzo[*h*]quinoline. The second run was scaled down so that a loading of 2 mol% of recovered **1** was maintained. Conversion to the product was determined by ^{1}H NMR spectroscopy by comparing the integration of H-2 and H-10 resonances in benzo[*h*]quinoline with the diagnostic ethoxy resonances in the product and also with an internal standard, 1,3,5-trimethoxybenzene. The recovery (isolated) yield was calculated with respect to the amount of **1** in the reaction mixture.

Run	Conversion to organic product (%)	Recovery yield of **1** (%)
1	97	75
2	98	72

2.3. Amination Reactions

The palladium-catalysed Buchwald–Hartwig amination reaction is an essential tool in organic synthesis and in the pharmaceutical industry in particular [4,35–39]. Pioneering work by Hartwig demonstrated that a range of amination reactions could be catalysed by PdCl$_2$(dppf) in the presence of additional dppf in an inert environment [28]. The similarities between PdCl$_2$(dppf) and the PdI$_2$(dppf) (**2**) generated from the [N^{n}Bu$_4$]$_2$[Pd$_2$I$_6$] (**1**) recovery product provided encouragement to explore the possibility of substituting current catalysts with more sustainable alternatives derived from waste. In addition, our aim was also to enhance the sustainability still further by exploring 'greener' solvents and more straightforward conditions that do not require the exclusion of oxygen and moisture. In order to explore these aspects, complexes **1** and **2** were applied to the benchmark cross-coupling reaction of *p*-bromobiphenyl and *p*-toluidine in a range of solvent systems in air (Table 2), following literature protocols (5 mol% Pd loading, 15 mol% dppf ancillary ligand and sodium *tert*-butoxide as a base for 3 h at 100 °C). The methyl resonances in *p*-toluidine (2.24 ppm) and the product (2.33 pm) were employed as characteristic ^{1}H NMR spectroscopic features to monitor the reaction through integration of the respective signals.

The experimental protocol in the original report [28] described heating at 100 °C in tetrahydrofuran for 3 h using 5 mol% Pd loading and 15 mol% dppf as an ancillary ligand. The runs using these conditions were performed in a high-pressure vial due to the significantly lower boiling point of the THF solvent (66 °C). Compounds **1** and **2** showed lower catalytic activity compared to the PdCl$_2$(dppf) in this initial set of experiments (Table 2). However, heating THF above its boiling point poses a safety risk due to increased pressure. Thus, solvents with higher boiling points were explored, such as toluene (b.p. 111 °C) and cyclopentyl methyl ether (CPME, b.p. 106 °C). Conducting the reaction in these solvents led to improved conversions (Table 2) for reactions catalysed by **1** and **2** as well as high conversions for the commercial palladium catalyst, PdCl$_2$(dppf). This was the first indication that PdI$_2$(dppf) (**2**), obtained from a palladium recovery pathway, was able to deliver comparable catalytic performance to the literature catalyst.

Table 2. Solvent screening for the benchmark reaction of *p*-bromobiphenyl and *p*-toluidine in the presence of 5 mol% Pd loading and 15 mol% dppf at 100 °C for 3 h using different catalysts. The conversion to the desired product was determined by ^{1}H NMR spectroscopy by comparing the integration of diagnostic methyl resonances in *p*-toluidine and the product, taking into account the excess of *p*-toluidine used. Values are the average of three independent experiments. CPME = cyclopentyl methyl ether.

Solvent	Conversion (%) with PdCl$_2$(dppf)	Conversion (%) with [N^{n}Bu$_4$]$_2$[Pd$_2$I$_6$] (1)	Conversion (%) with PdI$_2$(dppf) (2)
THF	90 ± 1	55 ± 1	81 ± 1
CPME	98 ± 1	92 ± 2	90 ± 1
Toluene	98 ± 1	80 ± 1	96 ± 2

In addition to its favourable properties, including low toxicity, high boiling point, low melting point, hydrophobicity and chemical stability, cyclopentyl methyl ether (CPME) can also be bio-derived from furfural [40]. It is known to broadly mimic the chemical properties of both THF and toluene, but with a lower environmental and health score [41]. On the basis of this improved sustainability and the high conversions it delivered for **1** and **2**, this solvent was chosen to be taken forward as a solvent for the remainder of this work. It was also determined that carrying out the reaction in air did not affect the conversion to the desired product when compared to an anhydrous CPME solution.

Ancillary ligands, such as electron-rich and bulky dppf ligands, are often required in amination reactions to increase the rate of the reaction, prevent the formation of palladium halide dimers after oxidative addition [27,28], and suppress β-hydride elimination by preventing an open coordination site [36]. Without additional equivalents of the dppf ligand, no product formation was observed under the conditions described above. Since PdI$_2$(dppf) (**2**) is readily obtained from [N^{n}Bu$_4$]$_2$[Pd$_2$I$_6$] (**1**) and dppf, the in situ generation of **2** was also investigated, as discussed below.

The catalytic performance of **1** and **2** towards the aryl iodide analogue, *p*-iodobiphenyl, was explored in CPME. In contrast to the aryl bromides, the catalytic activities of all catalysts appeared to be adversely affected by the switch to aryl iodide substrates, providing a lower conversion to the desired product (Table 3). This result agreed with the previously reported literature [35,37] suggesting that aryl iodides are less effective substrates than their aryl bromide counterparts in amination reactions. This has been suggested to be connected to the poisoning of the catalyst through the accumulation of iodide salts in the reaction medium [38].

Table 3. Reaction of *p*-iodobiphenyl and *p*-toluidine in the presence of 15 mol% dppf at 100 °C for 3 h using different catalysts at 2 and 5 mol% palladium loadings. The conversion to the desired product was determined by ^{1}H NMR spectroscopy by comparing the integration of diagnostic methyl resonances in *p*-toluidine and the product, taking into account the excess of *p*-toluidine used. Values are the average of three independent experiments.

Pd loading (mol%)	Conversion (%) with PdCl$_2$(dppf)	Conversion (%) with [N^{n}Bu$_4$]$_2$[Pd$_2$I$_6$] (1)	Conversion (%) with PdI$_2$(dppf) (2)
5	71 ± 3	59 ± 1	65 ± 3
2	80 ± 1	72 ± 1	57 ± 2

While encouraging, these results suggest that a lower catalyst loading than 5 mol% could actually be beneficial and that the reaction conditions still had room for improvement. Accordingly, we sought to optimise the conversion to the desired product for the benchmark reaction with catalyst **2** through an investigation of key variables using a 'Design of Experiments' (DoE) strategy, guided by JMP Statistical Discovery software. Definitive screening designs considered three factors at three levels, the minimum, centre and maximum, with a total of 20 test runs (ESI, Table S3-1). The results obtained (ESI, Table S5-1) were used to build a model to predict the outcome of the reaction and possible variable combinations of catalyst loading, dppf loading and reaction time. The results of this study (ESI, Table S3-2) indicate that the individual effect of time is the most significant parameter, followed by two-factor interactions of catalyst loading with time and dppf loading, in that order. Only slightly below this were the effects of catalyst loading and dppf amount. Figure 10A shows the relationships between the experimental and predicted conversions to the desired product. This graph helps to illustrate the error and the performance of the model. The data points should be split evenly by the 45-degree line to test the model's fit. The quality fit of the model equation was expressed by the coefficient regression (R^2) as 0.97, which is close to unity, signifying a good fit of experimental data to the model, providing confidence that it can predict the product yield.

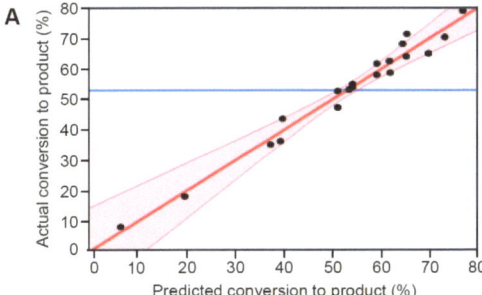 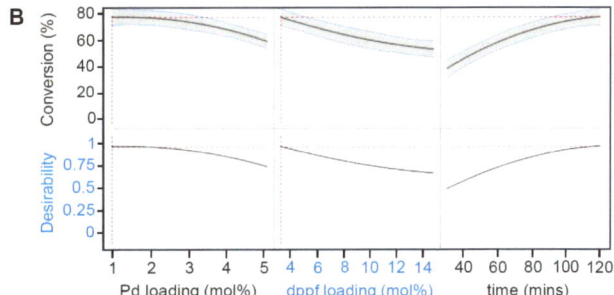

Figure 10. (**A**) Predicted and actual conversion to the desired product for the DoE model and (**B**) predicted optimal conditions (1 mol% 2, 3 mol% dppf, 120 min) for the benchmark reaction between *p*-bromobiphenyl and *p*-toluidine at 100 °C in CPME solution, as extracted from definitive screening design variable profiles.

The DoE analysis shown in Figure 10B indicates that the optimal predicted reaction yield for the reaction in CPME solution at 100 °C will be provided by 1 mol% PdI$_2$(dppf) (**2**) loading and 3 mol% dppf in 120 min. This confirmed that, while a 1:3 ratio of [Pd] to [dppf] was optimal, the palladium loading could be lowered significantly from the 5 mol% PdCl$_2$(dppf) typically used in the literature.

Kinetic experiments were conducted to investigate the conversion to the desired product over time and to better understand the behaviour of **2** as a catalyst in the model reaction (Figure 11 and ESI Table S5-2). This graph shows that no significant induction period was observed, with the fastest rate of reaction occurring at the start of the experiment. Extending the reaction time from 120 to 180 min led to only a slight increase in conversion to the product.

After establishing a set of catalytic conditions optimised for the reaction between *p*-toluidine and *p*-bromobiphenyl (1 mol% [Pd] loading, 3 mol% dppf at 100 °C for 3 h in CPME), the substrate scope was briefly explored with various aryl bromides (Table 4), keeping the toluidine constant. Both pre-formed PdI$_2$(dppf) (**2**) and **2** formed in situ from [N^{n}Bu$_4$]$_2$[Pd$_2$I$_6$] (**1**) and dppf were explored. For the in situ reactions, all reagents were combined and heating commenced without any catalyst pre-formation period. Overall, it was clear that the optimised catalytic conditions for the reaction between *p*-bromobiphenyl and *p*-toluidine (1 mol% [Pd]) proved insufficient to provide good conversions for all the substrates investigated in the subsequent substrate scope study. However, moving to 2 mol% [Pd] led to improved conversions to the desired product across the substrates explored both for pre-formed PdI$_2$(dppf) (**2**) and when **2** was generated in situ.

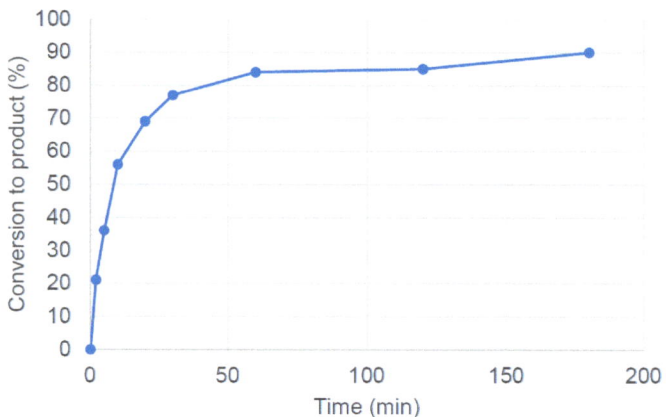

Figure 11. Reaction profile for the reaction between *p*-bromobiphenyl and *p*-toluidine at 100 °C in CPME in the presence of 1 mol% PdI$_2$(dppf) (**2**) and 3 mol% dppf. The conversion to the desired amination product was determined by ^{1}H NMR spectroscopy by comparing the integration of diagnostic methyl resonances in the product with an internal standard, 1,3,5-trimethoxybenzene. Values are the average of three independent experiments.

Table 4. Conversion to the products shown for various aromatic amines and aryl bromides after the reaction at 100 °C in CPME for 3 h in the presence of different [Pd] loadings of [N^{n}Bu$_4$]$_2$[Pd$_2$I$_6$] (**1**) or PdI$_2$(dppf) (**2**) and the stated amount of dppf. When using **1**, sufficient dppf was used to generate **2** in situ with the stated [Pd]:[dppf] ratio. The conversions to the desired products are an average of three independent experiments and were determined by ^{1}H NMR spectroscopy using relative integrations and 1,3,5-trimethoxybenzene as an internal standard (see ESI Section S3.3.7).

Product	Conversion to the Product Shown (%)				
	Catalyst **2** 1 mol% [Pd] 3 mol% [dppf]	Catalyst **1** 1 mol% [Pd] 3 mol% [dppf]	Catalyst **2** 2 mol% [Pd] 6 mol% [dppf]	Catalyst **1** 2 mol% [Pd] 6 mol% [dppf]	Catalyst **2** 2 mol% [Pd] 15 mol% [dppf]
Ph–C$_6$H$_4$–NH–C$_6$H$_4$–Me	85 ± 2	83 ± 1	99 ± 1	99 ± 1	99 ± 1
MeO–C$_6$H$_4$–NH–C$_6$H$_4$–Me	0	0	45 ± 2	60 ± 1	98 ± 2
Me–C$_6$H$_4$–NH–C$_6$H$_4$–Me	19 ± 3	26 ± 2	80 ± 1	96 ± 2	85 ± 2
OHC–C$_6$H$_4$–NH–C$_6$H$_4$–Me	17 ± 2	35 ± 3	17 ± 2	21 ± 4	45 ± 1
O$_2$N–C$_6$H$_4$–NH–C$_6$H$_4$–Me	62 ± 4	59 ± 1	81 ± 2	97 ± 1	87 ± 1
Ph–C$_6$H$_4$–NH–(m-tolyl)	15 ± 1	-	99 ± 1	86 ± 3	91 ± 1
Ph–C$_6$H$_4$–NH–(o-tolyl)	9 ± 2	-	82 ± 2	82 ± 2	95 ± 2

This higher loading helped to improve the conversion for more sterically hindered substrates, such as *ortho*-toluidine. However, *para*-bromobenzaldehyde proved a challenging substrate across all conditions employed, though a slightly improved conversion was obtained upon increasing the [Pd]:[dppf] ratio to 1:7.5 from 1:3. This change in conditions also benefitted the reaction with *para*-bromoanisole, resulting in a conversion to the desired product of 98%.

3. Materials and Methods

Synthesis and catalytic testing details are provided in the Supplementary Materials along with spectroscopic, crystallographic and analytical characterisation data. Tables are also included for the catalytic data presented in the plots in the main text.

4. Conclusions

Palladium catalysis continues to play a central role in metal-mediated transformations in synthetic chemistry, yet its widespread use is currently dependent on palladium sourced from mining, with all its attendant environmental and geopolitical issues. The low abundance of the metal and its limited geographical distribution encourages the greater recovery and re-use of the palladium-containing waste in our society, principally in the form of three-way catalysts (TWCs). This contribution draws on an effective and mild route to recover palladium in the form of the dimer $[N^nBu_4]_2[Pd_2I_6]$ (1) using inexpensive compounds, $[N^nBu_4]I$, I_2 and acetone.

In this work, compound **1** has been shown to be an effective oxidative C-H functionalisation catalyst, providing near-quantitative conversions to the desired product in the alkoxylation of benzo[*h*]quinoline and 8-methylquinoline, in most cases under significantly milder conditions, lower loading and shorter reaction times (50 °C, 1–2 mol% Pd loading, 2 h) than those reported in the literature [22]. The sustainability of the reaction was further enhanced by the demonstration that the palladium black formed after the reaction could be converted to $[N^nBu_4]_2[Pd_2I_6]$ (**1**) using $[N^nBu_4]I/I_2$ and re-used as a catalyst multiple times. The higher temperature typically used in previous reports (100 °C) was found to lead to the formation of Pd nanoparticles with both $Pd(OAc)_2$ (the literature catalyst) and **1**, which had a negative impact on the catalyst performance. The addition of dppf to compound **1** led to the formation of $PdI_2(dppf)$ (**2**) in high yield, allowing it to be used as a more sustainable alternative catalyst to the $PdCl_2(dppf)$ often used in Buchwald–Hartwig amination reactions. Optimisation of the conditions allowed a lower catalyst loading (1–2 mol% depending on the substrate) to be employed in a greener, bioderived solvent, cyclopentyl methyl ether (CPME). Catalyst **2** could also be generated in situ through the addition of dppf to $[N^nBu_4]_2[Pd_2I_6]$ (**1**), leading to similar conversions to the desired product.

In addition to sustainability improvements (milder conditions, greener solvents), this contribution illustrates the potential to replace palladium catalysts derived from mining with alternatives that can be obtained from waste under mild conditions. The use of **1** and **2** for two important palladium-catalysed transformations (C-H functionalisation and amination) bodes well for the same approach to be applied to other key Pd-mediated reactions, thus improving the sustainability of palladium catalysis.

Supplementary Materials: The following supporting information can be downloaded at: https://www.mdpi.com/article/10.3390/catal14050295/s1. Figure S1-1. Solid-state infrared spectrum of $[N^nBu_4]_2[Pd_2I_6]$ (**1**); Figure S1-2. Mass spectrum of $[N^nBu_4]_2[Pd_2I_6]$ (**1**); Figure S1-3. UV-Vis spectrum of $[N^nBu_4]_2[Pd_2I_6]$ (**1**) in MeCN; Figure S1-4; Solid-state infrared spectrum for $PdI_2(dppf)$ (**2**); Figure S1-5. ^{1}H NMR spectrum of $PdI_2(dppf)$ (**2**) in $CDCl_3$; Figure S1-6. $^{31}P\{^1H\}$ NMR spectrum of $PdI_2(dppf)$ (**2**) in $CDCl_3$; Figure S1-7. Mass spectrum of $PdI_2(dppf)$ (**2**); Figure S2-1. The crystal structure of **2** (50% probability ellipsoids); Figure S3-1. Reaction setup for catalytic reactions; Figure S3-2. ^{1}H NMR spectrum in $CDCl_3$ of the reaction mixture after the formation of 10-methoxybenzo[*h*]quinoline from benzo[*h*]quinoline showing the calculation of conversion using integration of the proton environments in the starting material and the product; Figure S3-3. ^{1}H NMR spectrum in $CDCl_3$ of isolated 10-methoxybenzo[*h*]quinoline; Figure S3-4. ^{1}H NMR spectrum in CDCl$_3$ of isolated 8-(methoxymethyl)quinoline; Figure S3-5. HPLC calibration curve of product against 1,3,5-trimethoxybenzene internal standard; Figure S3-6. HPLC calibration curve of *p*-bromobiphenyl against 1,3,5-trimethoxybenzene internal standard; Table S3-1. Definitive screening design parameters used in the Design of Experiments (DoE) model for the reaction of *p*-bromobiphenyl and p-toluidine at 100 °C in CPME solution; Table S3-2. Terms included in the definitive screening design model and their corresponding LogWorth values with larger values indicate a larger influence on the model. Terms with a LogWorth < 2 are not statistically significant and are not included; Table S4-1. Reaction profile for the conversion of benzo[*h*]quinoline to 10-methoxybenzo[*h*]quinoline with Pd(OAc)$_2$ (1.1 mol%) in the presence of 2 equivalents of PhI(OAc)$_2$ in MeOH at 50 °C and 100 °C; Table S4-2. Transformation of benzo[*h*]quinoline to 10-alkoxybenzo[*h*]quinoline catalysed by **1** in different solvents at 100 °C using 2 equivalents of PhI(OAc)$_2$; Table S4-3. Transformation of benzo[*h*]quinoline to 10-alkoxybenzo[*h*]quinoline catalysed by **1** (1–3 mol%) in different solvents at 50 °C using 2 equivalents of PhI(OAc)$_2$; Table S4-4. Transformation of benzo[h]quinoline to 10-alkoxybenzo[*h*]quinoline catalysed by PdI$_2$(dppf) **2** (1 or 2 mol%) in different solvents at 50 or 75 °C using 2 equivalents of PhI(OAc)$_2$; Table S4-5. Transformation of 8-methylquinoline to 8-(methoxymethyl)quinoline using **1** or **2** as catalysts (1 or 2 mol% [Pd]) in different solvents at 50 °C using 2 equivalents of PhI(OAc)$_2$; Table S5-1. Reaction conditions for the amination of *p*-bromobiphenyl with *p*-toluidine in CPME in the presence of ^{t}BuOK at 100 °C used to generate data for DoE analysis; Table S5-2. Reaction profile for the amination of *p*-bromobiphenyl with p-toluidine against time. Conversion to the desired product was determined by ^{1}H NMR spectroscopy. References [22,24,28,29,42–49] are cited in the Supplementary Materials.

Author Contributions: K.A.J., G.E. and S.M. carried out experimental work and performed characterisation. A.J.P.W. performed the crystallography. K.A.J. and J.D.E.T.W.-E. wrote the initial draft of the manuscript with guidance and input from D.C.B. and A.S. All authors have read and agreed to the published version of the manuscript.

Funding: This research was funded by the Ministry of Higher Education, Malaysia and Universiti Teknologi MARA, Malaysia, grant number KPT(BS)820710105709, the Centre for Doctoral training in Next Generation Synthesis and Reaction Technology (Imperial College London, EP/S023232/1) and the Public Scholarship, Development, Disability and Maintenance Fund of the Republic of Slovenia (11010-136/215).

Data Availability Statement: The original contributions presented in the study are included in the article/supplementary material, further inquiries can be directed to the corresponding author.

Acknowledgments: K.A.J. would like to thank the Ministry of Higher Education, Malaysia and Universiti Teknologi MARA, Malaysia, for a scholarship on the Post Doctoral Training Scheme. S.M. gratefully acknowledges the provision of a studentship from the Centre for Doctoral training in Next Generation Synthesis and Reaction Technology. G.E. thanks the Public Scholarship, Development, Disability and Maintenance Fund of the Republic of Slovenia for a scholarship. K.A.J. would like to thank Max King for contributions to the research on the amination reactions.

Conflicts of Interest: The authors declare no conflicts of interest.

References

1. Johansson Seechurn, C.C.; Kitching, M.O.; Colacot, T.J.; Snieckus, V. Palladium-catalysed cross-coupling: A historical contextual perspective to the 2010 Nobel Prize. *Angew. Chem. Int. Ed.* **2012**, *51*, 5062–5085. [CrossRef] [PubMed]
2. Gensch, T.; Hopkinson, M.N.; Glorius, F.; Wencel-Delord, J. Mild metal-catalysed C–H activation: Examples and concepts. *Chem. Soc. Rev.* **2016**, *45*, 2900–2936. [CrossRef] [PubMed]
3. Lyons, T.W.; Sanford, M.S. Palladium-catalyzed ligand-directed C-H functionalisation reactions. *Chem. Rev.* **2010**, *110*, 1147–1169. [CrossRef] [PubMed]
4. Muci, A.R.; Buchwald, S.L. Practical Palladium Catalysts for C-N and C-O Bond Formation. In *Cross-Coupling Reactions*; A Practical Guide (2002); Miyaura, N., Ed.; Springer: Berlin/Heidelberg, Germany, 2002; Volume 219, pp. 131–209.
5. Gazvoda, M.; Dhanjee, H.H.; Rodriguez, J.; Brown, J.S.; Farquhar, C.E.; Truex, N.L.; Loas, A.; Buchwald, S.L.; Pentelute, B.L. Palladium-mediated incorporation of carboranes into small molecules, peptides, and proteins. *J. Am. Chem. Soc.* **2022**, *144*, 7852–7860. [CrossRef] [PubMed]
6. Reichert, E.C.; Feng, K.; Sather, A.C.; Buchwald, S.L. Pd-Catalyzed Amination of Base-Sensitive Five-Membered Heteroaryl Halides with Aliphatic Amines. *J. Am. Chem. Soc.* **2023**, *145*, 3323–3329. [CrossRef] [PubMed]

7. ACS Green Chemistry Institute: Endangered Elements. 2020. Available online: https://www.acs.org/content/acs/en/greenchemistry/research-innovation/endangered-elements.html (accessed on 22 April 2024).
8. Johnson Matthey. PGM Market Report. 2019, pp. 1–48. Available online: http://www.platinum.matthey.com/services/market-research/pgm-market-reports (accessed on 22 April 2024).
9. Nuss, P.; Eckelman, M.J. Life cycle assessment of metals: A scientific synthesis. *PLoS ONE* **2014**, *9*, e101298. [CrossRef] [PubMed]
10. Michałek, T.; Hessel, V.; Wojnicki, M. Production, recycling and economy of palladium: A critical review. *Materials* **2023**, *17*, 45. [CrossRef]
11. Macklin, M.G.; Thomas, C.J.; Mudbhatkal, A.; Brewer, P.A.; Hudson-Edwards, K.A.; Lewin, J.; Scussolini, P.; Eilander, D.; Lechner, A.; Owen, J.; et al. Impacts of metal mining on river systems: A global assessment. *Science* **2023**, *381*, 1345–1350. [CrossRef]
12. Glaister, B.J.; Mudd, G.M. The environmental costs of platinum–PGM mining and sustainability: Is the glass half-full or half-empty? *Miner. Eng.* **2010**, *23*, 438–450. [CrossRef]
13. Yakoumis, I.; Panou, M.; Moschovi, A.M.; Panias, D. Recovery of platinum group metals from spent automotive catalysts: A review. *Cleaner Eng. Technol.* **2021**, *3*, 100112. [CrossRef]
14. Paiva, A.P.; Piedras, F.V.; Rodrigues, P.G.; Nogueira, C.A. Hydrometallurgical recovery of platinum-group metals from spent auto-catalysts—Focus on leaching and solvent extraction. *Sep. Purif. Technol.* **2022**, *286*, 120474. [CrossRef]
15. Wang, J.; Chen, H.; Hu, Z.; Yao, M.; Li, Y. A review on the Pd-based three-way catalyst. *Catal. Rev.* **2015**, *57*, 79–144. [CrossRef]
16. McCarthy, S.; Braddock, D.C.; Wilton-Ely, J.D.E.T. Strategies for sustainable palladium catalysis. *Coord. Chem. Rev.* **2021**, *442*, 213925. [CrossRef]
17. Hagelüken, B.C. Recycling the platinum group metals: A European perspective. *Platinum Met. Rev.* **2012**, *56*, 29–35. [CrossRef]
18. Dong, H.; Zhao, J.; Chen, J.; Wu, Y.; Li, B. Recovery of platinum group metals from spent catalysts: A review. *Int. J. Miner. Process.* **2015**, *145*, 108–113. [CrossRef]
19. Serpe, A.; Bigoli, F.; Cabras, M.C.; Fornasiero, P.; Graziani, M.; Mercuri, M.L.; Montini, T.; Pilia, L.; Trogu, E.F.; Deplano, P. Pd-dissolution through a mild and effective one-step reaction and its application for Pd-recovery from spent catalytic converters. *Chem. Commun.* **2005**, 1040–1042. [CrossRef] [PubMed]
20. Jantan, K.A.; Chan, K.W.; Melis, L.; White, A.J.P.; Marchiò, L.; Deplano, P.; Serpe, A.; Wilton-Ely, J.D.E.T. From recovered palladium to molecular and nanoscale catalysts. *ACS Sustain. Chem. Eng.* **2019**, *7*, 12389–12398. [CrossRef]
21. Jantan, K.A.; Kwok, C.Y.; Chan, K.W.; Marchiò, L.; White, A.J.P.; Deplano, P.; Serpe, A.; Wilton-Ely, J.D.E.T. From recovered metal waste to high-performance palladium catalysts. *Green Chem.* **2017**, *19*, 5846–5853. [CrossRef]
22. Dick, A.R.; Hull, K.L.; Sanford, M.S. A highly selective catalytic method for the oxidative functionalisation of C-H bonds. *J. Am. Chem. Soc.* **2004**, *126*, 2300–2301. [CrossRef]
23. McCarthy, S.; Desaunay, O.; Jie, A.L.; Hassatzky, M.; White, A.J.P.; Deplano, P.; Braddock, D.C.; Serpe, A.; Wilton-Ely, J.D.E.T. Homogeneous gold catalysis using complexes recovered from waste electronic equipment. *ACS Sustain. Chem. Eng.* **2022**, *10*, 15726–15734. [CrossRef]
24. Cuscusa, M.; Rigoldi, A.; Artizzu, F.; Cammi, R.; Fornasiero, P.; Deplano, P.; Marchiò, L.; Serpe, A. Ionic couple-driven palladium leaching by organic triiodide solutions. *ACS Sustain. Chem. Eng.* **2017**, *5*, 4359–4370. [CrossRef]
25. McCarthy, S.; Braddock, D.C.; Wilton-Ely, J.D.E.T. From waste to green applications: The use of recovered gold and palladium in catalysis. *Molecules* **2021**, *26*, 5217. [CrossRef] [PubMed]
26. Wilton-Ely, J.D.E.T. The use of recovered metal complexes in catalysis. *Johns. Matthey Technol. Rev.* **2023**, *67*, 300–302.
27. Wolfe, J.P.; Wagaw, S.; Buchwald, S.L. An improved catalyst system for aromatic carbon−nitrogen bond formation: The possible involvement of bis(phosphine) palladium complexes as key intermediates. *J. Am. Chem. Soc.* **1996**, *118*, 7215–7216. [CrossRef]
28. Driver, M.S.; Hartwig, J.F. A second-generation catalyst for aryl halide amination: Mixed secondary amines from aryl halides and primary amines catalysed by (DPPF)PdCl$_2$. *J. Am. Chem. Soc.* **1996**, *118*, 7217–7218. [CrossRef]
29. Colacot, T.J.; Qian, H.; Cea-Olivares, R.; Hernandez-Ortega, S. Synthesis, X-ray, spectroscopic and a preliminary Suzuki coupling screening studies of a complete series of dppfMX$_2$ (M = Pt, Pd; X = Cl, Br, I). *J. Organomet. Chem.* **2001**, *637*, 691–697. [CrossRef]
30. Topczewski, J.J.; Sanford, M.S. Carbon–hydrogen (C-H) bond activation at Pd IV: A Frontier in C-H functionalisation catalysis. *Chem. Sci.* **2015**, *6*, 70–76. [CrossRef] [PubMed]
31. Powers, D.C.; Ritter, T. Bimetallic Pd(III) complexes in palladium-catalysed carbon–heteroatom bond formation. *Nat. Chem.* **2009**, *1*, 302–309. [CrossRef] [PubMed]
32. Reetz, M.T.; Westermann, E. Phosphane-free palladium-catalysed coupling reactions: The decisive role of Pd nanoparticles. *Angew. Chem. Int. Ed.* **2000**, *39*, 165–168. [CrossRef]
33. Baumann, C.G.; De Ornellas, S.; Reeds, J.P.; Storr, T.E.; Williams, T.J.; Fairlamb, I.J.S. Formation and propagation of well-defined Pd nanoparticles (PdNPs) during C-H bond functionalisation of heteroarenes: Are nanoparticles a moribund form of Pd or an active catalytic species? *Tetrahedron* **2014**, *70*, 6174–6187. [CrossRef]
34. Reay, A.J.; Fairlamb, I.J.S. Catalytic C–H bond functionalisation chemistry: The case for quasi-heterogeneous catalysis. *Chem. Commun.* **2015**, *51*, 16289–16307. [CrossRef] [PubMed]
35. Guram, A.S.; Rennels, R.A.; Buchwald, S.L. A simple catalytic method for the conversion of aryl bromides to arylamines. *Angew. Chem. Int. Ed.* **1995**, *34*, 1348–1350. [CrossRef]
36. Wagaw, S.; Rennels, R.A.; Buchwald, S.L. Palladium-catalyzed coupling of optically active amines with aryl bromides. *J. Am. Chem. Soc.* **1997**, *119*, 8451–8458. [CrossRef]

37. Ali, M.H.; Buchwald, S.L. An improved method for the palladium-catalysed amination of aryl iodides. *J. Org. Chem.* **2001**, *66*, 2560–2565. [CrossRef] [PubMed]
38. Campeau, L.C.; Parisien, M.; Jean, A.; Fagnou, K. Catalytic direct arylation with aryl chlorides, bromides, and iodides: Intramolecular studies leading to new intermolecular reactions. *J. Am. Chem. Soc.* **2006**, *128*, 581–590. [CrossRef]
39. Forero-Cortés, P.A.; Haydl, A.M. The 25th anniversary of the Buchwald–Hartwig amination: Development, applications, and outlook. *Org. Process Res. Dev.* **2019**, *23*, 1478–1483. [CrossRef]
40. Azzena, U.; Carraro, M.; Pisano, L.; Monticelli, S.; Bartolotta, R.; Pace, V. Cyclopentyl methyl ether: An elective ecofriendly ethereal solvent in classical and modern organic chemistry. *ChemSusChem* **2019**, *12*, 40–70. [CrossRef] [PubMed]
41. Prat, D.; Wells, A.; Hayler, J.; Sneddon, H.; McElroy, C.R.; Abou-Shehada, S.; Dunn, P.J. CHEM21 selection guide of classical-and less classical-solvents. *Green Chem.* **2016**, *18*, 288–296. [CrossRef]
42. Dolomanov, O.V.; Bourhis, L.J.; Gildea, R.J.; Howard, J.A.K.; Puschmann, H. OLEX2: A Complete Structure Solution, Refinement and Analysis Program. *J. Appl. Cryst.* **2009**, *42*, 339–341. [CrossRef]
43. *SHELXTL v5.1*; Bruker AXS: Madison, WI, USA, 1998.
44. Sheldrick, G.M. Crystal structure refinement with SHELXL. *Acta Cryst.* **2015**, *C71*, 3–8.
45. Spek, A.L. (2003, 2009) PLATON, A Multipurpose Crystallographic Tool. *Acta. Cryst.* **2015**, *C71*, 9–18.
46. Old, D.W.; Wolfe, J.P.; Buchwald, S.L. A Highly Active Catalyst for Palladium-Catalyzed Cross-Coupling Reactions: Room-Temperature Suzuki Couplings and Amination of Unactivated Aryl Chlorides. *J. Am. Chem. Soc.* **1998**, *120*, 9722–9723. [CrossRef]
47. Wolfe, J.P.; Tomori, H.; Sadighi, J.P.; Yin, J.; Buchwald, S.J. Simple, Efficient Catalyst System for the Palladium-Catalyzed Amination of Aryl Chlorides, Bromides, and Triflates. *J. Org. Chem.* **2000**, *65*, 1158–1174. [CrossRef]
48. Semeniuchenko, V.; Braje, W.M.; Organ, M.G. Sodium Butylated Hydroxytoluene: A Functional Group Tolerant, Eco-Friendly Base for Solvent-Free, Pd-Catalysed Amination. *Chem. Eur. J.* **2021**, *27*, 12535–12539. [CrossRef] [PubMed]
49. Hajra, A.; Wei, Y.; Yoshikai, N. Palladium-Catalyzed Aerobic Dehydrogenative Aromatization of Cyclohexanone Imines to Arylamines. *Org. Lett.* **2012**, *14*, 5488–5491. [CrossRef] [PubMed]

Disclaimer/Publisher's Note: The statements, opinions and data contained in all publications are solely those of the individual author(s) and contributor(s) and not of MDPI and/or the editor(s). MDPI and/or the editor(s) disclaim responsibility for any injury to people or property resulting from any ideas, methods, instructions or products referred to in the content.

Article

Valorization of *Chlorella* Microalgae Residual Biomass via Catalytic Acid Hydrolysis/Dehydration and Hydrogenolysis/Hydrogenation

Antigoni G. Margellou [1], Stylianos A. Torofias [1], Georgios Iakovou [1] and Konstantinos S. Triantafyllidis [1,2,*]

[1] Department of Chemistry, Aristotle University of Thessaloniki, 54124 Thessaloniki, Greece; amargel@chem.auth.gr (A.G.M.); torofias@gmail.com (S.A.T.)
[2] Centre for Research and Technology Hellas, Chemical Process and Energy Resources Institute, 57001 Thessaloniki, Greece
* Correspondence: ktrianta@chem.auth.gr

Abstract: Microalgal biomass can be utilized for the production of value-added chemicals and fuels. Within this research, *Chlorella vulgaris* biomass left behind after the extraction of lipids and proteins was converted to valuable sugars, organic acids and furanic compounds via hydrolysis/dehydration using dilute aqueous sulfuric acid as a homogeneous catalyst. Under mild conditions, i.e., low temperature and low sulfuric acid concentration, the main products of hydrolysis/dehydration were monomeric sugars (glucose and xylose) and furanic compounds (HMF, furfural) while under more intense conditions (i.e., higher temperature and higher acid concentration), organic acids (propionic, formic, acetic, succinic, lactic, levulinic) were also produced either directly from sugar conversion or via intermediate furans. As a second valorization approach, the residual microalgal biomass was converted to value-added sugar alcohols (sorbitol, glycerol) via hydrogenation/hydrogenolysis reactions over metallic ruthenium catalysts supported on activated carbons (5%Ru/C). It was also shown that a low concentration of sulfuric acid facilitated the conversion of biomass to sugar alcohols by initiating the hydrolysis of carbohydrates to monomeric sugars. Overall, this work aims to propose valorization pathways for a rarely utilized residual biomass towards useful compounds utilized as platform chemicals and precursors for the production of a wide variety of solvents, polymers, fuels, food ingredients, pharmaceuticals and others.

Keywords: *Chlorella* microalgae; hydrolysis; hydrogenolysis; sugars; furans; sugar alcohols

Citation: Margellou, A.G.; Torofias, S.A.; Iakovou, G.; Triantafyllidis, K.S. Valorization of *Chlorella* Microalgae Residual Biomass via Catalytic Acid Hydrolysis/Dehydration and Hydrogenolysis/Hydrogenation. *Catalysts* **2024**, *14*, 286. https://doi.org/10.3390/catal14050286

Academic Editor: Carl Redshaw

Received: 13 March 2024
Revised: 21 April 2024
Accepted: 22 April 2024
Published: 23 April 2024

Copyright: © 2024 by the authors. Licensee MDPI, Basel, Switzerland. This article is an open access article distributed under the terms and conditions of the Creative Commons Attribution (CC BY) license (https://creativecommons.org/licenses/by/4.0/).

1. Introduction

The projected depletion of fossil fuels has spurred the development of emerging technologies for the conversion of renewable energy sources to value-added chemicals and biofuels. Lignocellulosic and microalgae biomass are recognized as the most promising renewable feedstocks for the production not only biofuels but also a wide variety of platform chemicals. Microalgae biomass has gained significant attention due to its fast growth rate and minimum growth demands without the need for chemicals and energy. Furthermore, their composition, enriched in lipids, proteins, carbohydrates and pigments, provides the potential for valorization towards many chemicals.

Lipids, accounting almost the half of microalgae weight, can be extracted via solvent extraction techniques supported by ultrasonication/microwaves and ionic liquids or solvent free techniques and supercritical carbon dioxide methods [1–3]. Regarding their composition and amount, the microalgae strains and the cultural conditions can determine both the fatty acid profile and the lipid content [4,5]. In general, high carbon-to-nitrogen ratios in cultivation or stress conditions such as nitrogen starvation, high salt concentration, high temperature and high pH favors lipid formation [4,6]. Microalgae lipids are composed of fatty acids with carbon numbers in the range C_{14}–C_{20} and polyunsaturated fatty acids with carbon numbers higher than C_{20} [5]. The former group of fatty acids is mainly utilized for third-generation biofuel production, as biodiesel or paraffinic hydrocarbons [7,8], while the latter group of polyunsaturated fatty acids, more specifically, eicosapentanenoic (EPA) and docosahexanoic acid (DHA), are used as health and nutritional supplements [9,10]. Taking into consideration that the human body cannot synthesize EPA and DHA, but both are essential fatty acids, oleaginous microalgae could be a good source [10,11].

With regard to sugar composition, the main sugars are glucose, identified in high concentrations, while rhamnose, fucose, ribose, xylose, arabinose, mannose and galactose were identified in lower amounts [12]. As in the case of lipids, carbohydrate content and composition exhibit differences between the microalgae species and can be controlled by finetuning the cultivation conditions [13]. In order to increase the carbohydrate content, two stage cultivation has been proposed, comprising a first stage where all the main nutrients are supplied to increase the production of biomass, and a second stage where specific nutrients are supplied to increase the formation of carbohydrates [14]. Microalgal carbohydrates are recognized as promising feedstocks for the production of biofuels and value-added chemicals [15]. After proper treatment, microalgal oligosaccharides can be utilized as a source of prebiotics or converted to bioethanol/biobutanol via fermentation and to biogas via anaerobic digestion [15–18]. Furthermore, microalgae-derived polysaccharides are widely used in the production of packaging materials [19,20], as animal feed providing antibiotic and antibacterial properties [13], and as promoters of plant growth and nutrient uptake [21]. Especially, delipidified microalgal biomass has been utilized in the synthesis of biopolymers, biomethane, biohydrogen and bioethanol [22].

Another major component of microalgae is proteins. Interestingly, some speciescan contain up to 80 wt.% proteins [23]. The proteins after extraction, separation and purification can be consumed by humans as alternative protein sources with valuable effects on human health [24–26]. Other value-added compounds which can be isolated from microalgae are pigments, such as chlorophylls, carotenoids and phycobilins [9]. Carotenoids are responsible for the yellow-red color of biomass and can be categorized in two main groups: xanthophylls and carotenes. Carotenes are linear hydrocarbons with 40 carbon atoms, and the most common compound is β-carotene, while xanthophylls are oxygenated derivatives of them [9]. Among the microalgae with the highest carotene production is *Dunaliella* which belongs to the Chlorophyta species and can be cultivated in highly saline environments [27]. The isolated β-carotene is used in food and pharmaceutical applications, and the foreseen global market is expected to be USD 380 billion by 2028 [28].

In addition to the valorization processes based on selective extraction of the various microalgae fractions, established thermochemical processes, such as pyrolysis, gasification and liquefaction, have also been studied towards the production of bio-oil/biocrude for further upgrading towards biofuels or other bio-based products. Pyrolysis is carried out at relatively high temperatures (400–550 °C) in an inert atmosphere towards bio-oil, char and gases, with typical yields being 50 wt.% bio-oil, 20 wt.% char and 30 wt.% gases [29,30]. With regard to the composition of bio-oil, complex mixtures are obtained that are rich in nitrogen-containing compounds, such as pyrroles, amines, amides and indoles, produced via proteins, carboxylic acids, phenolic compounds, deoxygenated aliphatics and aromatics [30,31]. Removal of nitrogen compounds can be achieved via downstream denitrogenation processes while removal of oxygen can be achieved via in-situ or ex-situ hydrodeoxygenation [31,32]. Heterogeneous catalysts with acid sites induce partial deoxygenation towards monoaromatics (benzene, toluene, etc.) [30,33]. Hydrothermal liquefaction is performed at lower temperatures (200–400 °C), higher residence times and higher pressures, using solvents and catalysts [34,35]. The main products are biocrude, gases and solids.

Within a biorefinery concept, the whole biomass needs to be converted to value-added chemicals and biofuels. Microalgae biorefining is usually based on the primary extraction of lipids and the valorization of delipidified biomass via pyrolysis, resulting, however, in slightly lower yields of bio-oils that do not containing lipid-derived compounds [33,36].

After lipid extraction and carbohydrate removal via saccharification, the residual biomass can yield bio-oil rich in phenolic and nitrogen-containing compounds [37]. Alternatively, after lipid extraction, proteins can be isolated, leaving carbohydrate- and pigment-enriched biomass [38,39]. Delipidified biomass can also be hydrolyzed towards the production of monomeric sugars, with potential for the production of biofuels. Usually, the hydrolysis of the carbohydrates is carried out using inorganic acids, such as sulfuric and hydrochloric acid, at 25–200 °C for between 5 min and 24 h, and the liquid product is rich in monomeric sugars, mainly glucose and xylose [40–42].

A very novel area of microalgal biomass valorization is the production of bio-based plastics. Microalgae-derived plastics can be produced via the extraction of lipids and carbohydrates and their downstream conversion to polymers. Alternatively, polyhydroxyalkanoates can be synthesized in microalgae cells under specified cultivation conditions [43,44].

The aim of this work was the valorization of *Chlorella* microalgae residual biomass via catalytic acid hydrolysis/dehydration and hydrogenolysis/hydrogenation. *Chlorella vulgaris* was subjected to solvent lipid extraction followed by protein removal and the remaining carbohydrate-enriched biomass was converted into sugars, organic acids and furans via hydrolysis/dehydration. Alternatively, hydrogenation/hydrogenolysis using heterogeneous catalysts was applied in-situ to convert sugars to sugar alcohols.

2. Results

2.1. Characterization of Chlorella vulgaris Strains

The composition of the *Chlorella vulgaris* biomass used in this study is shown in Table 1. The high nitrogen concentration in the cultivation of the LL (low lipid) sample led to higher protein (28.1 wt.%) and carbohydrate content (33.6 wt.%) compared to the ML (medium lipid) sample (see experimental section). The commercially available biomass (MF) exhibited the highest protein and lipid content (46.9 wt.% and 30.2 wt.%, respectively). The sample cultivated at pilot scale (AF) exhibited lower protein and lipid content (22.4 wt.% and 21.3 wt.%, respectively). On the other hand, this sample exhibited the lowest carbohydrate content (14.5 wt.%). The detailed analysis of the carbohydrate monomers is shown in Table S1. All the biomass samples contained mainly glucose and xylose, while the concentration of both sugars gradually increased after the extraction of lipids and proteins. It can be noted that the two lab-scale cultivated samples (LL and ML) exhibited significantly higher glucose content (16.8–26.8 wt.%) compared to the commercially available biomass (MF) and the pilot-scale cultivation (AF), which had glucose content of 8.9 and 5.7 wt.%, respectively. Furthermore, based on elemental analysis, all biomass samples exhibited high carbon (43.8–47.6 wt.%), hydrogen (6.6–7.8 wt.%) and nitrogen (6.1–11.7 wt.%) content and were almost sulfur-free. An exception to this trend is the AF samples, which showed the lowest carbon concentration (21.3 wt.%).

Table 1. Biochemical composition and elemental analysis of microalgae feedstocks.

Samples	Composition [a] (wt.%)				Elemental Analysis [a] (wt.%)				
	Lipids	Proteins	Carbohydrates	Ash	C	H	N	S	O
LL	15.0	28.1	33.6	19.3	47.6	6.7	11.7	0.0	14.7
ML	22.6	25.7	25.1	17.6	46.7	6.6	6.1	0.1	23.0
MF	30.2	46.9	14.8	6.2	43.8	7.8	9.8	0.0	32.4
AF	21.3	22.4	14.5	21.4	21.3	6.1	4.9	0.0	46.3

[a] Expressed as wt.% on dry biomass.

2.2. Catalytic Acid Hydrolysis/Dehydration

The hydrothermal hydrolysis of biomass was performed in aqueous solutions with sulfuric acid as an acidic homogeneous catalyst. Preliminary experiments were performed without a catalyst to study the effects of reaction temperature and feedstock composition. The hydrothermal treatment of the LL-Res sample (i.e., the remaining biomass after extraction of lipids from microalgae sample LL; see experimental section) at 175 °C for 15 min led to 63.7 wt.% solubilization while the extraction of proteins (sample LL-Re-P) enhanced the solubilization to 71.9 wt.%, as shown in Figure S1. The *Chlorella* strain with medium lipid content after lipid extraction (ML-Res sample) exhibited significantly higher solubilization (79.1 wt.%) compared to the strain of low lipid (LL) extracted biomass. An increase in the reaction temperature from 175 °C to 250 °C led to a slight increase in solubilization to 81.7 wt.%. The addition of dilute sulfuric acid (0.25% w/v) at 175 °C did not affect the solubilization (79.7 wt.%), as was shown in the case of ML-Res.

Regarding the composition of the hydrolysates, the main compounds identified via the HPLC analysis appertain to three main categories: sugars, organic acids and furanic compounds. The solubilization of the LL-Res biomass in neat water yielded a low concentration of monomeric sugars (5 mg/g) due to the partial depolymerization of carbohydrates, while the monomeric sugars were directly degraded to organic acids (Figure 1a). At 175 °C-15 min, LL-Res saw the formation of cellobiose (1.3 mg/g) and glucose (3.8 mg/g). The main organic acids formed were lactic acid (27.9 mg/g), acetic acid (4.6 mg/g) and propionic acid (7.9 mg/g), as can be observed in Figure 1b. A part of the formed organic acids, especially lactic acid and propionic acid, may also be attributed to protein conversion. More specifically, proteins can be converted to lactic acid via deamination, which is further dehydrated towards propionic acid [45]. Based on the analysis of the hydrolysate obtained via hydrothermal treatment at 190 °C-15 min (in neat water) of LL-Res-P (biomass derived after the extraction of lipids and proteins from the LL sample), it can be seen that the extraction of proteins led to a higher concentration of propionic acid (27.2 mg/g) but a significantly lower concentration of lactic acid (2.4 mg/g), proving that part of the lactic acid is derived from protein conversion but propionic acid is derived from carbohydrate conversion. Both LL-Res and LL-Res-P yielded low levels of furanic compounds.

The treatment of biomass remaining after the extraction of lipids from a microalgae sample with increased lipids (i.e., ML-Res) in neat water 175 °C-15 min) yielded a higher sugar concentration (15.5 mg/g), lower organic acids (29.9 mg/g) and an almost similar concentration of furans (0.9 mg/g). The high sugar content is mainly attributed to glucose, whose concentration is equal to 11.7 mg/g, and to a lesser extent to galactose, arabinose and mannose, whose concentrations add up to 3.9 mg/g. Also, more organic acids were identified, such as succinic acid (6.1 mg/g), formic acid (8.5 mg/g) and propionic acid (15.2 mg/g) compared to those from LL-Res. Both succinic acid and formic acid are formed as intermediate products during the dehydration of sugars into furans. Raising the reaction temperature from 175 °C to 250 °C induced more severe reaction conditions, which resulted in a slight increase in biomass solubilization and to a completely different composition of the hydrolysates. Under the more severe conditions, fewer sugars were obtained (8.4 mg/g) due to their conversion into organic acids (144 mg/g) and furanic compounds (2.6 mg/g), as can be observed in Figure 1a. Furthermore, the higher temperature yielded a lower glucose content (6.3 mg/g) but enhanced depolymerization of xylooligosaccharides towards xylose monomers (2.2 mg/g). In addition, the more severe conditions increased the concentration of formic acid to 22.4 mg/g and of propionic acid to 89.3 mg/g and further enhanced the formation of acetic acid (15.3 mg/g), lactic acid (11.3 mg/g) and levulinic acid (2.6 mg/g), either via direct conversion of sugars or via rehydration of HMF (i.e., in the case of formic and levulinic acids). The addition of sulfuric acid as a homogeneous acid catalyst in the hydrothermal treatment (175 °C-15 min, 0.25% w/v) of ML-Res, induced increased depolymerization of oligosaccharides to monomeric sugars (32.5 mg/g). i.e., containing glucose (26.3 mg/g) and xylose (6.3 mg/g). The acidic conditions facilitated the dehydration of sugars to furanic compounds (29.1 mg/g) and their further conversion to organic acids (98.5 mg/g). The dehydration of glucose led to the formation of HMF (21.3 mg/g), while the dehydration of xylose yielded furfural (7.9 mg/g), along with the subsequent formation of succinic acid (8.7 mg/g), lactic acid (7 mg/g), formic acid (23.1 mg/g), acetic acid (14.1 mg/g) and propionic acid (45.6 mg/g).

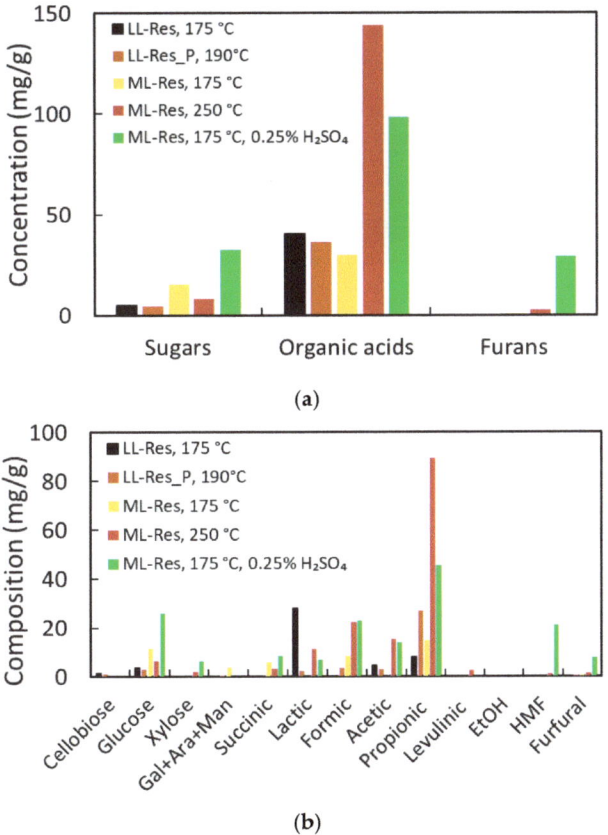

Figure 1. (a) Categories of the compounds and (b) composition of the hydrolysates of LL-Res and ML-Res *Chlorella vulgaris* hydrolysates.

A more detailed investigation of the effects of the process parameters (reaction temperature, time and acid catalyst concentration) was performed using the MF-Res biomass. Under the milder reaction conditions (190 °C, 15 min, 0.25% w/v H_2SO_4), the solubilization of the biomass was 84.0 wt.%, as shown in Figure S2. An increase in the reaction temperature from 190 °C to 230 °C enhanced the solubilization (90.0 wt.%), but a further increase to 250 °C led to slightly lower solubilization (87.0 wt.%), probably due to the formation of humins. Humins in the recovered solids were indirectly determined via TGA analysis. As can be observed in Table S2, the recovered solids exhibited two distinct weight-loss steps. The first step was in the range of 25–120 °C and corresponds to the evaporation of the remaining water, while the second, most dominant step was in the range 120–550 °C, corresponding to the decomposition of the remaining biomass. Above this temperature, the residual mass stabilized. An increase in the hydrolysis reaction temperature from 190 to 250 °C led to an increase in the residual mass from 41.6% to 50.6%, probably due to the formation of humins, which are more stable and resilient when heated in an inert atmosphere (TGA) and convert further to char. Unlike the reaction temperature, the reaction time did not significantly influence the solubilization of biomass. A prolonged reaction time (60 min) maintained the solubilization at 84.0 wt.%, equal to a shorter reaction time (15 min). The most profound impact was that of the concentration of sulfuric acid. A higher concentration (0.5% w/v) improved the solubilization from 84.0 wt.% to 93.0 wt.%. The sequential extraction of proteins (MF-Res-P) led to a partial decrease in the solubilization

(81.4 wt.% of the respective biomass sample). All the hydrolysates derived via the hydrolysis/dehydration of the MF-Res biomass are shown in Figure 2. All samples exhibited a light brown color, and changes in the severity of the treatment conditions did not influence the color of the product.

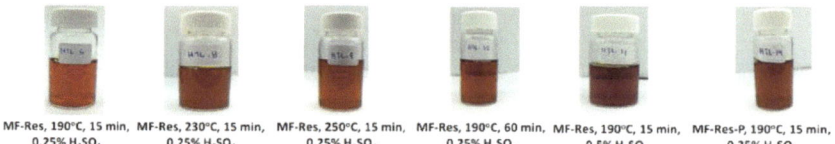

Figure 2. Hydrolysates obtained via the hydrolysis/dehydration of MF-Res *Chlorella vulgaris* under different reaction conditions.

Despite the similar levels of solubilization, reaction conditions did tailor the composition of the hydrolysates, as can be observed in Figure 3a,b. The least severe conditions (190 °C, 15 min, 0.25% w/v H_2SO_4) yielded the highest concentration of sugars (32.3 mg/g) and the lowest concentration of organic acids (91.4 mg/g). Surprisingly, these conditions also enhanced the dehydration of sugars towards furanic compounds, whose concentration was 27.8 mg/g. An increase in the reaction temperature from 190 °C to 230 °C yielded fewer sugars (23.0 mg/g), more organic acids (106.0 mg/g) and significantly lower furanic compounds (3.0 mg/g). A further increase in the reaction temperature to 250 °C proved to enhance further the formation of organic acids. This hydrolysate exhibited the lowest concentration of sugars (15.9 mg/g) and furanic compounds (3.1 mg/g), while organic acids exhibited the highest concentration (121 mg/g). To give more details regarding the content of individual compounds in the hydrolysates, it was observed that under the milder conditions, glucose was the main sugar formed (22.4 mg/g), with xylose and arabinose/galactose/mannose exhibiting lower concentrations (5.5 and 4.4 mg/g, respectively). The most abundant organic acid was propionic acid (46.8 mg/g), while succinic acid (9.2 mg/g), lactic acid (8.4 mg/g), formic acid (12.8 mg/g) and acetic acid (14.2 mg/g) were at substantially lower concentrations. The dehydration of glucose led to the formation of HMF (23.6 mg/g), while the dehydration of C_5 sugars led to the formation of furfural (4.3 mg/g). An increase in the reaction temperature from 190 °C to 230 °C yielded a lower glucose concentration (17.3 mg/g) as well as lower concentrations of xylose (3.5 mg/g) and other sugars (2.3 mg/g). A higher reaction temperature enhanced the formation of propionic acid, whose concentration increased to 80.5 mg/g, instead of the formation of other organic acids or furanic compounds. The highest reaction temperature (250 °C) induced further conversion of sugars towards organic acids. Glucose and xylose content were 12.3 mg/g and 3.5 mg/g, respectively, while none of the other sugars (arabinose, galactose and mannose) was identified. An increase in the reaction time from 15 to 60 min had a less profound effect on the hydrolysis/dehydration of sugars. The total sugar concentration decreased from 32.3 mg/g to 22.7 mg/g, mainly attributed to glucose, whose concentration was 18.6 mg/g, and xylose, with concentration 4.1 mg/g. The most profound effect was on the formation of organic acids, whose concentration increased from 91.4 mg/g to 121 mg/g. The main organic acid formed was propionic acid (75.4 mg/g). The acetic acid concentration also increased from 14.2 mg/g to 16.6 mg/g, while all the other acids were formed at lower concentrations. The concentration of the furanic compounds was low (2.0 mg/g), due to their subsequent conversion to organic acids as well as their condensation to humins. Characterization of the recovered solids via TGA showed that an increase in the hydrolysis reaction time from 15 to 60 min led to an increase in the residual mass from 41.6% to 50.7%, which may also be considered as indication of humin formation, as discussed above for the effect of higher hydrolysis temperature. Generally, harsher conditions in terms of reaction temperature and time are considered to decrease the carbohydrate yield in favor of furans and/or acids, while mild conditions can enhance the sugar yield [46].

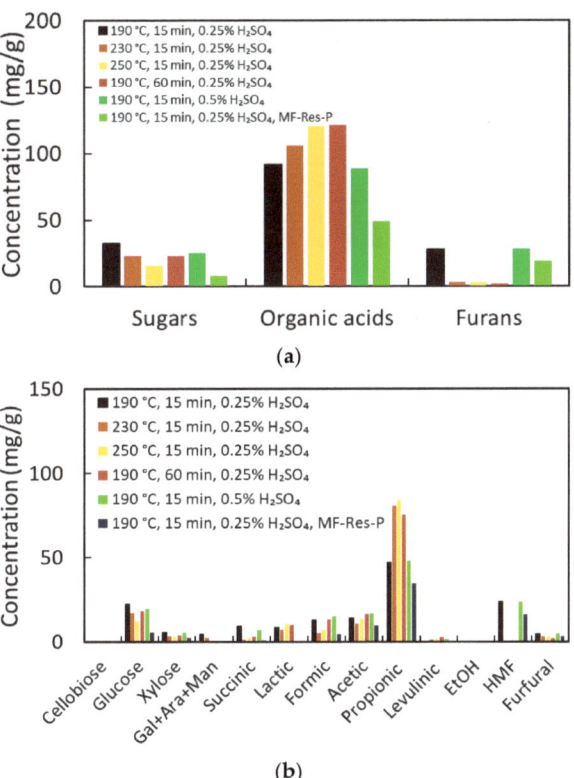

Figure 3. (**a**) Categories of the compounds and (**b**) composition of the MF-Res *Chlorella vulgaris* hydrolysates.

Regarding the effect of the acidic catalyst, a higher acid concentration, i.e., from 0.25 to 0.5% w/v H_2SO_4, did not significantly change the products' distribution (Figure 3a,b). More specifically, the higher acid concentration yielded slightly lower total sugars, 25.1 mg/g, compared to 32.3 mg/g formed when using 0.25% w/v H_2SO_4. Both glucose and xylose were formed at lower concentrations, 19.6 and 5.5 mg/g, respectively, and none of the other sugars was identified. Additionally, the total concentration of organic acids decreased from 91.4 mg/g to 88.9 mg/g. The higher acid concentration enhanced the formation of formic and acetic acids, whose content increased from 12.8 mg/g to 15.4 mg/g and from 14.2 mg/g to 16.7 mg/g, respectively. The succinic acid concentration decreased to 7.2 mg/g, while the propionic acid concentration remained almost the same at 48.0 mg/g. Furfural and HMF were measured at 4.9 and 23.7 mg/g, respectively. The lower formation of sugars and the higher concentration of acids and furanic compounds when using higher concentrations of sulfuric acid is in accordance with similar results from the literature for other microalgae [47].

A similar study was performed using the AF microalgae derived via pilot-scale cultivation. Regarding the solubilization of microalgae residual biomass (Figure S3), the hydrolysis of AF-Res (biomass obtained after lipid extraction from AF) at 190 °C, 15 min, 0.25% w/v H_2SO_4 yielded 91.7 wt.% solubilization, slightly higher than the 84 wt.% obtained during the hydrolysis of lab-scale cultivated samples (LL, ML) and the commercially (MF) available microalgae. Treatment of the AF-Res-P (lipid- and protein-extracted biomass) led to lower solubilization (83.4 wt.%) under the same hydrolysis conditions. An increase in the reaction time from 15 to 60 min led to slightly higher solubilization 85.7 wt.%, while higher sulfuric acid concentration further improved the solubilization to 90.2 wt.%. The obtained hydrolysates are shown in Figure 4. All hydrolysates exhibited a brown color, while increase of reaction time or of the sulfuric acid concentration turned the color to dark brown, possibly due to the formation of humins.

   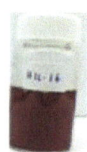

AF-Res, 190°C, 15 min, 0.25% H$_2$SO$_4$ AF-Res-P, 190°C, 15 min, 0.25% H$_2$SO$_4$ AF-Res-P, 190°C, 60 min, 0.25% H$_2$SO$_4$ AF-Res-P, 190°C, 15 min, 0.5% H$_2$SO$_4$

Figure 4. Hydrolysates obtained via the hydrolysis/dehydration of AF-Res and AF-Res-P *Chlorella vulgaris* under different reaction conditions.

The composition of the hydrolysates is shown in Figure 5a,b. The pilot-scale cultivation of *Chlorella* microalgae followed by lipid extraction (sample AF-Res) and hydrolysis at 190 °C, 15 min, 0.25% w/v H$_2$SO$_4$ yielded 27.4 mg/g sugars, 74.3 mg/g organic acids and 26.9 mg/g furanic compounds, which were slightly lower compared to the composition of the MF-Res hydrolysate under the same treatment conditions. Glucose (21.9 mg/g) and xylose (5.5 mg/g) were the only sugars formed. As can be observed in Figure 5b, propionic acid was the main organic acid formed (35.7 mg/g). Formic acid (17.8 mg/g), acetic acid (11.9 mg/g) and succinic acid (8.8 mg/g) were formed at lower concentrations, while appreciable amounts of furfural and HMF were also determined (ca. 7.6 mg/g and 19.4 mg/g, respectively). The subsequent extraction of proteins and the hydrothermal treatment of the respective sample (AF-Res-P) under the same conditions yielded a slightly lower concentration of sugars (18.7 mg/g) and organic acids (69.4 mg/g) but facilitated the in-situ dehydration of sugars to furanic compounds, whose concentrations amounted to 35.1 mg/g (HMF and furfural concentrations were 25.2 and 9.9 mg/g, respectively). Furthermore, the extraction of proteins did not affect the propionic acid and succinic acid concentrations but slightly reduced the concentrations of formic acid and acetic acid to 14.6 and 8.9 mg/g, respectively.

A higher residence time (60 min) at 190 °C for the AF-Res-P treatment enhanced the dehydration of sugars to furanic compounds as well as the conversion to organic acids. The concentration of sugars was found to be 13.4 mg/g, organic acids 85.9 mg/g and furanic compounds 39.5 mg/g. The prolonged reaction time yielded lower concentrations of glucose (9.4 mg/g) and xylose (6.7 mg/g). On the other hand, formic acid concentration increased from 14.6 to 17.4 mg/g, and acetic and propionic acid concentrations increased from 8.9 to 10.2 mg/g and from 37.0 to 55.5 mg/g, respectively. The main difference between the two reaction times is that levulinic acid (2.8 mg/g) was formed at 60 min, while succinic was not detected at all. Additionally, ethanol was produced at 60 min with a concentration of 35.7 mg/g.

On the other hand, higher sulfuric acid concentration facilitated the depolymerization of carbohydrates towards monomeric sugars, whose concentration increased to 23.5 mg/g (from 18.7 mg/g). Both glucose and xylose concentrations increased to 14.2 and 9.3 mg/g, respectively. Furthermore, furanic compounds increased to 42.0 mg/g while organic acid concentration decreased to 58.9 mg/g. Based on the above results, the higher sulfuric acid concentration enhanced the in-situ dehydration of sugars to furanic compounds without subsequent conversion to organic acids. Indeed, formic, acetic and succinic acid concentrations decreased to 6.7, 7.3 and 2.3 mg/g, respectively. On the other hand, HMF concentration increased from 25.2 to 29.8 mg/g and furfural concentration from 9.9 to 12.3 mg/g.

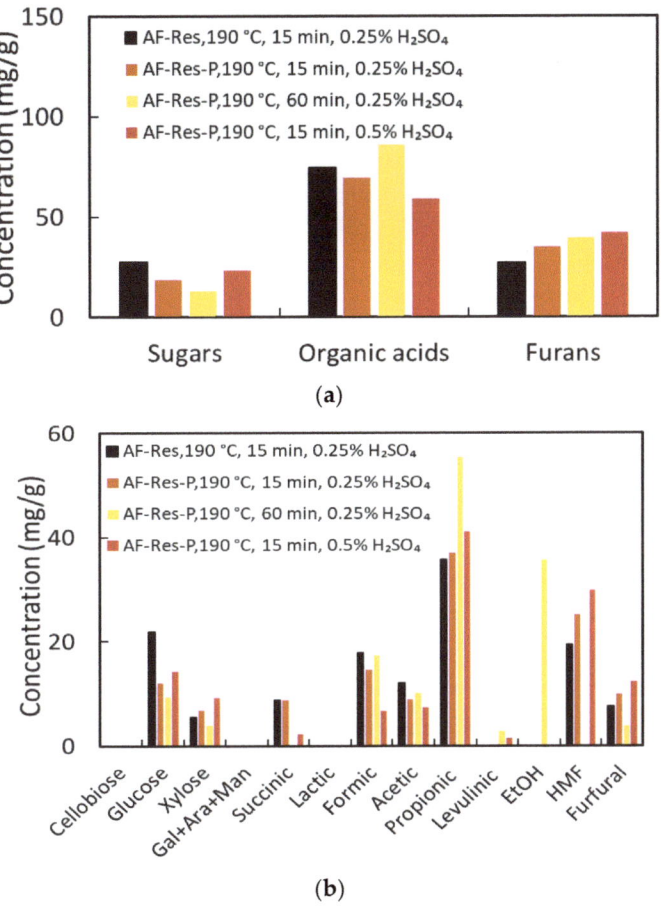

Figure 5. (**a**) Categories of the compounds and (**b**) composition of the AF-Res and AF-Res-P *Chlorella vulgaris* hydrolysates.

Based on the above results, it is shown that the careful selection of hydrolysis conditions (temperature, time and sulfuric acid concentration), which may also induce further/in situ conversion of the initially formed sugars to furans and organic acids, can finetune the composition of the final liquid hydrolysates. Usually, the boundaries between the initial hydrolysis of carbohydrates to monomer sugars, and their consequent conversion to furans and/or acids, are not distinct according to the different types of homogeneous or heterogeneous acid catalysts. As a result, it may be preferable to adjust and increase slightly the severity of the hydrolysis conditions towards the direct/in situ production of furans and/or organic acids. An additional benefit of doing this is that the tedious separation of sugars from the above mixture is avoided. As a general rule, relatively mild hydrolysis/reaction conditions select for sugars and some organic acids, while more severe conditions can shift the selectivity towards furanic compounds and other organic acids (from HMF hydrolysis). Humin formation under more intense conditions should also be taken into consideration. Another important outcome of the present results is the versatile character of the approach with respect to the type of feedstock used, i.e., lipid-extracted microalgae biomass or lipid-protein-extracted biomass. In addition, it is shown that similar results in terms of hydrolysis reactivity and product composition can be obtained from experimental/lab- and pilot-scale biomass cultivation and processing.

Another important point relates to the yields of the above-discussed targeted products (sugars, furans, acids) that are in the range of 7–20 wt.% of biomass. These yields are relatively low, mainly due to the relatively mild reaction conditions applied, which, however, restrict the formation of humins to low levels. Still, the initial sugar content of the biomass feedstocks (shown in Table S1) in the present work are not far from these product yields, i.e., they range from 5 to 40 wt.%, thus showing a relatively moderate/high yield of products when estimated on the basis of initial sugar content rather than the whole residual biomass. This further indicates that the residual biomass samples contain leftover lipids, proteins and ash.

2.3. Catalytic Hydrogenation/Hydrogenolysis

The catalytic hydrogenation/hydrogenolysis of microalgae residual biomass was performed using ethanol–water mixtures as solvents under increased hydrogen pressure, a sulfuric acid catalyst to induce the hydrolysis of carbohydrates towards sugar monomers, and ruthenium supported on activated carbon as the hydrogenation catalyst. The physicochemical characteristics of the 5%Ru/C catalyst are shown in Table 2 and Figure S4. The ruthenium was in its metallic phase; its crystallite size was calculated via Scherrer equation and found to be 8 nm. Regarding its porous properties, the catalyst exhibits mainly microporous characteristics, owing to the activated carbon support, alongside significant mesoporosity. The specific surface area, determined via the BET equation, was 1047 m^2/g, while the surface area corresponding to the micropores was 690 m^2/g, the rest being attributed to meso/macropores and external area. Taking into consideration the total pore volume, almost 37% corresponds to the micropore volume and 63% to the meso/macropore volume.

Table 2. Physicochemical characteristics of metallic catalysts supported on activated carbon (AC).

Catalyst	D_{XRD} (nm)	S_{BET} (m^2/g)	S_{micro} (m^2/g)	V_{total} (cm^3/g)	V_{micro} (cm^3/g)	$V_{meso/macro}$ (cm^3/g)
5%Ru/AC	8	1047	690	0.806	0.306	0.500

The biomass used as feed for the hydrogenation/hydrogenolysis experiments was the AF-Res-P, i.e., that derived from the pilot-scale microalgae AF after sequential extraction of lipids and proteins. The liquid products obtained via hydrogenation/hydrogenolysis exhibited a dark brown color, as can be observed in Figure 6.

0.1% w/v H_2SO_4
5%Ru/AC
190°C, 60 min

0.1% w/v H_2SO_4
5%Ru/AC
220°C, 60 min

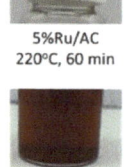
5%Ru/AC
220°C, 60 min

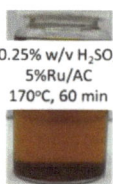
0.25% w/v H_2SO_4
5%Ru/AC
170°C, 60 min

Figure 6. Liquid products obtained via the hydrogenation/hydrogenolysis of AF-Res-P *Chlorella vulgaris* under different reaction conditions.

At 190 °C for 60 min, the solubilization of the AF-Res-P biomass was 80.9 wt.%, and by increasing the reaction temperature to 220 °C, the solubilization was raised to 87.6 wt.%, as shown in Figure S5. For purposes of comparison, the utilization of only the 5%Ru/AC as the catalyst, without the addition of H_2SO_4, resulted in almost the same solubilization, ca. 86.4 wt.%. Furthermore, a higher concentration of sulfuric acid, 0.25% w/v instead of 0.10% w/v, but under milder reaction conditions (170 °C), resulted in slightly lower solubilization, 74.3 wt.%, possibly due to humin formation.

The composition of the liquid products is shown in Figure 7a,b. Using 0.1% w/v H_2SO_4 + 5%Ru/AC at 190 °C, the concentration of sugars was 20.4 mg/g, organic acids 23.9 mg/g and furanic compounds 6.1 mg/g. Furthermore, under the hydrogenation/hydrogenolysis conditions applied, the in-situ production of sugar alcohols was achieved (21.9 mg/g), with sorbitol (13.6 mg/g) and glycerol (8.3 mg/g) being the main products. Via a generalized mechanism, the hydrogenation of glucose leads to the formation of sorbitol, which can then be converted to glycerol via hydrogenolysis [48,49]. An increase in the reaction temperature from 190 °C to 220 °C resulted in lower sugar concentration (11.1 mg/g), lower furans (5.4 mg/g) and significantly lower sugar alcohols (2.2 mg/g). On the other hand, the more intense conditions induced the conversion of sugars to organic acids, whose concentration increased to 32.2 mg/g, with the most abundant acids being propionic acid (21.5 mg/g), formic acid (3.6 mg/g) and levulinic acid (1.9 mg/g). This shift towards organic acid formation maybe attributed to the increased dehydration of sugars to furans (and their subsequent rehydration to organic acids) and/or the direct conversion of sugars to organic acids catalyzed by H_2SO_4 at higher temperatures, being faster than the hydrogenation reactivity of sugars towards sugar alcohols. Apart from the combination of sulfuric acid with 5%Ru/AC, the activity of neat 5%Ru/AC was examined at 220 °C for 60 min. As can be observed in Figure 7a, 5%Ru/AC enhanced the depolymerization of carbohydrates to monomeric sugars (13.8 mg/g), their dehydration to furanic compounds (4.7 mg/g), their conversion to organic acids (35.7 mg/g) and their hydrogenation/hydrogenolysis to sugar alcohols (18.9 mg/g). Regarding the sugars' composition, the presence of 5%Ru/AC enhanced the formation of glucose (5.6 mg/g) and galactose/arabinose/mannose (6.5 mg/g), while less xylose was produced (0.8 mg/g). Also, a narrower distribution of organic acids was observed with the formation of acetic acid (6.9 mg/g) and propionic acid (28.8 mg/g). The ruthenium-based catalyst enhanced the hydrogenolysis of the formed sorbitol to glycerol, whose concentration increased to 18.9 mg/g. Finally, aiming to achieve milder reaction conditions (mainly a lower temperature), a slightly higher concentration of sulfuric acid was used (0.25% w/v) at a lower reaction temperature (170 °C). Under these conditions, the selectivity towards sugars increased and their concentration amounted to 57.8 mg/g, along with increased concentrations of furanic compounds (11.8 mg/g) and sugar alcohols (40.6 mg/g). The higher concentration of sugars is attributed to the higher concentration of xylose (11.6 mg/g) and galactose/arabinose/mannose (39.1 mg/g). Interestingly, the only organic acids formed were succinic acid (8.1 mg/g) and acetic acid (5.1 mg/g). Regarding the sugar alcohols, sorbitol and glycerol were formed at concentrations of 13.5 mg/g and 27.1 mg/g, respectively.

The aim of this study was to provide a first proof of concept of the possible valorization of microalgae residual carbohydrates to sugar alcohols via hydrogenation/hydrogenolysis. The leaching of Ru was not detected in the liquid products, as the use of sulfuric acid was kept to very low levels, just to initiate the hydrolysis of carbohydrates to monomeric sugars that could then be hydrogenated using Ru. Certainly, an in-depth characterization of the spent catalysts and reuse/regeneration studies are needed to further investigate the exploitation potential of the proposed approach.

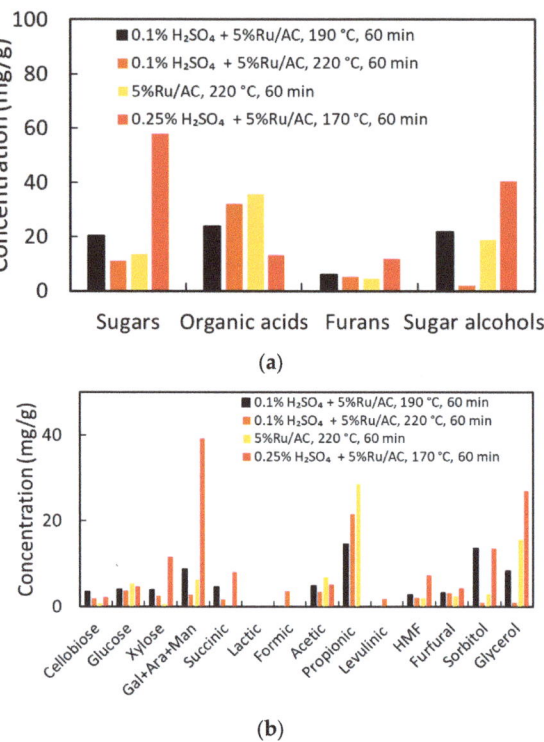

Figure 7. (**a**) Categories of the compounds and (**b**) composition of the AF-Res-P *Chlorella vulgaris* hydrolysates obtained via the hydrogenation/hydrogenolysis experiments.

3. Materials and Methods

3.1. Feedstocks

The microalgae samples used in this study are shown in Table 3. Two lab-scale cultures of *Chlorella vulgaris* were produced under different cultivation conditions, mainly nitrogen abundance, according to previously published procedures [33]. The high nitrogen concentration in the cultivation medium led to the *Chlorella vulgaris* LL (i.e., low lipid) sample, while the lower nitrogen concentration led to the production of the ML (i.e., moderate lipid) sample. Larger-scale amounts of *Chlorella vulgaris* biomass were produced by the Faculty of Agriculture, Aristotle University of Thessaloniki (photobioreactor, 2000L); this sample was named AF. For comparison purposes, a commercially available *Chlorella vulgaris* was also used (MF).

Table 3. *Chlorella vulgaris* residual biomass used in the hydrolysis/dehydration and hydrogenation/hydrogenolysis experiments.

Sample	Pretreatment
LL-Res	Lab-scale cultivation sample after lipid extraction of low lipid content strain
ML-Res	Lab-scale cultivation sample after lipid extraction of medium lipid content strain
MF-Res	Commercially available biomass after lipid extraction
AF-Res	Upscale cultivation sample after lipid extraction
LL-Res-P	Sample after protein extraction of LL-Res
ML-Res-P	Sample after protein extraction of ML-Res
MF-Res-P	Sample after protein extraction of MF-Res
AF-Res-P	Sample after protein extraction of AF-Res

3.2. Fractionation of Chlorella vulgaris Microalgae

The fractionation of the parent microalgae biomass was based on the extraction of lipids in the first step, followed by protein extraction in the second step. Thus, in the first step, 5 g of dry biomass was placed in a cellulose thimble. The extraction was performed in a Soxhlet apparatus, using a mixture of chloroform and methanol (2:1 v/v) as solvents, at 80 °C for 18 h. After the extraction, the solvent was recovered in a rotary evaporator (HB10, IKA) and the remaining lipid fraction was dried at 40 °C for 6 h. The remaining solid fraction, after lipid extraction, containing proteins and carbohydrates, was dried at room temperature overnight and at 40 °C for 6 h. The obtained samples were denoted as LL-Res, ML-Res, MF-Res and AF-Res.

In the second step, proteins were isolated according to the procedure described by Safi et al. [50]. Briefly, 0.5 g of dry biomass (after lipid extraction) was added to 25 mL 2 N NaOH solution (pH = 12) and left under stirring for 2 h at 40 °C. The supernatant was separated from the solid via centrifugation at 10,000× g, 40 °C, 10 min. The solid residue enriched in carbohydrates was washed several times with deionized water and centrifuged until pH = 6, dried at room temperature and at 40 °C for 6 h. Proteins were isolated after precipitation at pH = 3, with 0.1 M HCl and centrifugation at 10.000× g, 20 °C, 10 min. The obtained samples after protein extraction were denoted as LL-Res-P, ML-Res-P, MF-Res-P and AF-Res-P.

3.3. Characterization of Biomass Feedstocks

The characterization of the microalgae biomass feedstocks was performed according to NREL protocols. Moisture content and total solids were based on NREL/TP-5100-60956 protocol, according to which 100 mg of pulverized biomass was placed at 40 °C under vacuum for at least 18 h [51]. Afterwards, the ash content was determined after ignition at 575 °C for 3 h, according to the same protocol. The elemental composition (C/H/N/S) was determined in an elemental analyzer EA 3100 (EuroVector, Pavia, Italy). The samples were heated at 980 °C, under constant helium flow. Oxygen was determined via the equation O (wt.%) = 100 − C (wt.%) − H (wt.%) − N (wt.%) − S (wt.%) − Ash (wt.%). The carbohydrate content was determined via two-step sulfuric acid hydrolysis, according to NREL/TP-5100-60957 protocol [52].

3.4. Catalytic Hydrolysis/Dehydration

The hydrothermal hydrolysis/dehydration experiments were carried out in a batch stirred autoclave reactor. In each experiment, microalgal biomass was mixed with distilled water at liquid-to-solid ratio L/S = 32 and placed in the reactor, along with 0.25 or 0.50% w/v H_2SO_4 as an acid catalyst. After closing the reactor, 10 bar of nitrogen gas was purged into the reactor. The experiments were performed in the temperature range of 175–250 °C for 15–60 min, with a 400 rpm stirring rate to eliminate any mass transfer phenomena. The average heating rate to the targeted temperature was ~14 °C/min, thus limiting the conversion of biomass taking place during the heat-up of the experiment. After the reaction, the reactor was cooled to room temperature and vacuum filtration was applied to separate the liquid fraction from the biomass. The solid fraction, corresponding to the non-solubilized biomass, was dried at 80 °C for 6 h under vacuum, while the liquid fraction containing the solubilized biomass was analyzed by high-performance liquid chromatography (HPLC). Analysis was performed on an HPLC (LC-20AD, HPLC, Shimadzu, Tokyo, Japan) equipped with a refractive index detector (RID-10A, Shimadzu) and an oven (CTO-20A, Shimadzu). The identification and quantification of sugars, organic acids and furanic compounds was carried out in an SH-1011 column, at 45 °C and 0.01 N H_2SO_4 as the mobile phase with a flow of 0.7 mL/min. The recovered solids were characterized via thermogravimetric analysis (TGA). The measurements were performed in the Netzsch (Selb, Germany) STA 449 F5 Jupiter, at a temperature range of 25–950 °C, under N_2 atmosphere and a 10 °C/min heating rate. The experiments were performed in triplicate and the standard error was <5%.

3.5. Catalytic Hydrogenolysis/Hydrogenation

The hydrogenation/hydrogenolysis experiments were carried out in a batch autoclave reactor, using 0.4 g of biomass and a solvent mixture of 20 mL ethanol–water at a ratio of 90–10%. Dilute H_2SO_4 0.1 or 0.25% w/v was used as an acid catalyst, combined with the solid hydrogenation catalyst 5%Ru/AC. All experiments were performed with 30 bar hydrogen gas pressure (measured at room temperature), in the temperature range of 190–220 °C for 60 min and with a 400 rpm stirring rate to eliminate any mass transfer phenomena. The average heating rate to the targeted temperature was ~14 °C/min. Afterwards, the reactor was cooled to room temperature and vacuum filtration was applied to separate the liquid fraction from the biomass. The solid fraction was dried at 80 °C for 6 h under vacuum, while the liquid fraction was analyzed by high-performance liquid chromatography (HPLC), similarly to Section 3.4. The experiments were performed in triplicate and the standard error was <5%.

The catalyst 5%Ru/AC was synthesized via a wet impregnation method. Norit SX-Plus carbon was used as support. Prior to the impregnation, the support was thermally treated at 500 °C, 3 h under N_2 flow. Briefly, an appropriate amount of $RuCl_3$ was dissolved in 20 mL H_2O and then added dropwise and under stirring to the suspension of 10 g support which was suspended in 100 mL H_2O. Stirring continued for 1 h before the water was removed using a rotary evaporator. The final solid was dried overnight at 100 °C, calcined at 500 °C, 3 h under He (50 mL/min) and reduced under H_2 flow 50 mL/min. The catalyst was characterized via X-ray powder diffraction. The pattern was recorded using CuKa X-ray radiation, at 2θ = 5–85°, with 0.02°/step and 2 s/step. The porous properties were determined via nitrogen physisorption at -196 °C using an automatic volumetric sorption analyzer (Autosorb-1 MP, Quantachrome, Boynton Beac, FL, USA). Prior to taking the measurements, the sample was outgassed at 250 °C, 19 h under 5×10^{-9} Torr vacuum. Surface areas were determined via the multipoint BET method; the total pore volume was determined at $P/Po = 0.99$ while the microporous surface areas and volumes were determined via t-plot method.

4. Conclusions

This work focused on the potential valorization of various types of microalgal residual biomass towards the production of platform chemicals. The residual carbohydrate biomass that remained after the extraction of lipids and proteins can be converted to monomeric sugars, which can serve as precursors of a wide range of chemicals. The intensity of the mild acid-hydrothermal treatment of this residual biomass determined the extent of hydrolysis–dehydration–rehydration reactions that led to sugar monomers being produced at concentrations of 4.6–32.5 mg/g, furans at up to 42 mg/g and organic acids between 30 and 144 mg/g, respectively. Under more intense conditions, i.e., higher acid concentration and/or temperature, all the above products were suppressed due to further condensation reactions that lead to humin formation.

Each one of the above group of products, e.g., monomeric sugars, organic acids and furanic compounds, has its value in an integrated biorefinery scheme. Monomeric sugars can be used to produce bioethanol/butanol, a wide range of acids (succinic, lactic, etc.) as well as furans (furfural, HMF), all of them being high-added-value platform chemicals towards the production of fuels or biobased polymers. With regard to the furan/acid mixtures, these may be more easily separated, compared to when monomer sugars are also present, and utilized; for example, furfural can make furfuryl alcohol resins or HMF to produce furan dicarboxylic acid (FDCA) and then polyethylene furanoate (PEF), a potential replacement for PET. On the other hand, organic acids such as lactic and succinic are being used to produce valuable green plastics based on polylactic acid (PLA) or polybutylene succinates (PBS). Alternatively, the whole mixture of organic acids may find application in the production of phenol–formaldehyde resins (P–F) where a commercial/petroleum-derived, usually acetic acid-based, acidification medium is used to control/regulate the pH of the process.

A promising alternative pathway is the catalytic hydrogenation/hydrogenolysis of this residual carbohydrate biomass, with the assistance of homogeneous mild acid catalysts, towards the in-situ production of sugar alcohols, such as sorbitol and glycerol, whose concentrations add up to 2–41 mg/g. These products are also of high value and can be used as food ingredients, in polymer synthesis, in pharmaceuticals, etc., thus increasing the exploitation potential of microalgae residual biomass.

Supplementary Materials: The following supporting information can be downloaded at: https://www.mdpi.com/article/10.3390/catal14050286/s1, Figure S1: Solubilization degree of LL-Res and ML-Res *Chlorella vulgaris* biomass; Figure S2: Solubilization degree of MF-Res and MF-Res-P *Chlorella vulgaris* biomass; Figure S3: Solubilization degree of AF-Res and AF-Res-P *Chlorella vulgaris* biomass; Figure S4: (a) XRD and (b) nitrogen adsorption-desorption isotherms of 5%Ru/AC; Figure S5: Solubilization degree of AF-Res and AF-Res-P *Chlorella vulgaris* biomass; Table S1: Sugars composition of the initial microalgal biomass and the solids obtained after the extraction of lipids and proteins; Table S2: Thermal analysis of solids recovered after the hydrolysis experiments of MF-Res biomass.

Author Contributions: Conceptualization, K.S.T.; methodology, K.S.T. and A.G.M.; formal analysis, A.G.M., S.A.T. and G.I.; investigation, A.G.M., S.A.T. and G.I; data curation, A.G.M. and S.A.T.; writing—original draft preparation, A.G.M.; writing—review and editing, K.S.T. All authors have read and agreed to the published version of the manuscript.

Funding: This research has been co-financed by the European Regional Development Fund of the European Union and Greek national funds through the Operational Program Competitiveness, Entrepreneurship and Innovation (EPAnEK 2014-2020), under the Action "RESEARCH—CREATE—INNOVATE" B' CALL (ALGAFUELS project, code: T2EDK-00041).

Data Availability Statement: Data is contained within the article or Supplementary Materials.

Acknowledgments: The authors would like to thank the Faculty of Agriculture, Aristotle University of Thessaloniki, and the Zalidis group for providing the lab and upscale samples of microalgae *Chlorella vulgaris*.

Conflicts of Interest: The authors declare no conflicts of interest.

References

1. Lee, J.Y.; Yoo, C.; Jun, S.Y.; Ahn, C.Y.; Oh, H.M. Comparison of several methods for effective lipid extraction from microalgae. *Bioresour. Technol.* **2010**, *101* (Suppl. 1), S75–S77. [CrossRef] [PubMed]
2. Lee, S.Y.; Khoiroh, I.; Vo, D.-V.N.; Senthil Kumar, P.; Show, P.L. Techniques of lipid extraction from microalgae for biofuel production: A review. *Environ. Chem. Lett.* **2020**, *19*, 231–251. [CrossRef]
3. Jeevan Kumar, S.P.; Vijay Kumar, G.; Dash, A.; Scholz, P.; Banerjee, R. Sustainable green solvents and techniques for lipid extraction from microalgae: A review. *Algal Res.* **2017**, *21*, 138–147. [CrossRef]
4. Chew, K.W.; Yap, J.Y.; Show, P.L.; Suan, N.H.; Juan, J.C.; Ling, T.C.; Lee, D.J.; Chang, J.S. Microalgae biorefinery: High value products perspectives. *Bioresour. Technol.* **2017**, *229*, 53–62. [CrossRef] [PubMed]
5. Yen, H.W.; Hu, I.C.; Chen, C.Y.; Ho, S.H.; Lee, D.J.; Chang, J.S. Microalgae-based biorefinery—From biofuels to natural products. *Bioresour. Technol.* **2013**, *135*, 166–174. [CrossRef] [PubMed]
6. Kwak, H.S.; Kim, J.Y.H.; Woo, H.M.; Jin, E.; Min, B.K.; Sim, S.J. Synergistic effect of multiple stress conditions for improving microalgal lipid production. *Algal Res.* **2016**, *19*, 215–224. [CrossRef]
7. Yeletsky, M.P.; Kukushkin, G.R.; Stepanenko, A.S.; Koskin, P.A. Prospects in One-stage Conversion of Lipid-based Feedstocks into Biofuels Enriched with Branched Alkanes. *Curr. Org. Chem.* **2023**, *27*, 1114–1118. [CrossRef]
8. Koskin, A.P.; Zibareva, I.V.; Vedyagin, A.A. Conversion of Rice Husk and Nutshells into Gaseous, Liquid, and Solid Biofuels. In *Biorefinery of Alternative Resources: Targeting Green Fuels and Platform Chemicals*; Nanda, S., Vo, D.-V.N., Sarangi, P.K., Eds.; Springer: Singapore, 2020; pp. 171–194.
9. Levasseur, W.; Perre, P.; Pozzobon, V. A review of high value-added molecules production by microalgae in light of the classification. *Biotechnol. Adv.* **2020**, *41*, 107545. [CrossRef]
10. Karageorgou, D.; Rova, U.; Christakopoulos, P.; Katapodis, P.; Matsakas, L.; Patel, A. Benefits of supplementation with microbial omega-3 fatty acids on human health and the current market scenario for fish-free omega-3 fatty acid. *Trends Food Sci. Technol.* **2023**, *136*, 169–180. [CrossRef]
11. Patel, A.; Rova, U.; Christakopoulos, P.; Matsakas, L. Introduction to Essential Fatty Acids. In *Nutraceutical Fatty Acids from Oleaginous Microalgae*; Wiley: Hoboken, NJ, USA, 2020; pp. 1–22.

12. Brown, M.R. The amino-acid and sugar composition of 16 species of microalgae used in mariculture. *J. Exp. Mar. Biol. Ecol.* **1991**, *145*, 79–99. [CrossRef]
13. Moreira, J.B.; Vaz, B.d.S.; Cardias, B.B.; Cruz, C.G.; Almeida, A.C.A.d.; Costa, J.A.V.; Morais, M.G.d. Microalgae Polysaccharides: An Alternative Source for Food Production and Sustainable Agriculture. *Polysaccharides* **2022**, *3*, 441–457. [CrossRef]
14. Delattre, C.; Pierre, G.; Laroche, C.; Michaud, P. Production, extraction and characterization of microalgal and cyanobacterial exopolysaccharides. *Biotechnol. Adv.* **2016**, *34*, 1159–1179. [CrossRef] [PubMed]
15. de Carvalho Silvello, M.A.; Severo Goncalves, I.; Patricia Held Azambuja, S.; Silva Costa, S.; Garcia Pereira Silva, P.; Oliveira Santos, L.; Goldbeck, R. Microalgae-based carbohydrates: A green innovative source of bioenergy. *Bioresour. Technol.* **2022**, *344*, 126304. [CrossRef] [PubMed]
16. Gouda, M.; Tadda, M.A.; Zhao, Y.; Farmanullah, F.; Chu, B.; Li, X.; He, Y. Microalgae Bioactive Carbohydrates as a Novel Sustainable and Eco-Friendly Source of Prebiotics: Emerging Health Functionality and Recent Technologies for Extraction and Detection. *Front. Nutr.* **2022**, *9*, 806692. [CrossRef] [PubMed]
17. Smachetti, M.E.S.; Rizza, L.S.; Coronel, C.D.; Nascimento, M.D.; Curatti, L. Microalgal Biomass as an Alternative Source of Sugars for the Production of Bioethanol. In *Principles and Applications of Fermentation Technology*; Wiley: Hoboken, NJ, USA, 2018; pp. 351–386.
18. Markou, G.; Angelidaki, I.; Georgakakis, D. Microalgal carbohydrates: An overview of the factors influencing carbohydrates production, and of main bioconversion technologies for production of biofuels. *Appl. Microbiol. Biotechnol.* **2012**, *96*, 631–645. [CrossRef] [PubMed]
19. Morales-Jiménez, M.; Gouveia, L.; Yáñez-Fernández, J.; Castro-Muñoz, R.; Barragán-Huerta, B.E. Production, Preparation and Characterization of Microalgae-Based Biopolymer as a Potential Bioactive Film. *Coatings* **2020**, *10*, 120. [CrossRef]
20. Madadi, R.; Maljaee, H.; Serafim, L.S.; Ventura, S.P.M. Microalgae as Contributors to Produce Biopolymers. *Marine Drugs* **2021**, *19*, 466. [CrossRef] [PubMed]
21. Rachidi, F.; Benhima, R.; Sbabou, L.; El Arroussi, H. Microalgae polysaccharides bio-stimulating effect on tomato plants: Growth and metabolic distribution. *Biotechnol. Rep.* **2020**, *25*, e00426. [CrossRef] [PubMed]
22. Sarma, S.; Sharma, S.; Rudakiya, D.; Upadhyay, J.; Rathod, V.; Patel, A.; Narra, M. Valorization of microalgae biomass into bioproducts promoting circular bioeconomy: A holistic approach of bioremediation and biorefinery. *3 Biotech* **2021**, *11*, 378. [CrossRef] [PubMed]
23. Wild, K.J.; Steingass, H.; Rodehutscord, M. Variability in nutrient composition and in vitro crude protein digestibility of 16 microalgae products. *J. Anim. Physiol. Anim. Nutr.* **2018**, *102*, 1306–1319. [CrossRef]
24. Eilam, Y.; Khattib, H.; Pintel, N.; Avni, D. Microalgae-Sustainable Source for Alternative Proteins and Functional Ingredients Promoting Gut and Liver Health. *Glob. Chall.* **2023**, *7*, 2200177. [CrossRef] [PubMed]
25. Lucakova, S.; Branyikova, I.; Hayes, M. Microalgal Proteins and Bioactives for Food, Feed, and Other Applications. *Appl. Sci.* **2022**, *12*, 4402. [CrossRef]
26. Amorim, M.L.; Soares, J.; Coimbra, J.; Leite, M.O.; Albino, L.F.T.; Martins, M.A. Microalgae proteins: Production, separation, isolation, quantification, and application in food and feed. *Crit. Rev. Food Sci. Nutr.* **2021**, *61*, 1976–2002. [CrossRef] [PubMed]
27. Lortou, U.; Panou, M.; Papapanagiotou, G.; Florokapi, G.; Giannakopoulos, C.; Kavoukis, S.; Iakovou, G.; Zalidis, G.; Triantafyllidis, K.; Gkelis, S. Beneath the Aegean Sun: Investigating *Dunaliella* Strains' Diversity from Greek Saltworks. *Water* **2023**, *15*, 1037. [CrossRef]
28. Global Beta Carotene Market—Segmented by Application (Pharmaceuticals, D.S., Food & Beverage, Animal Feed, Others), By Source (Synthric, Algae, Fungi, Palm Oil, Others), & By Regional Analysis (North America, Europe, Asia Pacific, Latin America, and Middle East & Africa)—Global Industry Analysis, Size, Share, Growth, Trends, and Forecast (2023–2028). Available online: https://www.marketdataforecast.com/market-reports/beta-carotene-market (accessed on 5 January 2024).
29. Marcilla, A.; Catalá, L.; García-Quesada, J.C.; Valdés, F.J.; Hernández, M.R. A review of thermochemical conversion of microalgae. *Renew. Sustain. Energy Rev.* **2013**, *27*, 11–19. [CrossRef]
30. Yang, C.; Li, R.; Zhang, B.; Qiu, Q.; Wang, B.; Yang, H.; Ding, Y.; Wang, C. Pyrolysis of microalgae: A critical review. *Fuel Process. Technol.* **2019**, *186*, 53–72. [CrossRef]
31. Li, F.; Srivatsa, S.C.; Bhattacharya, S. A review on catalytic pyrolysis of microalgae to high-quality bio-oil with low oxygeneous and nitrogenous compounds. *Renew. Sustain. Energy Rev.* **2019**, *108*, 481–497. [CrossRef]
32. Liu, X.; Guo, Y.; Dasgupta, A.; He, H.; Xu, D.; Guan, Q. Algal bio-oil refinery: A review of heterogeneously catalyzed denitrogenation and demetallization reactions for renewable process. *Renew. Energy* **2022**, *183*, 627–650. [CrossRef]
33. Adamakis, I.D.; Lazaridis, P.A.; Terzopoulou, E.; Torofias, S.; Valari, M.; Kalaitzi, P.; Rousonikolos, V.; Gkoutzikostas, D.; Zouboulis, A.; Zalidis, G.; et al. Cultivation, characterization, and properties of Chlorella vulgaris microalgae with different lipid contents and effect on fast pyrolysis oil composition. *Environ. Sci. Pollut. Res. Int.* **2018**, *25*, 23018–23032. [CrossRef]
34. Sharma, N.; Jaiswal, K.K.; Kumar, V.; Vlaskin, M.S.; Nanda, M.; Rautela, I.; Tomar, M.S.; Ahmad, W. Effect of catalyst and temperature on the quality and productivity of HTL bio-oil from microalgae: A review. *Renew. Energy* **2021**, *174*, 810–822. [CrossRef]
35. Arun, J.; Varshini, P.; Prithvinath, P.K.; Priyadarshini, V.; Gopinath, K.P. Enrichment of bio-oil after hydrothermal liquefaction (HTL) of microalgae C. vulgaris grown in wastewater: Bio-char and post HTL wastewater utilization studies. *Bioresour. Technol.* **2018**, *261*, 182–187. [CrossRef] [PubMed]

36. Wang, K.; Brown, R.C.; Homsy, S.; Martinez, L.; Sidhu, S.S. Fast pyrolysis of microalgae remnants in a fluidized bed reactor for bio-oil and biochar production. *Bioresour. Technol.* **2013**, *127*, 494–499. [CrossRef] [PubMed]
37. Kim, S.-S.; Ly, H.V.; Kim, J.; Lee, E.Y.; Woo, H.C. Pyrolysis of microalgae residual biomass derived from *Dunaliella tertiolecta* after lipid extraction and carbohydrate saccharification. *Chem. Eng. J.* **2015**, *263*, 194–199. [CrossRef]
38. Obeid, S.; Beaufils, N.; Peydecastaing, J.; Camy, S.; Takache, H.; Ismail, A.; Pontalier, P.-Y. Microalgal fractionation for lipids, pigments and protein recovery. *Process Biochem.* **2022**, *121*, 240–247. [CrossRef]
39. Gerde, J.A.; Wang, T.; Yao, L.; Jung, S.; Johnson, L.A.; Lamsal, B. Optimizing protein isolation from defatted and non-defatted Nannochloropsis microalgae biomass. *Algal Res.* **2013**, *2*, 145–153. [CrossRef]
40. Harun, R.; Danquah, M.K. Influence of acid pre-treatment on microalgal biomass for bioethanol production. *Process Biochem.* **2011**, *46*, 304–309. [CrossRef]
41. Yang, Z.; Guo, R.; Xu, X.; Fan, X.; Luo, S. Fermentative hydrogen production from lipid-extracted microalgal biomass residues. *Appl. Energy* **2011**, *88*, 3468–3472. [CrossRef]
42. Naresh Kumar, A.; Min, B.; Venkata Mohan, S. Defatted algal biomass as feedstock for short chain carboxylic acids and biohydrogen production in the biorefinery format. *Bioresour. Technol.* **2018**, *269*, 408–416. [CrossRef] [PubMed]
43. Roy Chong, J.W.; Tan, X.; Khoo, K.S.; Ng, H.S.; Jonglertjunya, W.; Yew, G.Y.; Show, P.L. Microalgae-based bioplastics: Future solution towards mitigation of plastic wastes. *Environ. Res.* **2022**, *206*, 112620. [CrossRef]
44. Arora, Y.; Sharma, S.; Sharma, V. Microalgae in Bioplastic Production: A Comprehensive Review. *Arab. J. Sci. Eng.* **2023**, *48*, 7225–7241. [CrossRef]
45. Klingler, D.; Berg, J.; Vogel, H. Hydrothermal reactions of alanine and glycine in sub- and supercritical water. *J. Supercrit. Fluids* **2007**, *43*, 112–119. [CrossRef]
46. Martins, L.B.; Soares, J.; da Silveira, W.B.; Sousa, R.d.C.S.; Martins, M.A. Dilute sulfuric acid hydrolysis of *Chlorella vulgaris* biomass improves the multistage liquid-liquid extraction of lipids. *Biomass Convers. Biorefinery* **2021**, *11*, 2485–2497. [CrossRef]
47. Miranda, J.R.; Passarinho, P.C.; Gouveia, L. Pre-treatment optimization of Scenedesmus obliquus microalga for bioethanol production. *Bioresour. Technol.* **2012**, *104*, 342–348. [CrossRef] [PubMed]
48. Lazaridis, P.A.; Karakoulia, S.; Delimitis, A.; Coman, S.M.; Parvulescu, V.I.; Triantafyllidis, K.S. d-Glucose hydrogenation/hydrogenolysis reactions on noble metal (Ru, Pt)/activated carbon supported catalysts. *Catal. Today* **2015**, *257*, 281–290. [CrossRef]
49. Lazaridis, P.A.; Karakoulia, S.A.; Teodorescu, C.; Apostol, N.; Macovei, D.; Panteli, A.; Delimitis, A.; Coman, S.M.; Parvulescu, V.I.; Triantafyllidis, K.S. High hexitols selectivity in cellulose hydrolytic hydrogenation over platinum (Pt) vs. ruthenium (Ru) catalysts supported on micro/mesoporous carbon. *Appl. Catal. B Environ.* **2017**, *214*, 1–14. [CrossRef]
50. Safi, C.; Ursu, A.V.; Laroche, C.; Zebib, B.; Merah, O.; Pontalier, P.-Y.; Vaca-Garcia, C. Aqueous extraction of proteins from microalgae: Effect of different cell disruption methods. *Algal Res.* **2014**, *3*, 61–65. [CrossRef]
51. *NREL/TP-5100-60956*; Determination of Total Solids and Ash in Algal Biomass. NREL: Golden, CO, USA, 2015.
52. *NREL/TP-5100-60957*; Determination of Total Carbohydrates in Algal Biomass. NREL: Golden, CO, USA, 2015.

Disclaimer/Publisher's Note: The statements, opinions and data contained in all publications are solely those of the individual author(s) and contributor(s) and not of MDPI and/or the editor(s). MDPI and/or the editor(s) disclaim responsibility for any injury to people or property resulting from any ideas, methods, instructions or products referred to in the content.

Article

Non-Symmetrically Fused Bis(arylimino)pyridines with *para*-Phenyl Substitution: Exploring Their Use as N′,N,N″-Supports in Iron Ethylene Polymerization Catalysis

Yizhou Wang [1,2], Zheng Wang [1,3], Qiuyue Zhang [1], Yanping Ma [1], Gregory A. Solan [1,4,*], Yang Sun [1] and Wen-Hua Sun [1,2,*]

1. Key Laboratory of Engineering Plastics, Beijing National Laboratory for Molecular Sciences, Institute of Chemistry Chinese Academy of Sciences, Beijing 100190, China
2. CAS Research/Education Center for Excellence in Molecular Sciences, University of Chinese Academy of Sciences, Beijing 100049, China
3. College of Science, Hebei Agricultural University, Baoding 071001, China
4. Department of Chemistry, University of Leicester, University Road, Leicester LE1 7RH, UK
* Correspondence: gas8@leicester.ac.uk (G.A.S.); whsun@iccas.ac.cn (W.-H.S.); Tel.: +44-(0)116-2522096 (G.A.S.); +86-10-6255-7955 (W.-H.S.)

Citation: Wang, Y.; Wang, Z.; Zhang, Q.; Ma, Y.; Solan, G.A.; Sun, Y.; Sun, W.-H. Non-Symmetrically Fused Bis(arylimino)pyridines with *para*-Phenyl Substitution: Exploring Their Use as N′,N,N″-Supports in Iron Ethylene Polymerization Catalysis. *Catalysts* **2024**, *14*, 213. https://doi.org/10.3390/catal14030213

Academic Editors: Alfonso Grassi and Marc Visseaux

Received: 4 February 2024
Revised: 9 March 2024
Accepted: 18 March 2024
Published: 21 March 2024

Copyright: © 2024 by the authors. Licensee MDPI, Basel, Switzerland. This article is an open access article distributed under the terms and conditions of the Creative Commons Attribution (CC BY) license (https://creativecommons.org/licenses/by/4.0/).

Abstract: Through the implementation of a one-pot strategy, five examples of non-symmetrical [N,N-diaryl-11-phenyl-1,2,3,7,8,9,10-heptahydrocyclohepta[*b*]quinoline-4,6-diimine]iron(II) chloride complexes (aryl = 2,6-Me$_2$Ph **Fe1**, 2,6-Et$_2$Ph **Fe2**, 2,6-*i*-Pr$_2$Ph **Fe3**, 2,4,6-Me$_3$Ph **Fe4**, and 2,6-Et$_2$-4-MePh **Fe5**), incorporating fused six- and seven-membered carbocyclic rings and appended with a remote *para*-phenyl group, were readily prepared. The molecular structures of **Fe2** and **Fe3** emphasize the variation in fused ring size and the skewed disposition of the *para*-phenyl group present in the N′,N,N″-ligand support. Upon activation with MAO or MMAO, **Fe1–Fe5** all showed high catalytic activity for ethylene polymerization, with an exceptional level of 35.92 × 10^6 g (PE) mol^{-1} (Fe) h^{-1} seen for mesityl-substituted **Fe4**/MMAO operating at 60 °C. All catalysts generated highly linear polyethylene with good control of the polymer molecular weight achievable via straightforward manipulation of run temperature. Typically, low molecular weight polymers with narrow dispersity (M_w/M_n = 1.5) were produced at 80 °C (MMAO: 3.7 kg mol^{-1} and MAO: 4.9 kg mol^{-1}), while at temperatures between 40 °C and 50 °C, moderate molecular weight polymers were obtained (MMAO: 35.6–51.6 kg mol^{-1} and MAO: 72.4–95.5 kg mol^{-1}). Moreover, analysis of these polyethylenes by ^{1}H and ^{13}C NMR spectroscopy highlighted the role played by both β-H elimination and chain transfer to aluminum during chain termination, with the highest rate of β-H elimination seen at 60 °C for the MMAO-activated system and 70 °C for the MAO system.

Keywords: linear polyethylenes; high activity; molecular weight control; temperature effects; chain termination pathways

1. Introduction

Remarkable advances have been made in the design of iron and cobalt catalysts for olefin oligomerization and polymerization ever since the first disclosures in the late 1990s [1–4], which reflect the vital function played by the metal–ligand complex. With particular regard to the ligand, N,N,N-chelating 2,6-bis(arylimino)pyridines (**A**, Figure 1) have paved the way with a host of structural modifications reported over the last two and a half decades or so, culminating in notable effects on polymer structure and catalytic activity [5–11]. From an industrial perspective, 2-imino-1,10-phenanthroline complexes (**B**, Figure 1) have shown the greatest promise with a successful pilot-scale process for making linear α-olefins using iron derivatives now being scaled up into a 200,000-ton process in China (in construction since November 2021) [12–14].

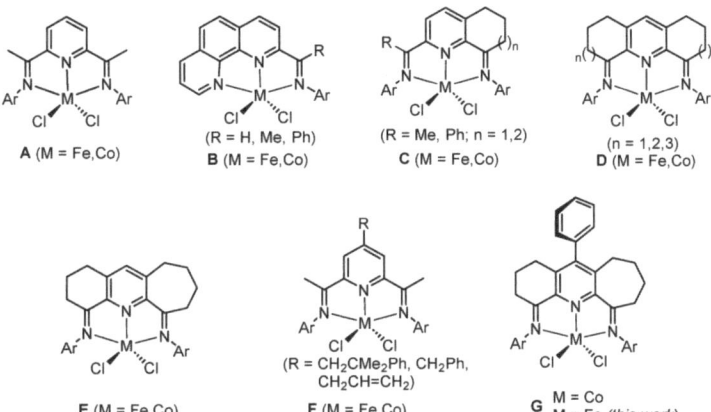

Figure 1. Bis(arylimino)pyridine iron(II)/cobalt(II) chlorides, **A**, and some key variations made to the *N,N,N*-ligand skeleton in **B–G**.

Elsewhere, the fusion of carbocyclic groups to the central pyridine in **A** has provided an effective means to not only enhance the thermostability and activity of the iron or cobalt catalyst [15–24] but also impart significant effects on the polymer products and catalytic activity. For example, cobalt-containing $\mathbf{C}_{n=1}$ (R = Me, Figure 1) incorporating a single six-membered fused ring [17] exhibited high activity for ethylene polymerization and, indeed, superior thermal stability to the prototypical cobalt catalyst **A** [1,2]. Similarly, iron and cobalt complexes containing a larger fused seven-membered ring, $\mathbf{C}_{n=2}$ (R = Ph, Figure 1) [18], showed improved thermal stability as well as a good catalytic lifetime. By contrast, doubly fused $\mathbf{D}_{2(n=1)}$-type iron and cobalt complexes (Figure 1) [15], bearing two six-membered carbocyclic rings, have a tendency towards generating mixtures of oligomers and polyethylenes that limit their usefulness. However, this limitation can be overcome by increasing the ring size of the two fused rings, with the seven-membered ring-containing iron complex $\mathbf{D}_{2(n=2)}$ (Figure 1) [20], affording uniquely polyethylene. Furthermore, the bis(eight-membered) iron examples of $\mathbf{D}_{2(n=3)}$ (Figure 1) afford higher molecular weight polymer with very high activity (up to 1.2×10^7 g (PE) mol^{-1} (Fe) h^{-1}) [17,22].

As a result of this fused ring strategy, bis(imino)pyridine complexes integrated with carbocyclic rings of different ring sizes have also started to emerge. For instance, $\mathbf{E}_{n=1/n=2}$-type cobalt complexes (Figure 1) [23], containing both six- and seven-membered carbocycles, allow access to low molecular-weight vinyl-terminated polyethylenes (M_w range: 1.5–22.8 kg mol^{-1}) with extremely narrow dispersity (M_w/M_n range: 1.5–2.1), materials that have received some attention as functional polymers for use in copolymerization [25–27]. In pursuit of better catalytic performance, iron examples of $\mathbf{E}_{n=1/n=2}$ have been shown to exhibit very high activity with levels reaching up to 1.6×10^7 g (PE) mol^{-1} (Fe) h^{-1} at 40 °C forming bimodal polyethylenes [24].

As an alternative direction in *N,N,N* ligand design, the introduction of hydrocarbyl groups (e.g., CH$_2$CH$_2$Ph, CH$_2$Ph, and CH$_2$CH=CH$_2$) to the *para*-position of the central pyridine in **A** has seen reports of iron and cobalt catalysts (**F**, Figure 1) that can influence molecular weight as well as the dispersity of the resulting polyethylene [28]. More recently, the introduction of a *para*-phenyl group to the central pyridine in cobalt-containing $\mathbf{E}_{n=1/n=2}$ to form **G** (Figure 1) has also seen unexpected enhancement effects on catalytic activity and molecular weight [29].

Given the superior catalytic performance generally seen for iron over cobalt in ethylene polymerization, we now target a series of *para*-phenyl-substituted ferrous examples of type **G** (Figure 1). In particular, five examples are disclosed that differ in the steric and electronic properties of the two *N*-aryl groups. To explore their use as catalysts in ethylene polymerization, a comprehensive evaluation has been undertaken that probes the type of aluminum alkyl activator as well as the effect of run temperature, Al:Fe molar ratio, reaction time, and ethylene pressure. To shed some light on any effects imparted by the *para*-phenyl group in the polymerizations, comparisons are made throughout with structurally related doubly fused iron catalysts such as **D** and **E** (Figure 1). Besides this polymerization study, the synthetic details and characterization of all five iron(II) complexes are also presented.

2. Results and Discussion

2.1. Synthesis and Characterization of Fe1–Fe5

To access the [*N,N*-diaryl-11-phenyl-1,2,3,7,8,9,10-heptahydrocyclohepta[*b*]quinoline-4,6-diimine]iron(II) chloride complexes (aryl = 2,6-Me$_2$Ph **Fe1**, 2,6-Et$_2$Ph **Fe2**, 2,6-i-Pr$_2$Ph **Fe3**, 2,4,6-Me$_3$Ph **Fe4**, and 2,6-Et$_2$-4-MePh **Fe5**), a straightforward one-pot approach proved effective. Specifically, by reacting 11-phenyl-1,2,3,7,8,9,10-heptahydrocyclohepta[*b*]quinoline-4,6-dione [24,29,30], the corresponding aniline and iron(II) chloride tetrahydrate in *n*-butanol at reflux under an atmosphere of nitrogen, in the presence of acetic acid as catalyst, **Fe1**–**Fe5** could be prepared as blue solids in yields of between 37% and 87% (Scheme 1). Attempts to make these complexes in a stepwise manner via the free ligand were hampered by difficulties in the purification of the diimine ligand due, in part, to the enamine tautomerism observed [15,24,31]. All five iron complexes proved stable when left to stand in the open air, but in solution, some evidence for oxidation could be detected. The characterization of these complexes was achieved using FT-IR spectroscopy, ESI mass spectrometry, and elemental analysis, while the crystals of **Fe2** and **Fe3** have been the subject of single X-ray diffraction studies.

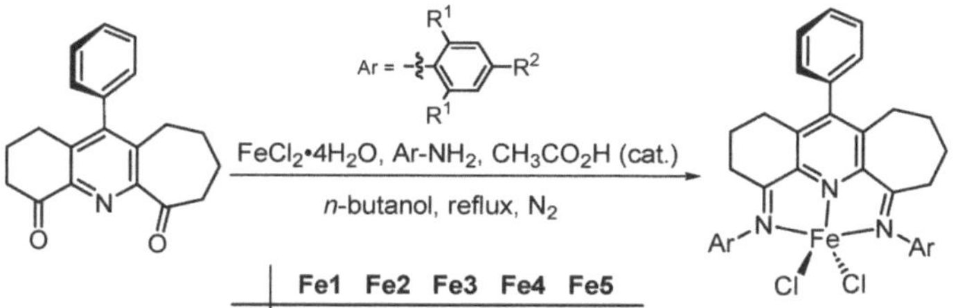

	Fe1	Fe2	Fe3	Fe4	Fe5
R^1	Me	Et	*i*-Pr	Me	Et
R^2	H	H	H	Me	Me
Yield %	85	66	65	87	37

Scheme 1. One-pot route to **Fe1**–**Fe5** from 11-phenyl-1,2,3,7,8,9,10-heptahydrocyclohepta[*b*]quinoline-4,6-dione.

Single crystals of **Fe2** and **Fe3** suitable for the X-ray determinations were grown in the glovebox under a nitrogen atmosphere by the slow diffusion of diethyl ether into a dichloromethane solution of the corresponding complex at ambient temperature. Perspective views of **Fe2** and **Fe3** are displayed in Figures 2 and 3; selected bond lengths and bond angles are presented in Table 1. For **Fe3**, the six- and seven-membered fused rings were disordered across both carbocyclic positions, an observation that has been noted previously [29]. Each structure consists of a single iron center coordinated by a tridentate *N,N*-diaryl-11-phenyl-1,2,3,7,8,9,10-heptahydrocyclohepta[*b*]quinoline-4,6-diimine ligand (aryl = 2,6-diethylphenyl **Fe2**, 2,6-diisopropylphenyl **Fe3**) and two chlorides to form a five-coordinate geometry, which can be best regarded as a distorted square pyramidal with Cl2 occupying the apex and the three nitrogen atoms and Cl1 the base. Some level of quantification of the degree of distortion can be demonstrated by using the tau value (τ), where τ = 0 represents an ideal square pyramidal and τ = 1 is an ideal trigonal bipyramid [32]. For **Fe2** and **Fe3**, τ = 0.11 and 0.21, respectively, which points towards modest distortions from an ideal square pyramidal, with the more sterically hindered 2,6-diisopropyl-containing **Fe3** causing a slightly larger distortion than in **Fe2**. The iron atom itself in each structure sits at a distance of 0.576 Å above the N1–N2–N3–Cl1 basal plane in **Fe2** and 0.536 Å in **Fe3**.

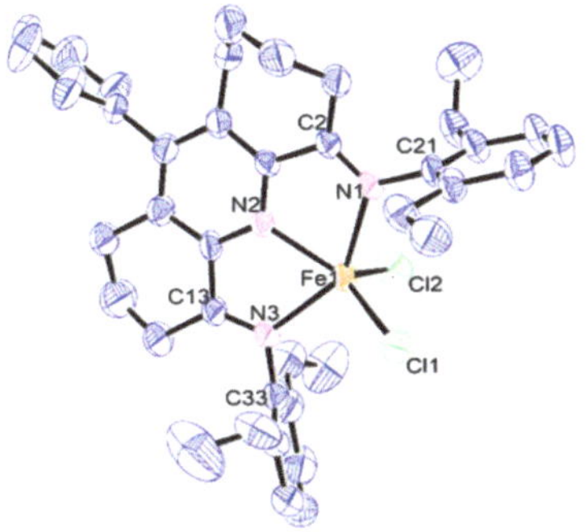

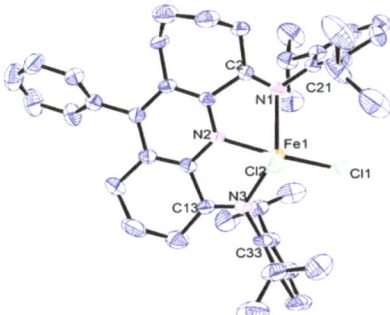

Figure 2. ORTEP representation of **Fe2** with the thermal ellipsoids shown at the 50% probability level; all the hydrogen atoms and solvents have been omitted for clarity.

Figure 3. ORTEP representation of **Fe3** with the thermal ellipsoids shown at the 50% probability level; all the hydrogen atoms have been omitted for clarity.

Table 1. Selected bond lengths and angles for **Fe2** and **Fe3**.

	Fe2	Fe3
Bond lengths (Å)		
Fe(1)–N(2)	2.142(5)	2.076(3)
Fe(1)–N(1)	2.241(5)	2.228(2)
Fe(1)–N(3)	2.240(5)	2.169(3)
Fe(1)–Cl(1)	2.2824(18)	2.2537(10)
Fe(1)–Cl(2)	2.3023(19)	2.3177(9)
N(1)–C(2)	1.277(8)	1.208(15)
N(1)–C(21)	1.441(7)	1.443(4)
N(3)–C(13)	1.286(9)	1.281(15)
N(3)–C(33)	1.436(8)	1.447(4)
Bond angles (°)		
Cl(1)–Fe(1)–Cl(2)	107.94(9)	116.77(4)
N(2)–Fe(1)–Cl(1)	137.74(15)	152.59(9)
N(2)–Fe(1)–Cl(2)	114.30(15)	90.63(8)
N(1)–Fe(1)–Cl(1)	103.03(14)	98.63(7)
N(1)–Fe(1)–Cl(2)	98.74(15)	102.38(7)
N(3)–Fe(1)–Cl(1)	99.50(15)	100.58(8)
N(3)–Fe(1)–Cl(2)	99.99(16)	99.99(8)
N(3)–Fe(1)–N(1)	144.49(19)	139.70(11)
N(2)–Fe(1)–N(1)	72.52(18)	73.13(9)
N(2)–Fe(1)–N(3)	72.32(18)	73.54(10)

With regard to N',N,N''–Fe chelation, the Fe–N distance involving the central Fe–$N_{pyridine}$ bond is the shortest (Fe1–N2: 2.142(5) Å **Fe2**, 2.076(3) Å **Fe3**), as is typical of previous analogs and reflects the binding properties of the tridentate ligand [20,22,24]. For the longer exterior Fe–N_{imino} bonds, some variation is evident between structure and those in **Fe2** comparable (2.241(5) and 2.240(5) Å), while in **Fe3** the Fe1–N1 distance (2.228(2) Å) is noticeably longer than the Fe1–N3 one (2.169(3) Å), likely reflecting the steric properties of the larger 7-membered carbocycle and the bulky *ortho*-isopropyl groups. Notably, in the non-phenyl substituted iron(II) counterpart of **Fe3**, **E** (M = Fe, Ar = diisopropylphenyl, Figure 1), the exterior Fe–N_{imine} distances are equivalent (2.216(1) Å) [24], which suggests the *para*-phenyl group in **Fe3** exerts some long-range influence on the binding properties of this particular N',N,N''-ligand. On the other hand, a similar comparison of the Fe–$N_{pyridine}$ bond distance in **Fe3** with that in **E** (M = Fe, Ar = diisopropylphenyl, Figure 1) shows only a modest lengthening in the former (2.076(3) vs. 2.060(1) Å). For both **Fe2** and **Fe3**, some deviation from coplanarity between the exterior imine groups and pyridine ring is seen, as is evidenced by the N2–C1–C2–N1 torsion angles (11.82° **Fe2**, 11.40° **Fe3**) and N2–C14–C13–N3 (8.12° **Fe2**, 10.90° **Fe3**), which can be credited to the strain imparted by the carbocyclic groups on the N',N,N''-ligand framework.

As is common to many bis(arylimino)pyridine-iron complexes, the two N-aryl groups are inclined essentially perpendicularly to the N',N,N''-coordination plane in both **Fe2** and **Fe3**, thereby positioning the *ortho*-substituents above and below this plane. The six- and seven-membered carbocycles in the chelating ligand adopt puckered arrangements, with the latter showing the most flexibility. The *para*-phenyl groups themselves in **Fe2** and **Fe3** adopt skewed dispositions with respect to the neighboring pyridine plane (tors. C9–C8–C15–C20 84.66° **Fe2**, 56.16° **Fe3**) due to the rotary flexibility of the *para*-phenyl group; similar flexibility has been noted in their cobalt analogs [29]. There are no intermolecular contacts of note.

In the FT-IR spectra of **Fe1**–**Fe5** (Figures S14–S18), characteristic peaks were identifiable for the imine double bonds between 1604 and 1617 cm^{-1}, a range that is quite typical for coordinated imine groups in related iron(II) complexes [20,22,24,33,34]. In addition, the microanalytical data for each iron complex lend good support for compositions of the type LFeCl$_2$.

2.2. Ethylene Polymerization Investigations Using **Fe1**–**Fe5**

As has been noted previously, methylaluminoxane (MAO) and modified methylaluminoxane (MMAO) are among the most effective activators for iron-based ethylene polymerization catalysis [20,22,24]. Accordingly, these two aluminoxanes were taken forward as part of parallel studies with the purpose of exploring not only structure–activity/polymer correlations in **Fe1**–**Fe5** but also the impact of the type of activator on these correlations. All polymerizations were conducted in toluene, with the ethylene pressure initially set at 10 atm.

2.2.1. Polymerization Studies Using **Fe1**–**Fe5** under Activation with MMAO

To allow a workable set of polymerization conditions using MMAO, **Fe4** was initially employed as the test precatalyst, and the most effective conditions identified were then extended to evaluate **Fe1**–**Fe3** and **Fe5**; the complete set of results is shown in Table 2.

To examine the effect of run temperature, the polymerizations using **Fe4**/MMAO were initially carried out at temperatures between 40 °C and 80 °C with the Al:Fe molar ratio fixed at 2500:1 (**Fe4**: 1 µmol) and a run time of 30 min (entries 1–5, Table 2). Inspection of the results reveals the catalytic activity reached a maximum of 19.14×10^6 g (PE) mol^{-1} (Fe) h^{-1} at 60 °C and still remained high with the temperature at 80 °C, with a level of 7.32×10^6 g (PE) mol^{-1} (Fe) h^{-1} noted (Figure 4). These performance characteristics reflect the appreciable thermal stability of **Fe4**. With regard to the polymer generated, the GPC traces indicate that the molecular weight rapidly decreased from 51.6 kg mol^{-1} to 3.7 kg mol^{-1} as the temperature was increased from 40 °C to 80 °C, respectively, in line with the higher rate of chain termination at higher temperatures [24]. Furthermore, the dispersity of the polymer generated significantly narrowed as the temperature was increased (M_w/M_n: from 14.2 to 1.5), with an initially bimodal distribution becoming unimodal at temperatures in excess of 60 °C. This evidence of bimodality at lower temperatures could be attributed to the different chain transfer pathways operational, namely β-H elimination and chain transfer to aluminum alkyl species (e.g., AlMe$_3$) in MMAO [4,10,24]. Alternatively, the bimodality could be caused by the presence of oxidized iron complexes, leading to additional active centers [22]. On the other hand, at higher run temperatures, it would seem plausible that either a single active center predominates or a single chain transfer pathway becomes prevalent [22].

Table 2. Ethylene polymerization screening using **Fe1–Fe5** with MMAO as activator [a].

Entry	Precat.	Al:Fe	T (°C)	t (min)	Mass of PE (g)	Activity [b]	M_w [c]	M_w/M_n [c]	T_m (°C) [d]
1	Fe4	2500	40	30	5.14	10.28	51.6	14.2	131.1
2	Fe4	2500	50	30	8.69	17.38	35.6	10.7	129.8
3	Fe4	2500	60	30	9.57	19.14	9.8	3.1	130.0
4	Fe4	2500	70	30	8.91	17.82	4.8	1.5	127.7
5	Fe4	2500	80	30	3.66	7.32	3.7	1.5	128.8
6	Fe4	1500	60	30	2.57	5.14	46.0	2.8	132.4
7	Fe4	1750	60	30	9.75	19.50	36.2	2.7	133.1
8	Fe4	2000	60	30	17.05	34.10	81.3	9.6	131.8
9	Fe4	2250	60	30	12.50	25.00	85.1	7.0	132.8
10	Fe4	2000	60	5	8.12	97.44	13.7	3.5	128.8
11	Fe4	2000	60	15	11.27	45.08	23.0	3.9	131.2
12	Fe4	2000	60	45	18.01	24.01	90.6	9.1	135.3
13	Fe4	2000	60	60	18.24	18.24	94.3	9.2	132.6
14 [e]	Fe4	2000	60	30	0.61	1.22	0.8	1.3	120.2
15 [f]	Fe4	2000	60	30	8.98	17.96	79.0	7.4	132.0
16	Fe1	2000	60	30	12.89	25.78	26.4	6.4	129.5
17	Fe2	2000	60	30	14.67	29.34	70.0	6.9	131.9
18	Fe3	2000	60	30	5.33	10.66	51.6	8.1	131.1
19	Fe5	2000	60	30	3.82	7.64	69.4	13.6	130.1

[a] Conditions: MMAO as an activator, 1.0 µmol of the iron precatalyst, 100 mL of toluene, and 10 atm of ethylene; [b] Values in units of 10^6 g (PE) mol^{-1} (Fe) h^{-1}; [c] M_w and M_n in units of kg mol^{-1}, determined by GPC; [d] Determined by DSC; [e] 1 atm of ethylene; [f] 5 atm of ethylene.

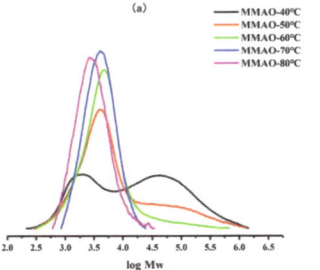 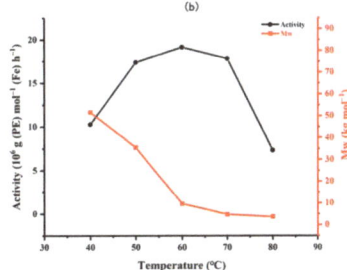

Figure 4. (**a**) GPC traces of the polyethylenes generated using **Fe4**/MMAO at different temperatures and (**b**) plots of catalytic activity and polymer molecular weight as a function of run temperature.

Thereafter, the polymerizations using **Fe4**/MMAO were undertaken at Al:Fe molar ratios between 1500:1 and 2500:1, with the temperature kept at 60 °C (entries 3 and 6–9, Table 2). As can be seen from the tabulated results, the highest activity of 34.10 × 10^6 g (PE) mol^{-1} (Fe) h^{-1} was recorded with the ratio at 2000:1, which was extremely high when compared with previously reported N,N,N-iron catalysts under comparable reaction conditions [20,22,24]. With the Al:Fe molar ratio gradually increasing from 1500:1 to 2500:1, the highest molecular weight polymer of 85.1 kg mol^{-1} was recorded at 2250:1, while the lowest of 9.8 kg mol^{-1} was obtained at 2500:1 (Figure 5). This dramatic decline in molecular weight with ratios exceeding 2250:1 reflects the critical value of aluminoxane necessary before chain transfer from the active species to aluminum in MMAO becomes significant [22,24].

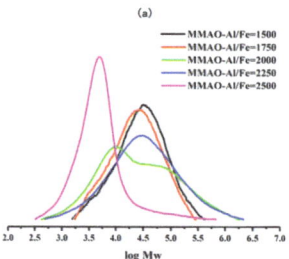
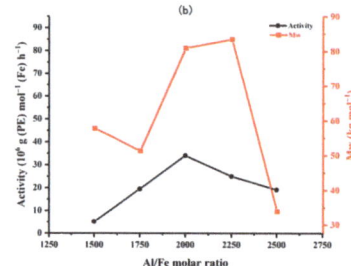

Figure 5. (a) GPC traces of the polyethylenes generated using **Fe4**/MMAO at various Al:Fe molar ratios and (b) plots of catalytic activity and polymer molecular weight as a function of Al:Fe molar ratio.

To explore the effect of the run time using **Fe4**/MMAO, five separate polymerizations were conducted over 5, 15, 30, 45, and 60 min with the Al:Fe molar ratio kept at 2000:1 and the temperature at 60 °C (entries 8 and 10–13, Table 2). As time elapsed, the activity rapidly decreased from 97.44×10^6 g (PE) mol^{-1} (Fe) h^{-1} after 5 min to 18.24×10^6 g (PE) mol^{-1} (Fe) h^{-1} by the one-hour mark (Figure 6), suggesting a short induction period for this catalyst and a relatively long lifetime for the active species [20,22,24]. With respect to the molecular weight data, a gradual increase was observed in the first 15 min, followed by a rapid increase between 15 and 30 min, and then assuming a more constant level of molecular weight as the run reached 60 min. In terms of the dispersity, a steady increase was noted over the first 15 min (M_w/M_n: from 3.5 to 3.9) before a significant broadening and the onset of some bimodality were seen between 30 and 60 min (M_w/M_n: from 9.1 to 9.6), suggesting the formation of another active species [22] or an additional chain transfer pathway becoming operative.

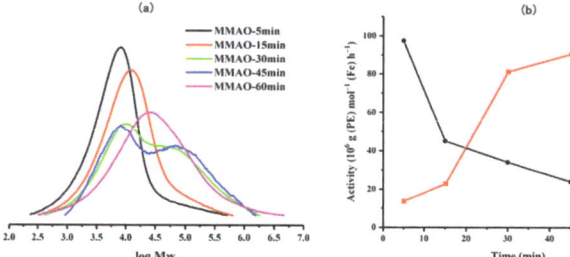

Figure 6. (a) GPC traces of the polyethylenes generated using **Fe4**/MMAO as the run time was varied, and (b) plots of catalytic activity and polymer molecular weight as a function of reaction time.

To investigate the effect of the ethylene pressure on the polymerizations using **Fe4**/MMAO, the runs were additionally conducted at pressures of 1 and 5 atm with the Al:Fe molar ratio fixed at 2000:1 and the temperature kept at 60 °C (entries 8, 14, and 15, Table 2). At 1 atm, the lowest activity of 1.22×10^6 g (PE) mol^{-1} (Fe) h^{-1} was observed, which can be credited to the relatively low solubility of ethylene in toluene at this pressure [24]. On the other hand, at 10 atm, the activity attained its highest value (34.10×10^6 g (PE) mol^{-1} (Fe) h^{-1}) and indeed was around twice that seen at 5 atm (17.96×10^6 g (PE) mol^{-1} (Fe) h^{-1}), underscoring the beneficial effect of ethylene pressure on productivity.

On the strength of the most effective set of conditions identified using **Fe4**/MMAO (run temperature = 60 °C, Al:Fe = 2000:1, run time = 30 min, P_{C2H4} = 10 atm), the remaining iron precatalysts, **Fe1**, **Fe2**, **Fe3**, and **Fe5**, were then studied accordingly (entries 16–19, Table 2). All catalysts displayed high activity (range: 7.64–34.10×10^6 g (PE) mol^{-1} (Fe) h^{-1}), with the relative order being: **Fe4** (2,4,6-trimethyl) > **Fe2** (2,6-diethyl) > **Fe1** (2,6-dimethyl) > **Fe3** (2,6-diisopropyl) > **Fe5** (2,6-diethyl-4-methyl) (Figure 7). These results suggest that less bulky *ortho*-R^1 groups are, in general, beneficial to the process of ethylene coordination/migratory insertion [24,30,35]. Nevertheless, it is evident that there are some anomalies in this structure/activity correlation, which could be attributable to the long-range effect of the *para*-phenyl group on carbocyclic ring flexibility and, in turn, N-aryl group binding/inclination [29].

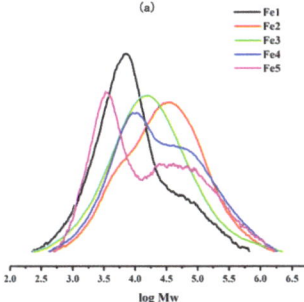 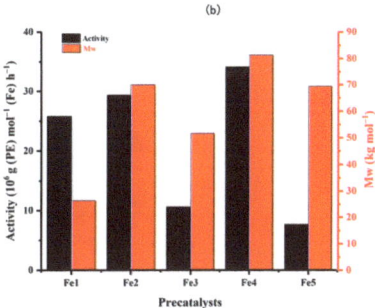

Figure 7. (**a**) GPC traces of the polyethylenes generated using **Fe1**–**Fe5** with MMAO as an activator and (**b**) a bar chart showing catalytic activity and polymer molecular weight with respect to the iron precatalyst employed.

As for the GPC traces, moderate molecular weight polymer (M_w: 26.4–81.3 kg mol^{-1}) was obtained for all precatalysts, while the dispersities were in all cases broad (M_w/M_n: 6.4–13.6). Indeed, the least hindered **Fe4**-formed polymer exhibited a higher molecular weight than that seen for diisopropyl-containing **Fe3**, which was quite different from previous studies where larger *ortho*-substituents tend to promote higher molecular weight material [20,22,24]. The reason for this observation remains uncertain but may conceivably be attributed to the *para*-phenyl group somehow influencing chain propagation, which is most apparent in the more sterically hindered **Fe3**. By contrast, the less bulky **Fe1** (2,6-dimethyl) and **Fe2** (2,6-diethyl) are less affected by the *para*-phenyl group, resulting in the molecular weight of the polymer formed by **Fe2** being greater than that of **Fe1**, an observation that can be credited to the bulkier *ortho*-ethyl groups better protecting the active center [20,22,24].

To understand the performance characteristics of the current catalysts in terms of activity and polymer molecular weight and dispersity, data for mesityl-containing **Fe4** (denoted in **G** as in Figure 1) are collected in Figure 8 alongside that for some previously reported iron systems incorporating doubly fused carbocycles. From this bar chart, it is evident that the highest activity of these iron complexes follows the order: **G** (**Fe4**: this work) >> $\mathbf{E}_{n=1/n=2}$ > $\mathbf{D}_{2(n=3)}$ ≈ $\mathbf{D}_{2(n=2)}$ (Figure 1) [20,22,24]. Evidently, the *para*-phenyl group in **G** exerts a positive effect on catalytic activity, which is most striking when compared to its non-phenyl substituted comparator $\mathbf{E}_{n=1/n=2}$, a finding that mirrors that seen with the cobalt counterparts [29]. While the origin of the effect remains uncertain, we consider the *para*-phenyl group to have a limited effect on the binding properties of this mesityl-substituted N',N,N''-ligand. Indeed, analysis of the X-ray structure of the 2,6-diethylphenyl-containing **Fe2** showed no evidence of uneven binding of the chelating ligand, a feature that was notably apparent in the bulkier derivative **Fe3** (see above). One plausible explanation for **G** (**Fe4**) being more active than $\mathbf{E}_{n=1/n=2}$ stems from the electron-withdrawing properties of the phenyl group and the impact this has on the ethylene coordination and insertion at the active iron center. Furthermore, the molecular weight of the polyethylene decreases in the following order: $\mathbf{D}_{2(n=3)}$ > **G** (**Fe4**: this work) > $\mathbf{E}_{n=1/n=2}$ > $\mathbf{D}_{2(n=2)}$. This finding shows that the presence of both the large eight-membered fused structure ($\mathbf{D}_{2(n=3)}$) and the *para*-phenyl group in **G** (**Fe4**) can have a favorable effect on the molecular weight of the polymer, observations that are similar to those observed for their cobalt analogs. In short, the introduction of the *para*-phenyl group to the N',N,N''-ligand periphery in **G** has the effect of promoting the catalytic activity and increasing the molecular weight of the polymer for both iron and cobalt complexes. In terms of the optimal run temperature, **G** (**Fe4**: this work) proved the most effective by operating effectively at 60 °C, which compares to 40 °C for the non-*para*-phenyl substituted analog **E**$_{n=1/n=2}$ [24], a result that suggests the *para*-phenyl group can also help provide a more stable active species at higher operating temperatures. As a final point, it can be seen from the figure that the dispersity is broad for all of these iron complexes, in line with the different chain transfer pathways that are operational.

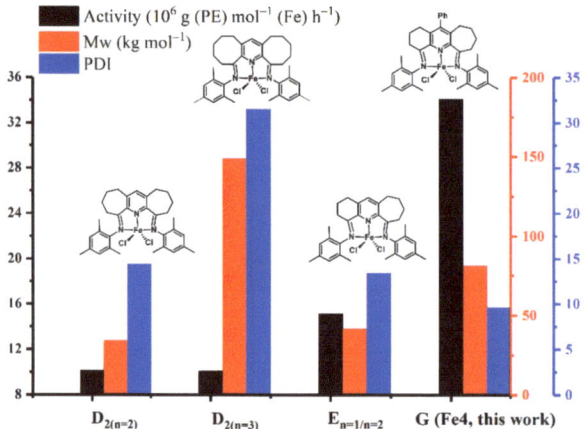

Figure 8. Comparison of the catalytic activity, polymer molecular weight, and dispersity obtained using **G** (**Fe4**, this work) with that for carbocyclic-fused iron precatalysts, **D**$_{2(n=2)}$, **D**$_{2(n=3)}$, and **E**$_{n=1/n=2}$; all polymerizations were conducted using MMAO at 10 atm of ethylene and under optimal reaction conditions.

2.2.2. Polymerization Studies Using **Fe1–Fe5** under Activation with MAO

To enable a comparison with the polymerization runs performed with MMAO, MAO was also employed in the activation of **Fe1–Fe5**. As before, **Fe4** was initially employed as the test precatalyst so as to identify an effective set of reaction conditions that could be used to screen the other iron precatalysts (Table 3).

In the first instance, the polymerization runs were performed using **Fe4**/MAO at run temperatures of between 40 and 80 °C (entries 1–5, Table 3) with the Al:Fe molar ratio fixed at 2500:1 and the run time at 30 min (entries 1–5, Table 3). The highest catalytic activity of 30.58×10^6 g (PE) mol^{-1} (Fe) h^{-1} was achieved at 60 °C, reflecting the outstanding thermal stability of this iron catalyst. Moreover, the molecular weight of the polymer dropped rapidly from 95.5 kg mol^{-1} at 40 °C to 4.9 kg mol^{-1} at 80 °C (Figure S1), which can be attributed to the higher rate of chain transfer at high temperatures [24]. As for the polymer dispersity, this decreased from 18.5 (40 °C) to 1.5 (80 °C) as the temperature increased, which manifests itself in the initially bimodal distribution becoming unimodal, findings that likely stem from the two chain transfer pathways becoming less competitive at higher temperatures [9,10,36].

Following this, the polymerizations were evaluated with Al:Fe molar ratios of 1000:1, 1500:1, 2000:1, 2500:1, and 3000:1 with the temperature fixed at 60 °C (entries 3 and 6–9, Table 3). As a result, the highest activity of 35.92×10^6 g (PE) mol^{-1} (Fe) h^{-1} was noted with the ratio of 2000:1 (Figure S2), which was marginally higher than the optimal activity for the MMAO runs (34.10×10^6 g (PE) mol^{-1} (Fe) h^{-1}). As for the molecular weight of the polymer, this was found to markedly decrease from 161.2 to 32.3 kg mol^{-1} as the Al:Fe molar ratio was increased, in line with a higher rate of chain transfer [24].

To shed light on the catalytic lifetime of the active center in **Fe4**/MAO, the polymerizations were carried out at different run times with the temperature and Al:Fe molar ratio kept at 60 °C and 2000:1, respectively. As expected, an exceptionally high activity of

84.96×10^6 g (PE) mol^{-1} (Fe) h^{-1} was observed after 5 min (entry 10, Table 3), which then progressively decreased to 18.65×10^6 g (PE) mol^{-1} (Fe) h^{-1} at 60 min (entry 13, Table 3), suggesting a short induction period and an excellent lifetime for the active center [20]. Meanwhile, the molecular weight of the polyethylene increased from 16.8 to 168.0 kg mol^{-1} over time (Figure S3), while the dispersity (M_w/M_n) increased from 3.8 to 16.6 in a similar manner to that seen for **Fe4**/MMAO.

Table 3. Ethylene polymerization screening using **Fe1–Fe5** with MAO as activator [a].

Entry	Precat.	Al:Fe	T (°C)	t (min)	Mass of PE (g)	Activity [b]	M_w [c]	M_w/M_n [c]	T_m (°C) [d]
1	Fe4	2500	40	30	6.04	12.08	95.5	18.5	130.2
2	Fe4	2500	50	30	10.03	20.06	72.4	11.7	130.7
3	Fe4	2500	60	30	15.29	30.58	65.0	8.8	131.7
4	Fe4	2500	70	30	9.85	19.70	18.4	3.0	130.3
5	Fe4	2500	80	30	5.05	10.10	4.9	1.5	132.7
6	Fe4	1000	60	30	10.54	21.08	161.2	8.4	132.1
7	Fe4	1500	60	30	12.52	25.04	152.5	13.8	132.3
8	Fe4	2000	60	30	17.96	35.92	129.0	11.4	132.3
9	Fe4	3000	60	30	14.80	29.60	32.3	4.3	130.7
10	Fe4	2000	60	5	7.08	84.96	16.8	3.8	129.4
11	Fe4	2000	60	15	11.05	44.20	29.7	6.1	130.0
12	Fe4	2000	60	45	18.13	24.17	142.1	15.4	131.5
13	Fe4	2000	60	60	18.65	18.65	168.0	16.6	133.0
14 [e]	Fe4	2000	60	30	0.84	1.68	3.4	4.0	124.8
15 [f]	Fe4	2000	60	30	7.04	14.08	130.7	13.0	132.7
16	Fe1	2000	60	30	14.10	28.20	48.9	8.8	131.1
17	Fe2	2000	60	30	10.18	20.36	59.2	8.5	131.4
18	Fe3	2000	60	30	6.52	13.04	35.2	7.0	131.4
19	Fe5	2000	60	30	8.73	17.46	50.2	8.8	132.2

[a] Conditions: MAO as an activator, 1.0 μmol of the iron precatalyst, 100 mL of toluene, 10 atm of ethylene; [b] Values in units of 10^6 g (PE) mol^{-1} (Fe) h^{-1}; [c] M_w and M_n in units of kg mol^{-1}, determined by GPC; [d] Determined by DSC; [e] 1 atm of ethylene; [f] 5 atm of ethylene.

On lowering the ethylene pressure from 10 atm to 1 atm using **Fe4**/MAO, a sharp decline in activity was observed from 35.92×10^6 g (PE) mol^{-1} (Fe) h^{-1} to 0.84×10^6 g (PE) mol^{-1} (Fe) h^{-1} (entries 8, 14, and 15, Table 3), which most likely derives from the lower solubility of ethylene in toluene at lower pressure [23,29]. Furthermore, the molecular weight of the polyethylene decreased noticeably from 129.0 to 3.4 kg mol^{-1} by dropping the pressure from 10 atm to 1 atm, an observation that supports the lower rates of chain propagation that are operational at lower pressure [24]. In addition, the molecular weight of the polyethylene formed during the **Fe4**/MAO runs was higher than that seen using **Fe4**/MMAO, reflecting the lower rate of chain transfer with MAO.

To gain further insight into the role of the *N*-aryl variations in **Fe1–Fe5**, the optimal conditions identified using **Fe4**/MAO (namely reaction temperature of 60 °C, Al:Fe molar ratio of 2000:1, ethylene pressure of 10 atm, and run time of 30 min) were deployed to evaluate the other iron precatalysts, **Fe1–Fe3** and **Fe5** (entries 16–19, Table 3). All iron complexes exhibited high activity (range: $13.04–35.92 \times 10^6$ g (PE) mol^{-1} (Fe) h^{-1}) and produced highly linear polyethylenes (T_m: 131–133 °C) with broad dispersities. On analysis of their relative performance, the catalytic activity decreased in the following order: **Fe4** (aryl = 2,4,6-trimethylphenyl) > **Fe1** (aryl = 2,6-dimethylphenyl) > **Fe2** (aryl = 2,6-diethylphenyl) > **Fe5** (aryl = 2,6-diethyl-4-methylphenyl) > **Fe3** (aryl = 2,6-diisopropylphenyl) (Figure S4). As is evident, the precatalysts bearing the least bulky substituents in the *ortho* positions of the *N*-aryl groups, **Fe4** and **Fe1**, exhibited higher activity due to the faster rates of ethylene insertion and chain propagation [24,30,35]. Moreover, mesityl-containing **Fe4** produced the highest molecular weight polyethylene of the series, which was also seen for the MMAO runs, but nonetheless unexpected when compared with previous work in which the most sterically hindered complex usually affords the highest molecular weight polymer. This unusual finding plausibly arose from the very high activity seen for **Fe4**, thereby inhibiting chain transfer.

2.3. Structural Analysis of the Polyethylene

The microstructural characteristics of polymers are of great significance to their mechanical properties and, in turn, their processing ability. As has already been highlighted, the DSC thermograms for the polyethylenes produced in this work typically display T_m values of >130 °C in accordance with highly linear polymers (see Tables 2 and 3) [16,18,20,23,24]. To obtain more information, selected polyethylene samples were further analyzed by ^{1}H and ^{13}C NMR spectroscopy with a view to understanding how run temperature and the type of aluminum activator impact the microstructure.

To begin with, we looked at the polymers generated using **Fe4**/MMAO at run temperatures of 40, 50, 60, 70, and 80 °C, where variations in molecular weight and dispersity were observed. To engender suitable solubility, these polymer samples were dissolved in the 1,1,2,2-tetrachloroethane-d_2 at 100 °C, and their ^{1}H NMR spectra were recorded at a similar temperature (entries 1–5, Table 2). For the ^{1}H NMR spectrum of the polyethylene obtained using **Fe4**/MMAO at 80 °C (entry 5, Table 2: M_w = 3.7 kg mol^{-1}, M_w/M_n = 1.5), a high-intensity resonance at δ 1.32 ppm for the –(CH$_2$)$_n$– repeat unit confirms the strictly linear nature of the polyethylene (Figure 9). Either side of this signal shows lower intensity peaks, which correspond to an n-propyl group (H$_f$, H$_e$, and H$_d$), while the more downfield region reveals multiplet peaks at δ 5.00 and 5.85 ppm, which can be assigned to a vinyl end-group (H$_a$ and H$_b$). Evidently, the latter observation highlights the involvement of β-H elimination to metal or monomer during chain termination [24]. Moreover, by considering the integral ratio of vinyl H$_b$ to methyl chain-end H$_f$ (1/18.82) [29,37,38], the molar ratio of the unsaturated polymer chain to the saturated polymer chain can be determined as 0.379 (see SI), a value that implies the additional presence of a fully saturated polymer in line with chain transfer to aluminum being also operational [4,5]. Further support for these assignments was provided by the ^{13}C NMR spectrum, which reveals a high-intensity peak around δ 29.43 ppm, which corresponds to the –(CH$_2$)$_n$– repeat unit, along with lower intensity peaks at δ 13.72, 22.38, and 31.66 ppm for an n-propyl end-group and weaker downfield peaks at δ 113.85 and 138.78 ppm for a terminal vinyl group (Figure 10) [24]. The absence of resonances corresponding to an isobutyl end-group rules out any chain transfer to aluminum-isobutyl species present within MMAO [30], which suggests that chain transfer to aluminum occurs solely with aluminum-methyl species.

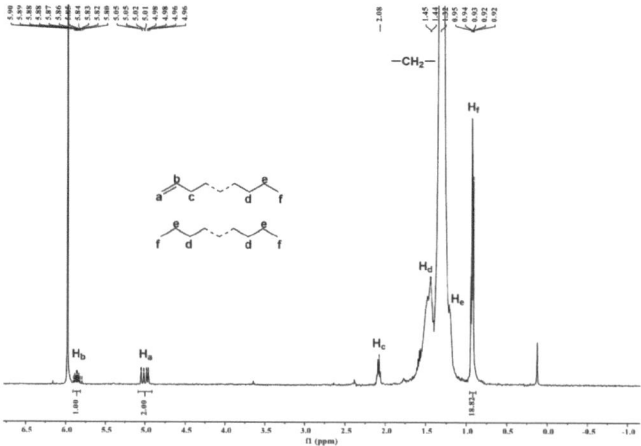

Figure 9. ^{1}H NMR spectrum of the polyethylene produced using **Fe4**/MMAO operating at 80 °C (entry 5, Table 2); recorded at 100 °C in 1,1,2,2-tetrachloroethane-d_2.

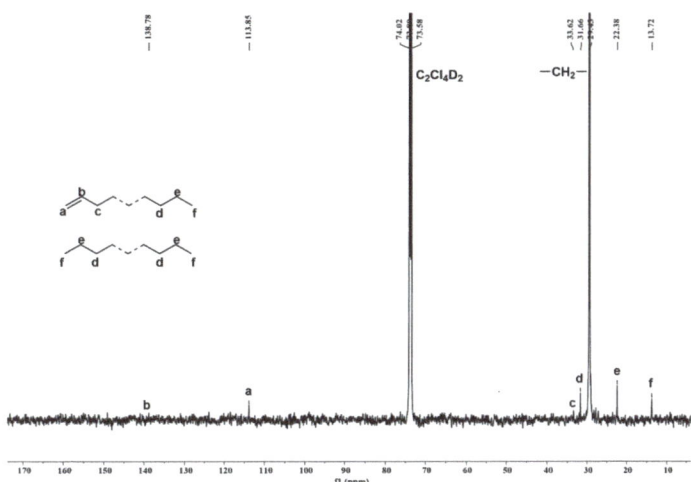

Figure 10. ^{13}C NMR spectrum of the polyethylene produced using **Fe4**/MMAO operating at 80 °C (entry 5, Table 2); recorded at 100 °C in 1,1,2,2-tetrachloroethane-d_2.

The ^{1}H NMR spectra for the polymers produced using **Fe4**/MMAO at the lower run temperatures (40–70 °C, entries 1–4, Table 2) revealed similar features, with signals characteristic of the -(CH$_2$)$_n$- repeat unit and weaker peaks for the *n*-propyl and vinyl chain-ends (Figures S5–S8). However, the molar ratio of unsaturated to saturated polymer chains showed some notable variations, with the value reaching a maximum of 1.98 at 60 °C (Figure 11), which implies that β-H elimination assumes the main chain termination pathway at this point. However, at temperatures below 60 °C, the ratio drops to 0.394 (50 °C) and then 0.108 (40 °C), reflecting the importance of chain transfer to aluminum-methyl species in MMAO at these temperatures. Evidently, the chain transfer pathways are affected by the run temperature, with the highest rate of β-H elimination occurring at 60 °C, which also corresponds to the optimal temperature in terms of catalytic activity.

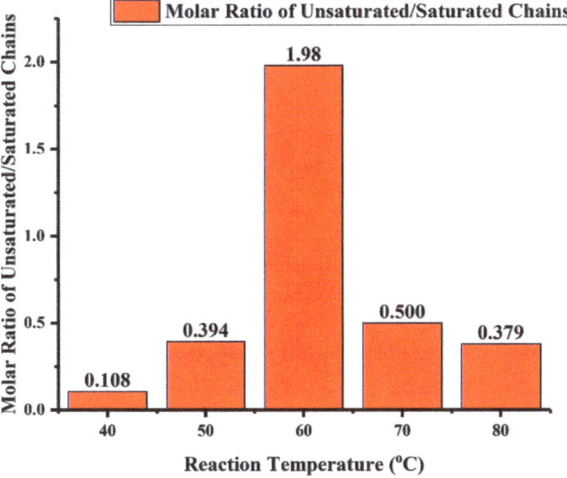

Figure 11. Bar chart showing the ratio of unsaturated to saturated chains for the polymer obtained as a function of run temperature using **Fe4**/MMAO as catalyst (entries 1–5, Table 2).

To appreciate the effect of the aluminum activator on these chain transfer pathways, the polyethylenes produced using **Fe4**/MAO over the same run temperature range (40–80 °C) as that used for **Fe4**/MMAO were similarly investigated by ^{1}H NMR spectroscopy. Once again, the ^{1}H NMR spectra of all five polymer samples revealed peaks characteristic of a highly linear polymer backbone along with varying amounts of chain end vinyl and *n*-propyl groups (Figures S9–S13). Interestingly, the highest rate of β-H elimination using **Fe4**/MAO was found at 70 °C (Figure 12), as evidenced by the highest molar ratio of unsaturated to saturated polymer chains in the polymer (1.05: H_b:H_f = 1/8.71). Nevertheless, this value was lower than that seen with the MMAO-activated system, an observation that reflects the differences in the type of aluminoxane and their impact on the competing chain transfer pathways.

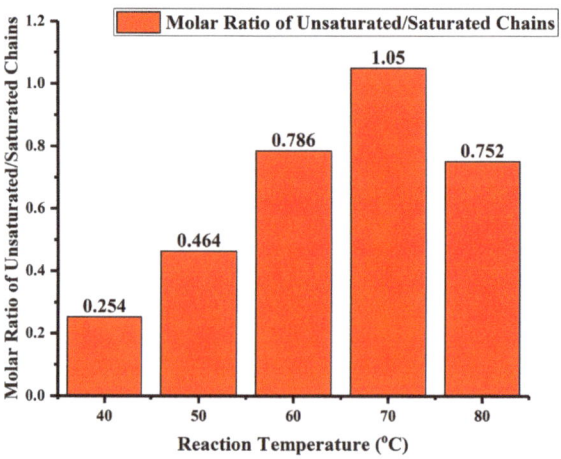

Figure 12. Bar chart showing the ratio of unsaturated to saturated chains for the polymer obtained as a function of run temperature using **Fe4**/MAO as catalyst (entries 1–5, Table 3).

3. Materials and Methods

3.1. General Considerations

All operations making use of moisture and/or air-sensitive compounds were conducted under a nitrogen atmosphere by using standard Schlenk techniques or in a nitrogen-filled glovebox. Toluene, the solvent used for the polymerization studies, was heated to reflux for more than 12 h over sodium benzophenone and distilled under nitrogen prior to use. The aluminum reagents, methylaluminoxane (MAO, 1.5 M in toluene) and modified methylaluminoxane (MMAO, 2.5 M in n-heptane, containing 20–25%Al(*i*-Bu)$_3$), were purchased from Anhui Botai Electronic Materials Co. (Chuzhou, China), while high-purity ethylene was procured from Beijing Yanshan Petrochemical Co. (Beijing, China) and used as received. Other chemical reagents were purchased from Concord Technology Co., Ltd. (Tianjin, China), Macklin Biochemical Technology Co., Ltd. (Shanghai, China), and Innochem Technology Co., Ltd. (Beijing, China) and used as received: diethyl ether (GR., Concord, Tianjin, China), *n*-hexane (GR., Concord, Tianjin, China), glacial acetic acid (AR., Concord, Tianjin, China), *n*-butanol (AR., Concord, Tianjin, China), ferrous chloride tetrahydrate (98% purity, Macklin, Shanghai, China), anilines (99% purity, Innochem, Beijing, China). The samples for FT-IR analysis were placed directly on the ATR attachment plane and pressed with a flat nut, and the FT-IR spectra were recorded on a PerkinElmer System 2000 FT-IR spectrometer (PerkinElmer Scientific, Waltham, MA, USA). The elemental analyses were conducted using a Flash EA 1112 microanalyzer (Thermo Fisher Scientific, Waltham, MA, USA). The ESI mass spectra were recorded using a Bruker solariX instrument (Bruker Corporation, Billerica, MA, USA) (Acquisition Mode: Single MS; Polarity: Positive; Broadband Low Mass: 57.7 *m/z*; Acquired Scans: 20; Broadband

High Mass: 1000.0 m/z). The melting points of the polyethylenes were determined on a PerkinElmer TA-Q2000 DSC analyzer (PerkinElmer Scientific, Waltham, MA, USA) under a nitrogen atmosphere according to the following procedure: around 5.0 mg of the polymer was heated to 160 °C at a rate of 20 °C min^{-1} and then maintained for 3 min at 160 °C to remove the thermal history. After cooling to −20 °C at a rate of 20 °C min^{-1}, the sample was then heated to 160 °C again at the rate of 10 °C min^{-1}. The polymer molecular weight (M_w) and dispersity (M_w/M_n) were measured using an Agilent PLGPC 220GPC system running (Agilent Technologies Inc., Santa Clara, CA, USA) at 150 °C using 1,2,4-trichlorobenzene as the mobile phase; sample preparation involved agitation of ca. 10 mg of the polyethylene sample in 1,2,4-trichlorobenzene (5 mL) at 170 °C for more than 8 h to allow dissolution. The ^{1}H NMR and ^{13}C NMR spectra of the polymers were measured at 100 °C using a Bruker AVANVE 500 MHz instrument (Bruker Corporation, Billerica, MA, USA); sample preparation involved dissolving the selected polyethylene (ca. 50 mg) in 1,1,2,2-tetrachloroethane-d_2 (0.6 mL), containing TMS as the internal standard, at 110 °C for more than 2 h. The 11-phenyl-1,2,3,7,8,9,10-heptahydrocyclohepta[b]quinoline-4,6-dione was prepared according to procedures reported elsewhere [29,39–41].

*3.2. Synthesis of [N,N-diaryl-11-phenyl-1,2,3,7,8,9,10-heptahydrocyclohepta[b]quinoline-4,6-diimine] iron(II) Chloride (**Fe1**–**Fe5**)*

3.2.1. Aryl = 2,6-Dimethylphenyl (**Fe1**)

Under an atmosphere of nitrogen, a mixture of 11-phenyl-1,2,3,7,8,9,10-heptahydrocyclohepta[b]quinoline-4,6-dione (0.153 g, 0.50 mmol), 2,6-dimethylaniline (0.242 g, 2.0 mmol), and ferrous chloride tetrahydrate (0.080 g, 0.40 mmol) was dissolved in n-butanol (2.0 mL), containing a catalytic amount of acetic acid (0.5 mL), and then stirred and heated to reflux for 2 h to give a blue-green solution. After cooling to room temperature, diethyl ether and n-hexane (32 mL, v:v = 1:1) were added to induce precipitation. This precipitate was then left to settle, and the supernatant solution was discarded. This process of adding diethyl ether/n-hexane (32 mL, v:v = 1:1) and discarding the supernatant solution was repeated two more times. Finally, the mixture was filtered and washed three times with diethyl ether and n-hexane (24 mL, v:v = 2:1), affording **Fe1** as a blue powder (0.217, 85%). FT-IR (cm^{-1}): 2950 (m), 1678 (m), 1617 (m, $\nu_{C=N}$), 1570 (m), 1441 (s), 1376 (m), 1218 (s), 1073 (s), 920 (m), 844 (m), 770 (s), 704 (s). ESI-MS (m/z): Calculated for $[C_{36}H_{37}ClFeN_3]^+$ (602.20254). Found: 602.20232 (Figure S19). Elemental analysis: Calculated for $C_{36}H_{37}Cl_2N_3Fe$ (638.46) C, 67.73, H, 5.84, N, 6.58. Found: C, 67.31, H, 6.25, N, 6.68%.

3.2.2. Aryl = 2,6-Diethylphenyl (**Fe2**)

By using a similar procedure to that outlined for **Fe1**, but with 2,6-diethylaniline as the arylamine, **Fe2** was isolated as a blue powder (0.171 g, 66%). FT-IR (cm^{-1}): 2965 (m), 2932 (m), 2870 (m), 1678 (m), 1613 (m, $\nu_{C=N}$), 1567 (m), 1492 (m), 1447 (s), 1330 (m), 1216 (m), 1189 (m), 1106 (m), 1071 (m), 1030 (m), 951 (m), 848 (m), 809 (m), 776 (s), 703 (s). ESI-MS (m/z): Calculated for $[C_{40}H_{45}ClFeN_3]^+$ (658.26514). Found: 658.26496 (Figure S20). Elemental analysis: Calculated for $C_{40}H_{45}Cl_2N_3Fe$ (694.57) C, 69.17, H, 6.53, N, 6.05. Found: C, 68.80, H, 6.71, N, 6.25%.

3.2.3. Aryl = 2,6-Diisopropylphenyl (**Fe3**)

By using a similar procedure to that outlined for **Fe1**, but with 2,6-diisopropylaniline as the arylamine, **Fe3** was isolated as a blue powder (0.196 g, 65%). FT-IR (cm^{-1}): 2962 (m), 2866 (m), 2361 (w), 1688 (w), 1611 (w, $\nu_{C=N}$), 1568 (m), 1458 (s), 1383 (m), 1360 (m), 1324 (m), 1245 (m), 1214 (m), 1183 (m), 1124 (m), 1103 (m), 1051 (m), 943 (m), 848 (m), 804 (s), 774 (s), 709 (s). ESI-MS (m/z): Calculated for $[C_{44}H_{53}ClFeN_3]^+$ (714.32774). Found: 714.32762 (Figure S21). Elemental analysis: Calculated for $C_{44}H_{53}Cl_2N_3Fe$ (750.67) C, 70.40, H, 7.12, N, 5.60. Found: C, 70.19, H, 7.42, N, 5.73%.

3.2.4. Aryl = 2,4,6-Trimethylphenyl (**Fe4**)

By using a similar procedure to that outlined for **Fe1**, but with 2,4,6-trimethylaniline as the arylamine, **Fe4** was isolated as a blue powder (0.232 g, 87%). ESI-MS (m/z): Calculated for $[C_{38}H_{41}ClFeN_3]^+$ (630.23384). Found: 630.23370 (Figure S22). FT-IR (cm^{-1}): 3336 (m), 2912 (s), 2108 (w), 1984 (w), 1683 (w), 1604 (w, $\nu_{C=N}$), 1561 (m), 1472 (s), 1440 (s), 1376 (m), 1325 (m), 1215 (s), 1148 (m), 1071 (m), 1033 (m), 952 (w), 853 (s), 780 (w), 757 (w), 702 (s). Elemental analysis: Calculated for $C_{38}H_{41}Cl_2N_3Fe$ (666.51) C, 68.84, H, 6.20, N, 6.30. Found: C, 68.71, H, 6.55, N, 6.17%.

3.2.5. Aryl = 2,6-Diethyl-4-methylphenyl (**Fe5**)

By using a similar procedure to that outlined for **Fe1**, but with 2,6-diethyl-4-methylaniline as the arylamine, **Fe5** was isolated as a blue powder (0.106 g, 37%). ESI-MS (m/z): Calculated for $[C_{42}H_{49}ClFeN_3]^+$ (686.29644). Found: 686.29592 (Figure S23). FT-IR (cm^{-1}): 3363 (m), 2964 (m), 2930 (m), 2870 (m), 1676 (w), 1612 (m, $\nu_{C=N}$), 1566 (m), 1456 (s), 1373 (m), 1335 (m), 1257 (m), 1215 (m), 1150 (m), 1070 (m), 950 (m), 857 (s), 774 (m), 704 (s). Elemental analysis: Calculated for $C_{42}H_{49}Cl_2N_3Fe$ (722.62) C, 69.81, H, 6.84, N, 5.82. Found: C, 69.57, H, 7.03, N, 5.61%.

3.3. Polymerization Studies

3.3.1. Ethylene Polymerization at 1 Atm Ethylene Pressure

A 250 mL Schlenk vessel, equipped with a stirrer bar, was loaded with **Fe4** (0.7 mg, 1.0 µmol), and the vessel was subjected to three cycles of evacuation and back-filling with nitrogen before the atmosphere was replaced with ethylene (1 atm). Freshly distilled toluene (30 mL) was then injected to dissolve the iron complex, and the mixture was stirred and heated to 60 °C. The activator (MAO or MMAO) was immediately injected via a syringe, and the run commenced. After 30 min, the supply of ethylene was stopped and the pressure within the Schlenk flask vented. The reaction mixture was then quenched with 10% hydrochloric acid in ethanol (30 mL). The polymer was washed with ethanol and dried under reduced pressure at 100 °C until it reached a constant weight.

3.3.2. Ethylene Polymerization at 5 Atm or 10 Atm Ethylene Pressure

A 250 mL stainless steel autoclave, fitted with a mechanical stirrer, an ethylene pressure control device, and a temperature controller, was employed for the higher-pressure (5 atm or 10 atm) ethylene polymerization evaluations. The autoclave was evacuated and back-filled with high-purity nitrogen (2 cycles) and then pressurized with 5 atm of ethylene to check the gas tightness. With the ethylene pressure returning to 1 atm, the iron precatalyst (1.0 µmol), pre-mixed with toluene (25 mL) in a Schlenk tube, was immediately injected into the autoclave. Any remaining iron complex was washed into the autoclave with more toluene (25 mL), and then a further volume of toluene (25 mL) was added. The required amount of activator (MAO or MMAO) and additional toluene (25 mL) were injected successively into the autoclave to take the total volume of solvent to 100 mL. The ethylene pressure was then set at either 5 atm or 10 atm, and the stirring immediately commenced. After the requisite reaction time, the ethylene gas supplying the reactor was stopped, and the autoclave was allowed to cool to room temperature by surrounding it with an ice/water bath. The reactor was then slowly vented, and the contents of the autoclave were quenched with a 5% hydrochloric ethanol solution (50 mL). Finally, the polymer was filtered, washed three times with ethanol (30 mL), and dried under reduced pressure at 100 °C until it reached a constant weight.

3.4. Single Crystal X-ray Diffraction Studies

X-ray diffraction studies were conducted on **Fe2** and **Fe3**, using single crystals of each iron complex that had been grown in the glove box by the slow diffusion of diethyl ether into a dichloromethane solution at ambient temperature. A crystal of each was selected and mounted on an XtaLAB Synergy R HyPix diffractometer (Rigaku Corporation, Tokyo, Japan) incorporating graphite-monochromated Cu-Kα radiation (λ = 1.54184 Å) at 170(2) K for **Fe2** and 169.99(10) K for **Fe3**, and data collection commenced. Using Olex2, the structures were solved by employing the ShelXT structure solution program using intrinsic phasing and refined with the ShelXL refinement package using least squares minimization [42,43].

4. Conclusions

In summary, a straightforward one-pot strategy has been successfully implemented to prepare five examples of iron(II) chloride complexes, **Fe1**–**Fe5**, bearing bis(imino)pyridines fused with six- and seven-membered carbocyclic rings and further appended with a *para*-phenyl group. All these iron complexes have been characterized via FT-IR spectroscopy, ESI mass spectrometry, elemental analysis, and, in the cases of **Fe2** and **Fe3**, by single crystal X-ray diffraction. On activation with MMAO or MAO, **Fe1**–**Fe5** exhibited extremely high activity, with a maximum of 35.92×10^6 g (PE) mol^{-1} (Fe) h^{-1} seen for **Fe4**/MAO operating at 60 °C. All catalysts generated highly linear polymers with varying levels of vinyl end-groups that were dependent on run temperature. Interestingly, analysis of ^{1}H NMR spectra of the polymers revealed higher rates of β-H elimination at 60 °C for the MMAO-activated system and 70 °C for the MAO-activated system, highlighting the key role played by temperature in influencing the competition between β-H elimination and chain transfer to aluminum. Moreover, considerable control of the polymer molecular weight could be achieved with **Fe4** found to form low molecular weight polymer with narrow dispersity at high temperatures (70 or 80 °C), while higher molecular weight material was formed at lower run temperatures.

Perhaps more importantly, **Fe4**/MMAO exhibited outstanding thermal stability and was indeed superior to the non-phenyl substituted iron analog **E**/MMAO (Ar = 2,4,6-trimethlyphenyl, M = Fe, Scheme 1), which is evidenced by the optimal run temperature of **Fe4**/MMAO being 60 °C (*cf.* 40 °C for **E**/MMAO) and activity of 34.10×10^6 g (PE) mol^{-1} (Fe) h^{-1} (*cf.* 15.15×10^6 g (PE) mol^{-1} (Fe) h^{-1} for **E**/MMAO). Given the remoteness of the *para*-phenyl group from the active iron center, it would seem unlikely that it has a direct steric influence on the active site, though an indirect effect by impacting on the flexibility/steric properties of the neighboring fused carbocycles and, in turn, the binding properties of the N',N,N''-ligand may be at play; an electronic effect cannot, however, be ruled out. More generally, these *para*-phenyl substituted carbocyclic-fused bis(imino)pyridine iron complexes exhibit excellent performance for ethylene polymerization when compared with a series of fused ring iron counterparts, including **D**$_{2(n=2)}$ and **D**$_{2(n=3)}$ (Figure 1), especially with respect to catalytic activity. As a similar effect on catalytic activity and molecular weight has been seen with their *para*-phenyl substituted cobalt counterparts (**G**, Figure 1), we aim in future work to further probe the generality of this structural variation in catalyst design.

Supplementary Materials: The following supporting information can be downloaded at: https://www.mdpi.com/article/10.3390/catal14030213/s1. Figure S1: (a) GPC traces for the polyethylenes generated using **Fe4**/MAO at different run temperatures and (b) plots of catalytic activity and polymer molecular weight as a function of run temperature; Figure S2: (a) GPC traces for the polyethylenes generated using **Fe4**/MAO at various Al:Fe molar ratios and (b) plots of catalytic activity and polymer molecular weight as a function of Al:Fe molar ratio; Figure S3: (a) GPC traces for the polyethylenes generated using **Fe4**/MAO over reaction time and (b) plots of catalytic activity and polymer molecular weight as a function of reaction time; Figure S4: (a) GPC traces for the polyethylenes generated using **Fe1**–**Fe5** under MAO activation and (b) a bar chart showing the effects of variation in *N*-aryl group in **Fe1**–**Fe5** on catalytic activity and polymer molecular weight; Figure S5: ^{1}H NMR spectrum of the polyethylene produced using **Fe4**/MMAO operating at 70 °C (entry 4, Table 2); recorded at 100 °C in 1,1,2,2-tetrachloroethane-d_2; Figure S6: ^{1}H NMR spectrum of the polyethylene produced using **Fe4**/MMAO operating at 60 °C (entry 3, Table 2); recorded at 100 °C in 1,1,2,2-tetrachloroethane-d_2; Figure S7: ^{1}H NMR spectrum of the polyethylene produced using **Fe4**/MMAO operating at 50 °C (entry 2, Table 2); recorded at 100 °C in 1,1,2,2-tetrachloroethane-d_2; Figure S8: ^{1}H NMR spectrum of the polyethylene produced using **Fe4**/MMAO operating at 40 °C (entry 1, Table 2); recorded at 100 °C in 1,1,2,2-tetrachloroethane-d_2; Figure S9: ^{1}H NMR spectrum of the polyethylene produced using **Fe4**/MAO operating at 40 °C (entry 1, Table 3); recorded at 100 °C in 1,1,2,2-tetrachloroethane-d_2; Figure S10: ^{1}H NMR spectrum of the polyethylene produced using **Fe4**/MAO operating at 50 °C (entry 2, Table 3); recorded at 100 °C in 1,1,2,2-tetrachloroethane-d_2; Figure S11: ^{1}H NMR spectrum of the polyethylene produced using **Fe4**/MAO operating at 60 °C (entry 3, Table 3); recorded at 100 °C in 1,1,2,2-tetrachloroethane-d_2; Figure S12: ^{1}H NMR spectrum of the polyethylene produced using **Fe4**/MAO operating at 70 °C (entry 4, Table 3); recorded at 100 °C in 1,1,2,2-tetrachloroethane-d_2; Figure S13: ^{1}H NMR spectrum of the polyethylene produced using **Fe4**/MAO operating at 80 °C (entry 5, Table 3); recorded at 100 °C in 1,1,2,2-tetrachloroethane-d_2; Table S1: Determining the molar ratios of *a* (chain A, unsaturated) to *b* (chain B, fully saturated) for the PE's produced using **Fe4**/MMAO at various temperatures; Table S2: Determining the molar ratios of *a* (chain A, unsaturated) to *b* (chain B, fully saturated) for the PE's produced using **Fe4**/MAO at various temperatures; Table S3: Crystal data and structure refinement for **Fe2** and **Fe3**. Figure S14: FT-IR spectrum of **Fe1**. Figure S15: FT-IR spectrum of **Fe2**. Figure S16: FT-IR spectrum of **Fe3**. Figure S17: FT-IR spectrum of **Fe4**. Figure S18: FT-IR spectrum of **Fe5**. Figure S19: ESI-MS spectrum of **Fe1**. Figure S20: ESI-MS spectrum of **Fe2**. Figure S21: ESI-MS spectrum of **Fe3**. Figure S22: ESI-MS spectrum of **Fe4**. Figure S23: ESI-MS spectrum of **Fe5**. References [24,29,38,44] are cited in the Supplementary Materials.

Author Contributions: Y.W.: methodology, investigation, data curation, writing—original draft preparation, and writing—review and editing; Z.W.: methodology, validation, supervision, and writing—review and editing; Q.Z.: data curation; Y.M.: software; G.A.S.: visualization, writing—review and editing; Y.S.: X-ray diffraction; W.-H.S.: conceptualization, resources, supervision, funding acquisition, and writing—review and editing. All authors have read and agreed to the published version of the manuscript.

Funding: This research was funded by the National Natural Science Foundation of China (21871275).

Data Availability Statement: CCDC reference numbers 2325145–2325146 contain Supplementary crystallographic Information for **Fe2** and **Fe3**, and the data presented in this study are available in this article.

Acknowledgments: G.A.S. thanks the Chinese Academy of Sciences for a President's International Fellowship for Visiting Scientists.

Conflicts of Interest: The authors declare no conflicts of interest.

References

1. Small, B.L.; Brookhart, M.; Bennett, A.M.A. Highly Active Iron and Cobalt Catalysts for the Polymerization of Ethylene. *J. Am. Chem. Soc.* **1998**, *120*, 4049–4050. [CrossRef]
2. Britovsek, G.J.P.; Gibson, V.C.; McTavish, S.J.; Solan, G.A.; White, A.J.P.; Williams, D.J.; Britovsek, G.J.P.; Kimberley, B.S.; Maddox, P.J. Novel olefin polymerization catalysts based on iron and cobalt. *Chem. Commun.* **1998**, 849–850. [CrossRef]
3. Small, B.L.; Brookhart, M. Iron-Based Catalysts with Exceptionally High Activities and Selectivities for Oligomerization of Ethylene to Linear α-Olefins. *J. Am. Chem. Soc.* **1998**, *120*, 7143–7144. [CrossRef]
4. Britovsek, G.J.P.; Mastroianni, S.; Solan, G.A.; Baugh, S.P.D.; Redshaw, C.; Gibson, V.C.; White, A.J.P.; Williams, D.J.; Elsegood, M.R.J. Oligomerisation of Ethylene by Bis(imino)pyridyliron and -cobalt Complexes. *Chem. Eur. J.* **2000**, *6*, 2221–2231. [CrossRef] [PubMed]
5. Wang, Z.; Solan, G.A.; Zhang, W.; Sun, W.-H. Carbocyclic-fused N,N,N-pincer ligands as ring-strain adjustable supports for iron and cobalt catalysts in ethylene oligo-/polymerization. *Coord. Chem. Rev.* **2018**, *363*, 92–108. [CrossRef]
6. Gibson, V.C.; Redshaw, C.; Solan, G.A. Bis(imino)pyridines: Surprisingly Reactive Ligands and a Gateway to New Families of Catalysts. *Chem. Rev.* **2007**, *107*, 1745–1776. [CrossRef]
7. Bianchini, C.; Giambastiani, G.; Luconi, L.; Meli, A. Olefin oligomerization, homopolymerization and copolymerization by late transition metals supported by (imino)pyridine ligands. *Coord. Chem. Rev.* **2010**, *254*, 431–455. [CrossRef]
8. Britovsek, G.J.P.; Bruce, M.; Gibson, V.C.; Kimberley, B.S.; Maddox, P.J.; Mastroianni, S.; McTavish, S.J.; Redshaw, C.; Solan, G.A.; Stromberg, S.; et al. Iron and Cobalt Ethylene Polymerization Catalysts Bearing 2,6-Bis(Imino)Pyridyl Ligands: Synthesis, Structures, and Polymerization Studies. *J. Am. Chem. Soc.* **1999**, *121*, 8728–8740. [CrossRef]
9. Ittel, S.D.; Johnson, L.K.; Brookhart, M. Late-Metal Catalysts for Ethylene Homo- and Copolymerization. *Chem. Rev.* **2000**, *100*, 1169–1204. [CrossRef]
10. Gibson, V.C.; Solan, G.A. Olefin Oligomerizations and Polymerizations Catalyzed by Iron and Cobalt Complexes Bearing Bis(imino)pyridine Ligands. In *Catalysis without Precious Metals*; Bullock, R.M., Ed.; Wiley-VCH: Weinheim, Germany, 2010; pp. 111–141.

11. Gibson, V.C.; Solan, G.A. Iron-Based and Cobalt-Based Olefin Polymerisation Catalysts. In *Metal Catalysts in Olefin Polymerization*; Guan, Z., Ed.; Springer: Berlin/Heidelberg, Germany, 2009; pp. 107–158.
12. Sun, W.-H.; Jie, S.; Zhang, S.; Zhang, W.; Song, Y.; Ma, H. Iron Complexes Bearing 2-Imino-1,10-phenanthrolinyl Ligands as Highly Active Catalysts for Ethylene Oligomerization. *Organometallics* **2006**, *25*, 666–677. [CrossRef]
13. Pelletier, J.D.A.; Champouret, Y.D.M.; Cadarso, J.; Clowes, L.; Gañete, M.; Singh, K.; Thanarajasingham, V.; Solan, G.A. Electronically variable imino-phenanthrolinyl-cobalt complexes; synthesis, structures and ethylene oligomerisation studies. *J. Organomet. Chem.* **2006**, *691*, 4114–4123. [CrossRef]
14. Jie, S.; Zhang, S.; Sun, W.-H.; Kuang, X.; Liu, T.; Guo, J. Iron(II) complexes ligated by 2-imino-1,10-phenanthrolines: Preparation and catalytic behavior toward ethylene oligomerization. *J. Mol. Catal. A Chem.* **2007**, *269*, 85–96. [CrossRef]
15. Appukuttan, V.K.; Liu, Y.; Son, B.C.; Ha, C.-S.; Suh, H.; Kim, I. Iron and Cobalt Complexes of 2,3,7,8-Tetrahydroacridine-4,5(1H,6H)-diimine Sterically Modulated by Substituted Aryl Rings for the Selective Oligomerization to Polymerization of Ethylene. *Organometallics* **2011**, *30*, 2285–2294. [CrossRef]
16. Zhang, W.; Chai, W.; Sun, W.-H.; Hu, X.; Redshaw, C.; Hao, X. 2-(1-(Arylimino)ethyl)-8-arylimino-5,6,7-trihydroquinoline Iron(II) Chloride Complexes: Synthesis, Characterization, and Ethylene Polymerization Behavior. *Organometallics* **2012**, *31*, 5039–5048. [CrossRef]
17. Sun, W.-H.; Kong, S.; Chai, W.; Shiono, T.; Redshaw, C.; Hu, X.; Guo, C.; Hao, X. 2-(1-(Arylimino)ethyl)-8-arylimino-5,6,7-trihydroquinolylcobalt dichloride: Synthesis and polyethylene wax formation. *Appl. Catal. A Gen.* **2012**, *447–448*, 67–73. [CrossRef]
18. Zhang, Y.; Suo, H.; Huang, F.; Liang, T.; Hu, X.; Sun, W.H. Thermo-stable 2-(arylimino)benzylidene-9-arylimino-5,6,7,8-tetrahydro cyclohepta[b]pyridyliron(II) precatalysts toward ethylene polymerization and highly linear polyethylenes. *J. Polym. Sci. Part A Polym. Chem.* **2016**, *55*, 830–842. [CrossRef]
19. Du, S.; Zhang, W.; Yue, E.; Huang, F.; Liang, T.; Sun, W.H. α,α′-Bis(arylimino)-2,3:5,6-bis(pentamethylene)pyridylcobalt Chlorides: Synthesis, Characterization, and Ethylene Polymerization Behavior. *Eur. J. Inorg. Chem.* **2016**, *2016*, 1748–1755. [CrossRef]
20. Du, S.; Wang, X.; Zhang, W.; Flisak, Z.; Sun, Y.; Sun, W.-H. A practical ethylene polymerization for vinyl-polyethylenes: Synthesis, characterization and catalytic behavior of α,α′-bisimino-2,3:5,6-bis(pentamethylene)pyridyliron chlorides. *Polym. Chem.* **2016**, *7*, 4188–4197. [CrossRef]
21. Wang, Z.; Solan, G.A.; Mahmood, Q.; Liu, Q.; Ma, Y.; Hao, X.; Sun, W.-H. Bis(imino)pyridines Incorporating Doubly Fused Eight-Membered Rings as Conformationally Flexible Supports for Cobalt Ethylene Polymerization Catalysts. *Organometallics* **2018**, *37*, 380–389. [CrossRef]
22. Wang, Z.; Zhang, R.; Zhang, W.; Solan, G.A.; Liu, Q.; Liang, T.; Sun, W.-H. Enhancing thermostability of iron ethylene polymerization catalysts through N,N,N-chelation of doubly fused α,α′-bis(arylimino)-2,3:5,6-bis(hexamethylene)pyridines. *Catal. Sci. Technol.* **2019**, *9*, 1933–1943. [CrossRef]
23. Wang, Z.; Ma, Y.; Guo, J.; Liu, Q.; Solan, G.A.; Liang, T.; Sun, W.-H. Bis(imino)pyridines fused with 6- and 7-membered carbocylic rings as N,N,N-scaffolds for cobalt ethylene polymerization catalysts. *Dalton Trans.* **2019**, *48*, 2582–2591. [CrossRef] [PubMed]
24. Wang, Z.; Solan, G.A.; Ma, Y.; Liu, Q.; Liang, T.; Sun, W.-H. Fusing Carbocycles of Inequivalent Ring Size to a Bis(imino)pyridine-Iron Ethylene Polymerization Catalyst: Distinctive Effects on Activity, PE Molecular Weight, and Dispersity. *Research* **2019**, *2019*, 9426063. [CrossRef] [PubMed]
25. Mazzolini, J.; Boyron, O.; Monteil, V.; Gigmes, D.; Bertin, D.; D'Agosto, F.; Boisson, C. Polyethylene End Functionalization Using Radical-Mediated Thiol–Ene Chemistry: Use of Polyethylenes Containing Alkene End Functionality. *Macromolecules* **2011**, *44*, 3381–3387. [CrossRef]
26. Wang, X.; Nozaki, K. Selective Chain-End Functionalization of Polar Polyethylenes: Orthogonal Reactivity of Carbene and Polar Vinyl Monomers in Their Copolymerization with Ethylene. *J. Am. Chem. Soc.* **2018**, *140*, 15635–15640. [CrossRef] [PubMed]
27. Amin, S.B.; Marks, T.J. Versatile Pathways for In Situ Polyolefin Functionalization with Heteroatoms: Catalytic Chain Transfer. *Angew. Chem. Int. Ed.* **2008**, *47*, 2006–2025. [CrossRef] [PubMed]
28. Cámpora, J.; Naz, A.M.; Palma, P.; Rodríguez-Delgado, A.; Álvarez, E.; Tritto, I.; Boggioni, L. Iron and Cobalt Complexes of 4-Alkyl-2,6-diiminopyridine Ligands: Synthesis and Ethylene Polymerization Catalysis. *Eur. J. Inorg. Chem.* **2008**, *2008*, 1871–1879. [CrossRef]
29. Wang, Y.; Wang, Z.; Zhang, Q.; Zou, S.; Ma, Y.; Solan, G.A.; Zhang, W.; Sun, W.-H. Exploring Long Range para-Phenyl Effects in Unsymmetrically Fused bis(imino)pyridine-Cobalt Ethylene Polymerization Catalysts. *Catalysts* **2023**, *13*, 1387. [CrossRef]
30. Zhang, Q.; Yang, W.; Wang, Z.; Solan, G.A.; Liang, T.; Sun, W.-H. Doubly fused N,N,N-iron ethylene polymerization catalysts appended with fluoride substituents; probing catalytic performance via a combined experimental and MLR study. *Catal. Sci. Technol.* **2021**, *11*, 4605–4618. [CrossRef]
31. Masuda, J.D.; Wei, P.; Stephan, D.W. Nickel and palladium phosphinimine-imine ligand complexes. *Dalton Trans.* **2003**, *18*, 3500–3505. [CrossRef]
32. Addison, A.W.; Rao, T.N. Synthesis, structure, and spectroscopic properties of copper(II) compounds containing nitrogen–sulphur donor ligands; the crystal and molecular structure of aqua[1,7-bis(N-methylbenzimidazol-2′-yl)-2,6-dithiaheptane]copper(II) perchlorate. *J. Chem. Soc. Dalton Trans.* **1984**, 1349–1356. [CrossRef]

33. Long, Z.; Wu, B.; Yang, P.; Li, G.; Liu, Y.; Yang, X.-J. Synthesis and characterization of para-nitro substituted 2,6-bis(phenylimino)pyridyl Fe(II) and Co(II) complexes and their ethylene polymerization properties. *J. Organomet. Chem.* **2009**, *694*, 3793–3799. [CrossRef]
34. Guo, L.; Gao, H.; Zhang, L.; Zhu, F.; Wu, Q. An Unsymmetrical Iron(II) Bis(imino)pyridyl Catalyst for Ethylene Polymerization: Effect of a Bulky Ortho Substituent on the Thermostability and Molecular Weight of Polyethylene. *Organometallics* **2010**, *29*, 2118–2125. [CrossRef]
35. Liu, T.; Ma, Y.; Solan, G.A.; Sun, Y.; Sun, W.-H. Unimodal polyethylenes of high linearity and narrow dispersity by using ortho-4,4′-dichlorobenzhydryl-modified bis(imino)pyridyl-iron catalysts. *New J. Chem.* **2023**, *47*, 5786–5795. [CrossRef]
36. Bianchini, C.; Giambastiani, G.; Rios, I.G.; Mantovani, G.; Meli, A.; Segarra, A.M. Ethylene oligomerization, homopolymerization and copolymerization by iron and cobalt catalysts with 2,6-(bis-organylimino)pyridyl ligands. *Coord. Chem. Rev.* **2006**, *250*, 1391–1418. [CrossRef]
37. Han, M.; Oleynik, I.I.; Liu, M.; Ma, Y.; Oleynik, I.V.; Solan, G.A.; Liang, T.; Sun, W.H. Ring size enlargement in an ortho-cycloalkyl-substituted bis(imino)pyridine-cobalt ethylene polymerization catalyst and its impact on performance and polymer properties. *Appl. Organomet. Chem.* **2022**, *36*, e6529. [CrossRef]
38. Zhang, Q.; Zuo, Z.; Ma, Y.; Liang, T.; Yang, X.; Sun, W.-H. Fluorinated 2,6-bis(arylimino)pyridyl iron complexes targeting bimodal dispersive polyethylenes: Probing chain termination pathways via a combined experimental and DFT study. *Dalton Trans.* **2022**, *51*, 8290–8302. [CrossRef] [PubMed]
39. Dabiri, M.; Baghbanzadeh, M.; Nikcheh, M.S. Oxalic Acid: An Efficient and Cost-Effective Organic Catalyst for the Friedländer Quinoline Synthesis under Solvent-Free Conditions. *Monatsh. Chem.* **2007**, *138*, 1249–1252. [CrossRef]
40. Bell, T.W.; Khasanov, A.B.; Drew, M.G.B. Role of Pyridine Hydrogen-Bonding Sites in Recognition of Basic Amino Acid Side Chains. *J. Am. Chem. Soc.* **2002**, *124*, 14092–14103. [CrossRef]
41. Vierhapper, F.W.; Eliel, E.L. Selective Hydrogenation of Quinoline and Its Homologs, Isoquinoline, and Phenyl-Substituted Pyridines in the Benzene Ring. *J. Org. Chem. Vol.* **1975**, *40*, 2729–2734. [CrossRef]
42. Sheldrick, G.M. Crystal structure refinement with SHELXL. *Acta Crystallogr. C* **2015**, *71*, 3–8. [CrossRef]
43. Sheldrick, G. SHELXT—Integrated space-group and crystal-structure determination. *Acta Crystallogr. A* **2015**, *71*, 3–8. [CrossRef]
44. Fulmer, G.R.; Miller, A.J.M.; Sherden, N.H.; Gottlieb, H.E.; Nudelman, A.; Stoltz, B.M.; Bercaw, J.E.; Goldberg, K.I. NMR Chemical Shifts of Trace Impurities: Common Laboratory Solvents, Organics, and Gases in Deuterated Solvents Relevant to the Organometallic Chemist. *Organometallics* **2010**, *29*, 2176–2179. [CrossRef]

Disclaimer/Publisher's Note: The statements, opinions and data contained in all publications are solely those of the individual author(s) and contributor(s) and not of MDPI and/or the editor(s). MDPI and/or the editor(s) disclaim responsibility for any injury to people or property resulting from any ideas, methods, instructions or products referred to in the content.

Article

Extended Hydrogen-Bonded Molybdenum Arrays Derived from Carboxylic Acids and Dianilines: ROP Capability of the Complexes and Parent Acids and Dianilines

William Clegg [1], Mark R. J. Elsegood [2] and Carl Redshaw [3,*]

[1] Chemistry, School of Natural and Environmental Sciences, Newcastle University, Newcastle upon Tyne NE1 7RU, UK; bill.clegg@ncl.ac.uk
[2] Chemistry Department, Loughborough University, Loughborough LE11 3TU, UK; m.r.j.elsegood@lboro.ac.uk
[3] Plastics Collaboratory, Chemistry, School of Natural Sciences, University of Hull, Hull HU6 7RX, UK
* Correspondence: c.redshaw@hull.ac.uk

Citation: Clegg, W.; Elsegood, M.R.J.; Redshaw, C. Extended Hydrogen-Bonded Molybdenum Arrays Derived from Carboxylic Acids and Dianilines: ROP Capability of the Complexes and Parent Acids and Dianilines. *Catalysts* 2024, *14*, 214. https://doi.org/10.3390/catal14030214

Academic Editor: Mauro Bassetti

Received: 23 January 2024
Revised: 13 March 2024
Accepted: 15 March 2024
Published: 21 March 2024

Copyright: © 2024 by the authors. Licensee MDPI, Basel, Switzerland. This article is an open access article distributed under the terms and conditions of the Creative Commons Attribution (CC BY) license (https://creativecommons.org/licenses/by/4.0/).

Abstract: From reactions involving sodium molybdate and dianilines [2,2'-(NH_2)C_6H_4]$_2$(CH_2)$_n$ (n = 0, 1, 2) and amino-functionalized carboxylic acids 1,2-(NH_2)(CO_2H)C_6H_4 or 2-$H_2NC_6H_3$-1,4-(CO_2H)$_2$, in the presence of Et_3N and Me_3SiCl, products adopting H-bonded networks have been characterized. In particular, the reaction of 2,2'-diaminobiphenyl, [2,2'-NH_2(C_6H_4)]$_2$, and 2-aminoterephthalic acid, $H_2NC_6H_3$-1,4-(CO_2H)$_2$, led to the isolation of [($MoCl_3$[2,2'-N(C_6H_4)]$_2$]$_2${HNC_6H_3-1-(CO_2),4-(CO_2H)]·2[2,2'-NH_2(C_6H_4)]$_2$·3.5MeCN (**1**·3.5MeCN), which contains intra-molecular N–H···Cl H-bonds and slipped $\pi \cdots \pi$ interactions. Similar use of 2,2'-methylenedianiline, [2,2'-(NH_2)C_6H_4]$_2$$CH_2$, in combination with 2-aminoterephthalic acid led to the isolation of [$MoCl_2$($O_2CC_6H_3NHCO_2SiMe_3$)($NC_6H_4CH_2C_6H_4NH_2$)]·3MeCN(**2**·3MeCN). Complex **2** contains extensive H-bonds between pairs of centrosymmetrically-related molecules. In the case of 2,2'ethylenedianiline, [2,2'-(NH_2)C_6H_4]$_2$$CH_2CH_2$, and anthranilic acid, 1,2-(NH_2)(CO_2H)C_6H_4, reaction with Na_2MoO_4 in the presence of Et_3N and Me_3SiCl in refluxing 1,2-dimethoxyethane afforded the complex [$MoCl_3${1,2-(NH)(CO_2)C_6H_4}{$NC_6H_4$$CH_2CH_2C_6H_4NH_3$}]·MeCN (**3**·MeCN). In **3**, there are intra-molecular bifurcated H-bonds between NH_3 H atoms and chlorides, whilst pairs of molecules H-bond further via the NH_3 groups to the non-coordinated carboxylate oxygen, resulting in H-bonded chains. Complexes **1** to **3** have been screened for the ring opening polymerization (ROP) of both ε-caprolactone (ε-CL) and δ-valerolactone (δ-VL) using solvent-free conditions under N_2 and air. The products were of moderate to high molecular weight, with wide Đ values, and comprised several types of polymer families, including OH-terminated, OBn-terminated (for PCL only), and cyclic polymers. The results of metal-free ROP using the dianilines [2,2'-(NH_2)C_6H_4]$_2$(CH_2)$_n$ (n = 0, 1, 2) and the amino-functionalized carboxylic acids 1,2-(NH_2)(CO_2H)C_6H_4 or 2-$H_2NC_6H_3$-1,4-(CO_2H)$_2$ under similar conditions (no BnOH) are also reported. The dianilines were found to be capable of the ROP of δ-VL (but not ε-CL), whilst anthranilic acid outperformed 2-aminoterephthalic acid for both ε-Cl and δ-VL.

Keywords: hydrogen-bonded network; amino-functionalized carboxylic acids; dianilines; molybdenum complexes; solid-state structures; ring opening polymerization (ROP); metal-free ROP; cyclic esters

1. Introduction

Despite the intensive research on imido compounds over the last few decades, driven partly by their presence in metathesis catalysts [1], examples of complexes bearing highly functionalized imido ligation remain somewhat limited [2]. With this in mind, we have been interested in exploiting amino-functionalized carboxylic acids, and our early work focused on the use of anthranilic acid [3–5], and in separate studies we have reported the use of potentially chelating dianilines, to access new structural motifs [6,7]. Our entry into such chemistry is via sodium molybdate, which is combined with anilines in the presence of triethylamine and trimethylsilyl chloride, using refluxing dimethoxyethane as the solvent [8–11]. The use of dianilines or amino-functionalized carboxylic acids in combination with molybdate under such conditions has the potential to afford products with extensive intra- and/or inter-molecular hydrogen bonding. For example, use of [(2-NH$_2$C$_6$H$_4$)$_2$O] led to the salt [Et$_3$NH][MoCl$_4$[(2-NH$_2$C$_6$H$_4$)(2-NC$_6$H$_4$)O] comprising discrete units of two anions and two cations linked by a series of H-bonds [11]. We also note that chiral diimido complexes (of Mo and W) have been reported by Sundermeyer et al. [12]. Herein, we combine the two sets of 'ligands' in one pot, i.e., an amino-functionalized carboxylic acid and a chelating dianiline, with the aim of further increasing the degree of hydrogen bonding present. In particular, use of the dianilines [2,2′-(NH$_2$)C$_6$H$_4$]$_2$(CH$_2$)$_n$ (n = 0, 1, 2), in combination with either anthranilic acid or 2-aminoterephthalic acid, in the 'molybdate' preparation led to products (Scheme 1) with extensive intra- and inter-molecular interactions resulting in the formation of H-bonded pairs, chains, or a 3D network.

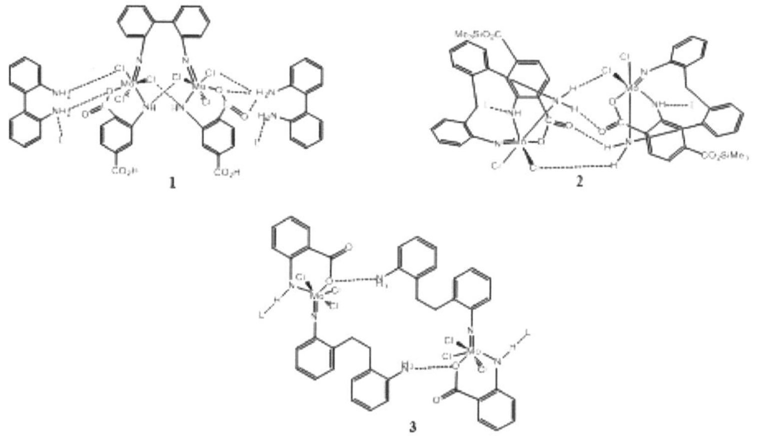

Scheme 1. Complexes **1–3** isolated in this work (L = MeCN).

Given the interest in alternatives to petroleum-derived plastics [13], there is currently much research devoted to seeking catalysts that are capable of producing greener polymers via ring opening polymerization (ROP) of cyclic esters. Both transition metal and metal-free ROP systems are attracting interest [14–24]. Ideally, such catalysts should be highly active, inexpensive, and non-toxic. In the case of metals, the use of earth-abundant metals is now attracting attention [25]. In terms of molybdenum-based systems, the limited number that have thus far been reported tend to exhibit activity only at elevated temperatures, with early examples based on ammonium decamolybdate (melt at 150 °C) [26] or bis-(salicylaldhydato)dioxomolybdenum (110 °C in mesitylene) [27]. More recent examples include oxo and imido complexes derived from chelating phenols [28,29], or the oxydianiline [(2-NH$_2$C$_6$H$_4$)$_2$O] [11], or a family of iodoanilines [30]. For many of these molybdenum-based systems, trans-esterification is evident. Given the somewhat limited arsenal of molybdenum-based ROP catalysts, we have investigated the three complexes herein (shown in Scheme 1) for their potential in the ROP of ε-caprolactone and δ-valerolactone. Moreover, given the aforementioned interest in metal-free catalysts for ROP, we have also examined the use of the precursor anilines and acids (Scheme 2) employed in the synthesis of **1–3** as ROP catalysts. We also note that there is growing interest in the ability of hydrogen bonding to promote ROP, particularly in carboxylic acid-based systems [31] and amine/amide-containing systems [32–39].

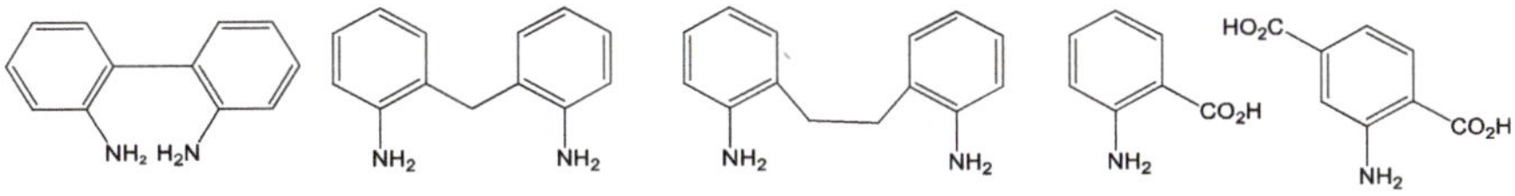

Scheme 2. Dianilines and acids employed herein.

2. Results and Discussion

2.1. Synthesis and Characterization of Hydrogen-Bonded 2,2′-Diaminobiphenyl-Derived Molybdenum Complex

Use of the established sodium molybdate/aniline/Et$_3$N/Me$_3$SiCl preparation [8–11] but employing a combination of 2,2′-diaminobiphenyl [2,2′-NH$_2$(C$_6$H$_4$)]$_2$ and 2-aminoterephthalic acid H$_2$NC$_6$H$_3$-1,4-(CO$_2$H)$_2$ as the amine sources, following work-up (MeCN), led to the isolation of small orange/red prisms in moderate yield. In the IR spectrum, a broad peak at 3444 cm^{-1} is assigned to vNH (Figure S1, SI). Single crystals suitable for X-ray diffraction were obtained from a saturated acetonitrile solution at 0 °C. The molecular structure is shown in Figure 1, with selected bond lengths and angles given in the caption; alternative views are given in Figure S2, SI. The asymmetric unit comprises one molybdenum-containing complex, two 2,2′-diaminobiphenyl molecules, and three and a half acetonitrile solvent molecules of crystallization, i.e., [(MoCl$_3$[2,2′-N(C$_6$H$_4$)]$_2$){HNC$_6$H$_3$-1-(CO$_2$),4-(CO$_2$H)]·2[2,2′-NH$_2$(C$_6$H$_4$)]$_2$·3.5MeCN **1**·2[2,2′-NH$_2$(C$_6$H$_4$)]$_2$·3.5MeCN. The metal complex ligand conformations are supported by intra-molecular N–H⋯Cl H-bonds (Table 1) and the two amino-terephthalate ligands adopt a slipped π⋯π interaction with C(16) and C(23) overlaying the centroid of the aromatic ring of the opposite ligand at a distance of ca. 3.5 Å. The closest atom⋯atom contacts are: C(15)⋯C(22) = 3.334 Å; C(20)⋯C(26) = 3.487 Å; and C(17)⋯C(24) = 3.535 Å. Thus, the rings are closer at the N end.

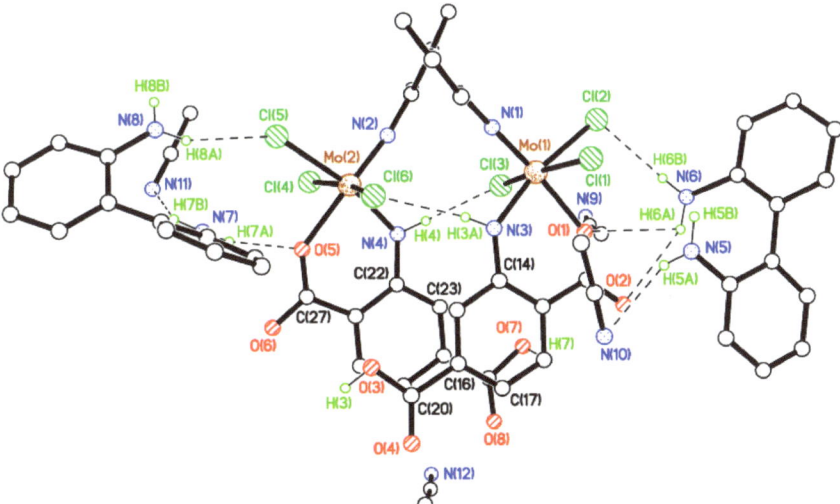

Figure 1. Molecular structure of **1**·2[2,2′-NH$_2$(C$_6$H$_4$)]$_2$·3.5MeCN. Selected bond lengths (Å) and angles (°): Mo(1)–N(1) 1.715(9), Mo(1)–N(3) 1.977(8), Mo(1)–O(1) 2.092(6), Mo(1)–Cl(1) 2.386(3), Mo(1)–Cl(2) 2.455(2), Mo(1)–Cl(3) 2.421(7); O(1)–Mo(1)–N(3) 81.7(3), Cl(2)–Mo(1)–N(3) 167.5(2), N(1)–Mo(1)–O(1) 170.9(3). Elements are shown in different representations (C black, H pale green, N blue, O red, Cl green, Mo brown) and hydrogen bonds are shown as dashed lines (in all structural figures).

Table 1. Hydrogen-bond geometry (Å, °) for **1**·2[2,2′-NH$_2$(C$_6$H$_4$)]$_2$·3.5MeCN.

D—H···A	D—H	H···A	D···A	D—H···A
N3—H3A···Cl6	0.88	2.42	3.278 (8)	164
O3—H3···O2 [i]	0.84	1.73	2.564 (9)	170
N4—H4···Cl3	0.88	2.32	3.190 (8)	172
O7—H7···O6 [ii]	0.84	1.89	2.707 (10)	165
N5—H5A···N10	0.88	2.53	3.37 (2)	159
N5—H5B···Cl4 [iii]	0.88	2.74	3.413 (11)	135
N6—H6A···O2	0.88	2.56	3.319 (12)	145
N6—H6A···N5	0.88	2.28	2.736 (13)	112
N6—H6B···Cl2	0.88	2.44	3.307 (9)	170
N7—H7A···O5	0.88	2.13	2.990 (10)	165
N7—H7B···N11	0.88	2.34	3.02 (2)	135
N8—H8A···Cl5	0.88	2.95	3.432 (11)	116

Symmetry codes: (i) $x + 1/2, -y + 3/2, z + 1/2$; (ii) $x - 1/2, -y + 3/2, z - 1/2$; (iii) $x - 1, y, z$.

The Mo$_2$ complex is capped at each end with a neutral, co-crystallized C$_{12}$H$_{12}$N$_2$ molecule via 2 or 3 N–H···Cl or N–H···O H-bonds. The NH$_2$ H atoms also H-bond to MeCN molecules of crystallization or symmetry-related Mo$_2$ complexes. The carboxylic acid OH groups H-bond to the carboxylate oxygen atoms of neighboring Mo$_2$ complexes via charge-assisted H-bonds, so they do not form the often-observed carboxylic acid head-to-tail $R^2_2(8)$ motif (Figure 2) [40,41].

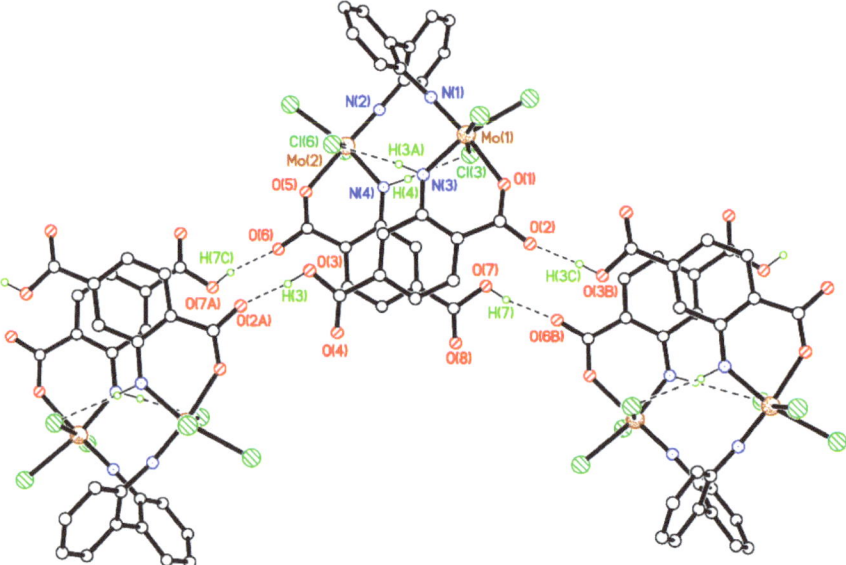

Figure 2. View of H-bonded chains running parallel to *c*.

The only NH hydrogen atom that does not act as an H-bond donor is H(8B), presumably due to the lack of a suitably positioned acceptor. The structure adopts a 3D H-bonded network overall (Figure 3).

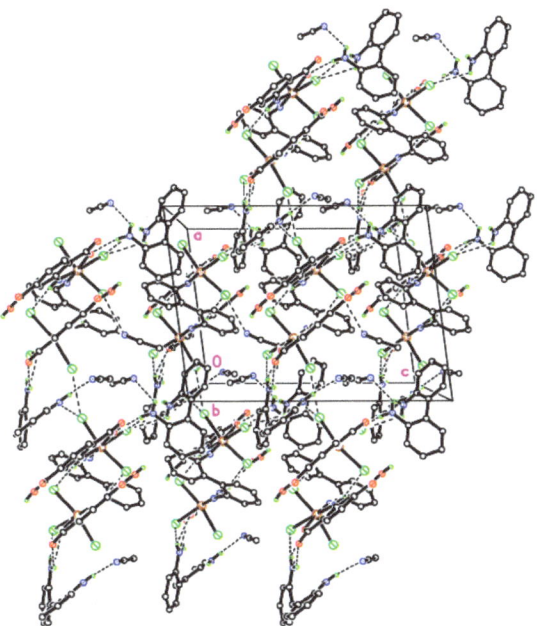

Figure 3. H-bonded 3D network of $\mathbf{1} \cdot 2[2,2'\text{-}NH_2(C_6H_4)]_2 \cdot 3.5MeCN$.

The MALDI-ToF spectrum of $\mathbf{1} \cdot 2[2,2'\text{-}NH_2(C_6H_4)]_2 \cdot 3.5MeCN$ is given in Figure S3, SI; for assignments see experimental Section 4.1. Complex **1** (and **2** and **3** below) despite being recrystallized from MeCN proved to be insoluble in common organic solvents, which not only hampered characterization in solution (e.g., only a weak ^{1}H NMR spectrum of **1** was obtained, Figure S4, SI for the aromatic region), but also restricted our ROP studies to the melt phase, vide infra.

2.2. Synthesis and Characterization of the Methylene-Bridged Dianiline-Derived Molybdenum Complex

We then investigated the use of the methylene-bridged dianiline $[2,2'\text{-}(NH_2)C_6H_4]_2CH_2$ in combination with 2-aminoterephthalic acid $H_2NC_6H_3\text{-}1,4\text{-}(CO_2H)_2$. Following work-up (extraction into MeCN), orange crystals were isolated, which were subjected to a single-crystal X-ray determination. The molecular structure is shown in Figure 4, with selected bond lengths and angles given in the caption. The asymmetric unit comprises one molybdenum complex and three molecules of non-coordinated acetonitrile. The dianiline-derived ligand binds in imido/amine fashion about the octahedral metal center, whilst the 2-aminoterephthalic acid binds in carboxylate/amide fashion. In the IR spectrum (Figure S5, SI), there are peaks at 3501, 3446, and 3385 cm^{-1}, which are assigned to vNH. Interestingly, a trimethylsilylation has occurred at the 2-aminoterephthalic acid, as has been observed previously in the system $[Et_3NH][MoCl_3\{2\text{-}(HN)C_6H_4CO_2\}\{2\text{-}Me_3SiO_2CC_6H_4N\}]$ [2]. Moreover, a search of the Cambridge Structural Database (CSD) reveals four other occurrences of this type of PhCO$_2$SiMe$_3$ motif [42–45]. The coordination around molybdenum is completed by two chloride ligands, which are trans to the amine and amide functions. The formula can thus be written as $[MoCl_2(O_2CC_6H_3NHCO_2SiMe_3)(NC_6H_4CH_2C_6H_4NH_2)] \cdot 3MeCN$ (**2**·3MeCN).

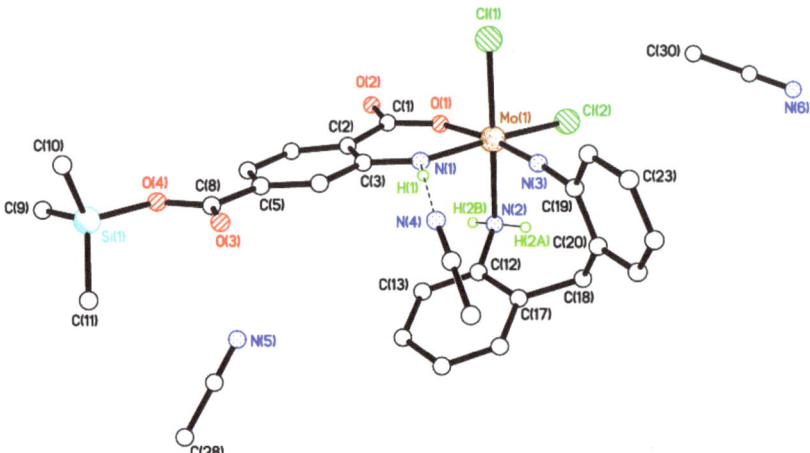

Figure 4. Molecular structure of [MoCl$_2$(O$_2$CC$_6$H$_3$NHCO$_2$SiMe$_3$)(NC$_6$H$_4$CH$_2$C$_6$H$_4$NH$_2$)]·3MeCN (**2**·3MeCN). Selected bond lengths (Å) and angles (°): Mo(1)–O(1) 2.0644(9), Mo(1)–N(1) 1.9428(11), Mo(1)–N(2) 2.2375(11), Mo(1)–N(3) 1.7370(11), Mo(1)–Cl(1) 2.3643(3), Mo(1)–Cl(2) 2.4267(3); N(1)–Mo(1)–O(1) 82.81(4), N(2)–Mo(1)–N(3) 90.85(4).

Pairs of centrosymmetrically-related molecules are connected via H-bonds utilizing both NH$_2$ hydrogen atoms as donors and a chloride and non-coordinated carboxylate oxygen as acceptors (Figure 5 and Table 2); alternative views are given in Figure S6, SI. The angle between the two aromatic rings in the imido ligand is 75.99 (4)°. The N–H group forms an H-bond to one of the three molecules of crystallization. The MALDI-ToF spectrum of [MoCl$_2$(O$_2$CC$_6$H$_3$NHCO$_2$SiMe$_3$)(NC$_6$H$_4$CH$_2$C$_6$H$_4$NH$_2$)]·3MeCN (**2**·3MeCN) is given in Figure S7, SI; for assignments see experimental Section 4.2.

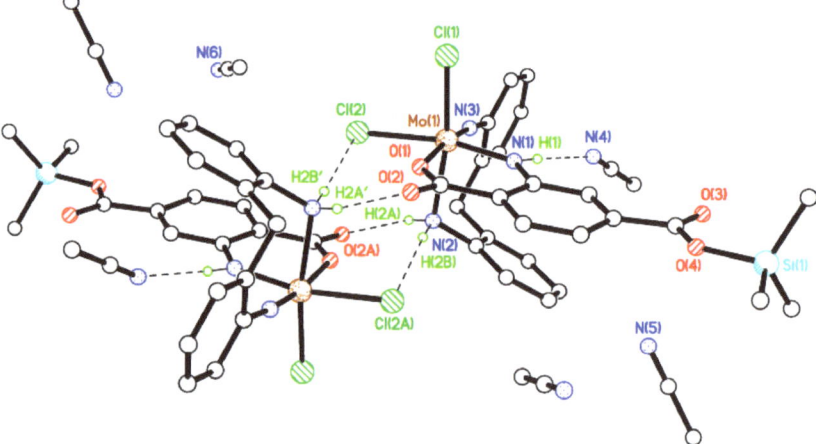

Figure 5. H-bonds between pairs of centrosymmetrically-related molecules in **2**·3MeCN.

Table 2. Hydrogen-bond geometry (Å, °) for **2**·3MeCN.

D—H⋯A	D—H	H⋯A	D⋯A	D—H⋯A
N1—H1⋯N4	0.82 (1)	2.13 (1)	2.9518 (17)	176 (2)
N2—H2A⋯O1 [i]	0.81 (1)	2.47 (2)	2.9215 (14)	116 (2)
N2—H2A⋯O2 [i]	0.81 (1)	2.28 (1)	3.0736 (15)	168 (2)
N2—H2B⋯Cl2 [i]	0.84 (1)	2.68 (1)	3.5139 (11)	174 (2)

Symmetry codes: (i) $-x + 1, -y, -z + 1$.

2.3. Synthesis and Characterization of the Ethylene-Bridged Dianiline-Derived Molybdenum Complex

The length of the bridge in the dianiline was then extended to an ethylene linkage, namely [2,2′-(NH$_2$)C$_6$H$_4$]$_2$CH$_2$CH$_2$, and, to avoid the previous silylation issue, anthranilic acid 1,2-(NH$_2$)(CO$_2$H)C$_6$H$_4$ was employed rather than 2-aminoterephthalic acid. Work-up as before led to the isolation of purple crystals of X-ray diffraction quality. In the IR spectrum (Figure S8, SI), there are peaks at 3452/3348 cm^{-1}, which are assigned to vNH. The molecular structure is shown in Figure 6, with selected bond lengths and angles given in the caption; alternative views are given in Figure S9, SI. This purple complex was identified as [MoCl$_3${1,2-(NH)(CO$_2$)C$_6$H$_4$}{NC$_6$H$_4$CH$_2$CH$_2$C$_6$H$_4$NH$_3$}]·MeCN (**3**·MeCN) and formed in good yield (ca. 65%). Single crystals suitable for X-ray diffraction were obtained from a saturated acetonitrile solution at 0 °C. In the asymmetric unit, there are two independent Mo complexes and two MeCN molecules of crystallization. The two molybdenum-containing zwitterionic molecules differ slightly in their conformations in the solid state. For example, the two rings in the N—N groups are twisted relative to each other by different amounts: C(8) > C(13) vs. C(16) > C(21) = 49.81 (13)° and C(29) > C(34) vs. C(37) > C(42) = 30.87 (13)°. In both unique complexes, there are intra-molecular bifurcated H-bonds between one of the NH$_3$ H atoms and the coordinated chlorides Cl(1) and Cl(2) or Cl(5) and Cl(6). The two MeCN solvent molecules of crystallization both act as H-bond acceptors from the N–H groups at N(1) and N(4) (Table 3).

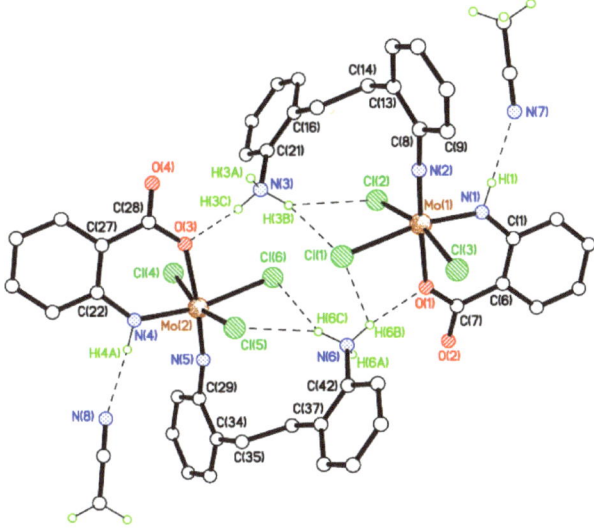

Figure 6. Molecular structure of **3**·MeCN. Selected bond lengths (Å) and angles (°): Mo(1)–N(1) 1.939(3), Mo(1)–N(2) 1.730(2), Mo(1)–O(1) 2.1137(19), Mo(1)–Cl(1) 2.4535(8), Mo(1)–Cl(2) 2.3963(8), Mo(1)–Cl(3) 2.4071(9); N(1)–Mo(1)–O(1) 82.72(9), O(1)–Mo(1)–N(2) 177.09(10).

Table 3. Hydrogen-bond geometry (Å, °) for **3**·MeCN.

D—H⋯A	D—H	H⋯A	D⋯A	D—H⋯A
N1—H1⋯N7	0.81 (2)	2.19 (2)	2.995 (4)	172 (3)
N3—H3A⋯O2 [i]	0.87 (2)	1.96 (2)	2.828 (3)	172 (3)
N3—H3B⋯Cl1	0.87 (2)	2.54 (2)	3.323 (3)	151 (3)
N3—H3B⋯Cl2	0.87 (2)	2.85 (3)	3.502 (3)	133 (3)
N3—H3C⋯O3	0.88 (2)	1.93 (2)	2.803 (3)	174 (3)
N3—H3C⋯O4	0.88 (2)	2.55 (3)	3.054 (3)	117 (2)
N4—H4⋯N8	0.81 (2)	2.21 (2)	3.010 (4)	170 (3)
N6—H6A⋯O4 [ii]	0.88 (2)	1.98 (2)	2.847 (3)	171 (3)
N6—H6B⋯Cl1	0.87 (2)	2.99 (3)	3.446 (3)	115 (2)
N6—H6B⋯O1	0.87 (2)	2.00 (2)	2.858 (3)	171 (3)
N6—H6B⋯O2	0.87 (2)	2.59 (3)	3.059 (3)	115 (2)
N6—H6C⋯Cl5	0.87 (2)	2.70 (2)	3.416 (3)	141 (3)
N6—H6C⋯Cl6	0.87 (2)	2.59 (2)	3.326 (3)	143 (3)

Symmetry codes: (i) $x, -y + 1/2, z + 1/2$; (ii) $x, -y + 1/2, z - 1/2$.

In the packing of **3**·MeCN, the two unique Mo complexes H-bond in a head-to-tail fashion via the NH$_3$ groups to a coordinated carboxylate oxygen atom in the case of H(3C) to O(3), or via a bifurcated pair of interactions between H(6B) and O(1) and Cl(1); this is not over a crystallographic inversion center. These pairs of molecules then H-bond further to other molecules via the NH$_3$ groups to the non-coordinated carboxylate, resulting in H-bonded chains running parallel to *c* (Figure 7). These diamond-shaped interactions again occur in head-to-tail pairs not dictated by crystallographic inversion symmetry, involving O(2) and O(4).

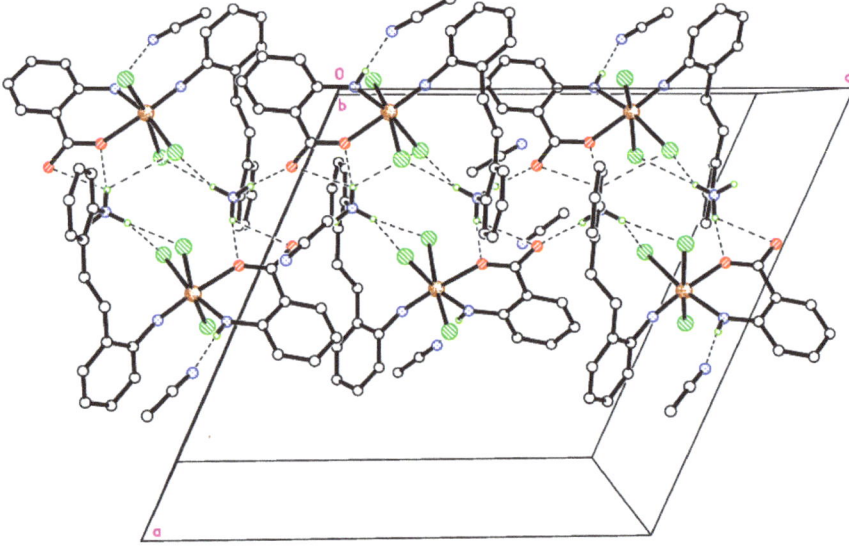

Figure 7. View of H-bonded chains running parallel to *c* for **3**·MeCN.

The MALDI-ToF spectrum of [MoCl$_3$\{1,2-(NH)(CO$_2$)C$_6$H$_4$\}\{NC$_6$H$_4$CH$_2$CH$_2$C$_6$H$_4$NH$_3$\}]· MeCN (**3**·MeCN) is given in Figure S10, SI; for assignments see experimental Section 4.3.

3. Ring Opening Polymerization (ROP)

3.1. Ring Opening Polymerization of ε-Caprolactone (ε-CL)

Complexes **1**–**3** have been screened for their ability to act as catalysts in the presence of benzyl alcohol (BnOH) for the ROP of ε-caprolactone, and the results are presented in

Table 4. Results for **1–3** are compared with the related molybdenum-containing complexes **I** and **II** [11] (see Figure 8). Based on previous molybdenum ROP studies in our group [11,28–30], we selected the conditions of 130 °C with a ratio of ε-CL to complex of 500:1 in the presence of one (for **2** and **3**) or two (for **1**) equivalents of benzyl alcohol per metal over 24 h, i.e., a [CL]:[catalyst]:[BnOH] ratio of 500:1:1 or 500:1:2. However, given the aforementioned problematic issues with solubility, the complexes herein were only screened as melts under either N_2 or air (see Table 4). All complexes were found to be active under these polymerization conditions with similar monomer conversions (>95%, e.g., Figures S11–S14, SI), affording relatively high molecular weight polymers, with **3** under N_2 (Entry 5, Table 4) affording the highest at ca. 31,650 Da, albeit with poor control (Đ = 3.66); selected gpc traces are given in Figures S15–S21. End group analysis by ^{1}H NMR spectroscopy reveals signals at 3.63, 5.09, and 7.34 ppm consistent with the presence of a BnO end group, which indicates that the polymerization proceeds through a coordination-insertion mechanism, whereby the monomer coordinates to the metal followed by the acyl oxygen bond cleavage of the monomer and chain propagation. Interestingly, consistent with the wide Đ values, the MALDI-TOF spectra revealed several families of products including OH terminated polymers, OBn terminated polymers, and cyclic polymers (e.g., Figures 9 and 10; expansions are given in the SI, Figures S22–S25). There was evidence of trans-esterification, and all observed M_n values were significantly lower than the calculated values.

Table 4. The ROP of ε-CL catalyzed by **1–3** and **I, II**.

Entry	Catalyst	[CL]:[Cat]:BnOH	Conversion [a] (%) [b]	M_n (obsd) [b]	M_n Corrected [c]	$M_n calc$ [d]	Đ [b]
1	**1**	500:1:2	100	14,820	8300	57,180	2.78
2 [e]	**1**	500:1:2	100	16,590	9290	57,180	3.00
							3.00
3	**2**	500:1:1	100	10,430	5840	57,180	1.12
4 [e]	**2**	500:1:1	99	22,880	12,810	56,610	1.85
5	**3**	500:1:1	86	56,520	31,650	49,190	3.66
6 [e]	**3**	500:1:1	99	34,200	19,150	56,610	2.84
7	**I**	500:1:1	100	64,160	35,930	57,180	3.83
8 [e]	**I**	500:1:1	100	34,110	19,100	57,180	3.61
9	**II**	500:1:1	100	21,040	11,780	57,180	2.26
10 [e]	**II**	500:1:1	97	15,620	8750	55,460	3.69
11 [e]	[2,2'-$NH_2(C_6H_4)]_2$	500:1	-	-	-	-	-
12 [e]	[2,2'-$(NH_2)C_6H_4]_2CH_2$	500:1	-	-	-	-	-
13 [e]	[2,2'-$(NH_2)C_6H_4]_2CH_2CH_2$	500:1	-	-	-	-	-
14 [e]	1,2-$(NH_2)(CO_2H)C_6H_4$	500:1	98	14,090	7890	56,070	1.37
15 [e]	$H_2NC_6H_3$-1,4-$(CO_2H)_2$	500:1	28	-	-	-	-

[a] Determined by ^{1}H NMR spectroscopy; [b] Measured by GPC in THF relative to polystyrene standards; [c] M_n calculated after Mark–Houwink correction [44,45]; M_n corrected = 0.56 × M_n obsd; [d] Calculated from ([monomer]$_0$/[cat]$_0$) × conv (%) × monomer molecular weight (M_{CL} = 114.14) + end groups (OBn/OH used in this case); [e] Conducted in air.

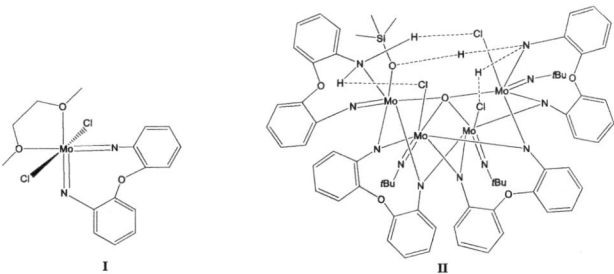

Figure 8. Known complexes **I** and **II**.

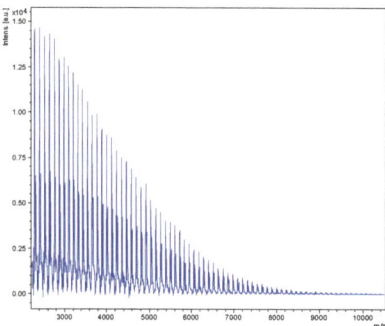
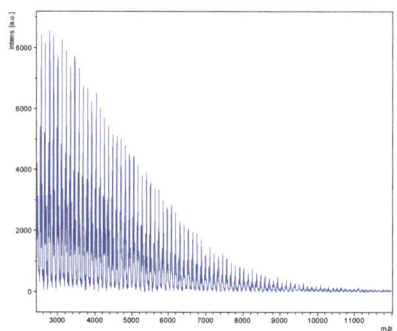

Figure 9. MALDI-ToF spectrum of PCL using **1** left: (Entry 1, Table 4); right (Entry 2, Table 4). For Entry 1, Table 4: The main families are (i) chain polymer (terminated by 2 OH) as sodium adducts [M = 17 (OH) + 1(H) + n × 114.14 (CL) + 22.99 (Na$^+$)] (e.g., for n = 23, calc. 2666.2 obsv. 2668.4; n = 25 calc. 2894.5 obsv. 2895.9 with peaks offset by about 1.2 Da; (ii) with BnO end groups [M = n × 114.14 (CL) + 108.05 (BnOH) + 22.99 (Na$^+$)] (e.g., for n = 50 calc. 5838, obsv. 5836.4; for n = 55, calc. 6408.7 obsv. 6406.9) with peaks offset by about 1.7 Da; (iii) cyclic polymers as the potassium adducts, e.g., n = 50 calc. 5746.1 obsv. 5745.7. For Entry 2, Table 4: The main families are (i) chain polymer (terminated by 2 OH) as sodium adducts [M = 17 (OH) + 1(H) + n × 114.14 (CL) + 22.99 (Na$^+$)], e.g., for n = 30, calc. 3465.2 obsv. 3465.6; n = 40 calc. 4606.6 obsv. 4605.2; (ii) cyclic polymers as the potassium adducts, e.g., n = 50 calc. 5746.1 obsv. 5745.7, n = 60 calc. 6887.5 obsv. 6886.4. Expansions of all the peaks highlighted above can be found in SI (Figures S37 and S38).

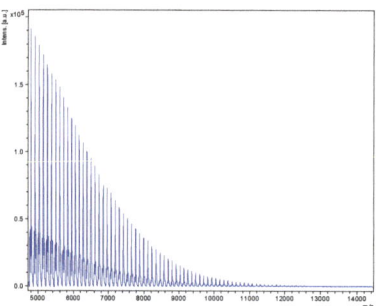
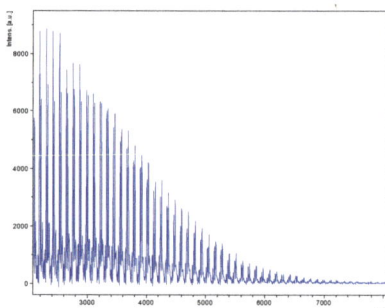

Figure 10. MALDI-ToF spectrum of PCL left using **2** (Entry 3, Table 4); right using **3** (Entry 5, Table 4). For Entry 3, Table 4: The main families are (i) chain polymer (terminated by 2 OH) as sodium adducts [M = 17 (OH) + 1(H) + n × 114.14 (CL) + 22.99 (Na$^+$)] (e.g., for n = 30, calc. 3465.2 obsv. 3466.3; n = 40 calc. 4606.6 obsv. 4606.2; (ii) cyclic polymers as the potassium adducts e.g., n = 40 calc. 4604.7 obsv. 4606.0; n = 50 calc. 5746.1 obsv. 5747.1. For Entry 5, Table 4: The main family are polymers with BnO end groups as the potassium adducts [M = n × 114.14 (CL) + 108.05 (BnOH) + 39.1 (K$^+$)] (e.g., for n = 50 calc. 5854.2, obsv. 5854.0). Expansions of all the peaks highlighted above can be found in the SI (Figures S39 and S40).

Given that metal-free catalysts are known [19], the dianilines [2,2'-(NH$_2$)C$_6$H$_4$]$_2$(CH$_2$)$_n$ (n = 0, 1, 2) and the amino-functionalized carboxylic acids 1,2-(NH$_2$)(CO$_2$H)C$_6$H$_4$ or 2-H$_2$NC$_6$H$_3$-1,4-(CO$_2$H)$_2$ employed in this work (Scheme 2) to prepare **1**–**3** were screened under similar conditions, albeit in the absence on BnOH for the ROP of ε-CL. In the case of the dianilines, little or no conversion was observed. However, both amino-functionalized carboxylic acids exhibited conversion, with 1,2-(NH$_2$)(CO$_2$H)C$_6$H$_4$ achieving 98%; the MALDI-ToF spectrum (Figure S26, SI) indicated the presence of anthranilic acid-derived end groups; the ^{1}H NMR spectrum is given in Figure S27, SI. We note that benzoic acid has been shown to be an efficient ROP catalyst for ε-Cl [32].

3.2. Ring Opening Polymerization of δ-Valerolactone (δ-VL)

As in the case of ε-CL, the complexes herein were only screened for the ROP of δ-VL as melts under either N_2 or air (see Table 5) using a ratio of [VL]:[catalyst]:[BnOH] of 500:1:2 (for **1**) or 500:1:1 (for **2**, **3**, and **I**, **II**). All complexes were found to be active under these polymerization conditions with monomer conversions (>75%, Figures S28–S35, SI), affording relatively high molecular weight polymers; selected gpc traces are given in Figures S36–S42, SI. As for CL, the conversions observed for **3** were somewhat lower than for the other systems. The highest molecular weights were afforded by **2** under air (Entry 4, Table 5) and **II** under N_2 (Entry 9, Table 5), although in each case there was evidence of bimodal behavior. ^{1}H NMR spectra of the PVL again indicated that the products were catenulate. The MALDI-TOF spectra revealed several families of products including OH-terminated polymers and cyclic polymers (e.g., Figures 11 and 12; expansions are given in Figures S43–S46, SI). As for PCL, there was evidence of trans-esterification, and all observed PVL M_n values were significantly lower than the calculated values.

Table 5. The ROP of δ-VL catalyzed by **1–3** and **I**, **II**.

Entry	Catalyst	[VL]:[Cat]:BnOH	Conversion [a] (%) [a]	M_n (obsd) [b]	M_n Corrected [c]	M_n calc [d]	Đ [b]
1	**1**	500:1:2	93	22,160	12,630	46,570	2.38
2 [e]	**1**	500:1:2	98	29,320	16,710	49,080	1.38
3	**2**	500:1:1	83	24,390/10,910	13,900/6220	41,570	1.24/1.06
4 [e]	**2**	500:1:1	96	27,470	15,660	48,080	2.71
5	**3**	500:1:1	76	23,460	13,370	38,060	2.75
6 [e]	**3**	500:1:1	86	16,770	9560	43,070	1.90
7	**I**	500:1:1	94	10,560	6020	47,070	1.69
8 [e]	**I**	500:1:1	94	24,460	13,940	47,070	2.06
9	**II**	500:1:1	98	29,400/5930	16,760/3380	49,080	1.69/1.47
10 [e]	**II**	500:1:1	97	20,440/4720	11,650/2690	48,580	1.39/1.23
11 [e]	[2,2'-$NH_2(C_6H_4)$]$_2$	500:1	95	15,910	9070	47,580	2.29
12 [e]	[2,2'-$(NH_2)C_6H_4$]$_2CH_2$	500:1	94	29,490	16,810	47,070	1.21
13 [e]	[2,2'-$(NH_2)C_6H_4$]$_2CH_2CH_2$	500:1	94	29,860	17,020	47,070	1.32
14 [e]	1,2-$(NH_2)(CO_2H)C_6H_4$	500:1	99	8140	4640	49,580	2.69
15 [e]	$H_2NC_6H_3$-1,4-$(CO_2H)_2$	500:1	13	-	-	-	-

[a] Determined by ^{1}H NMR spectroscopy. [b] Measured by GPC in THF relative to polystyrene standards; [c] M_n calculated after Mark–Houwink correction [44,45]: M_n corrected = 0.57 × M_n obsd; [d] Calculated from ([monomer]$_0$/[cat]$_0$) × conv (%) × monomer molecular weight (M_{VL} = 100.12) + end groups (OH used in this case). [e] Conducted in air.

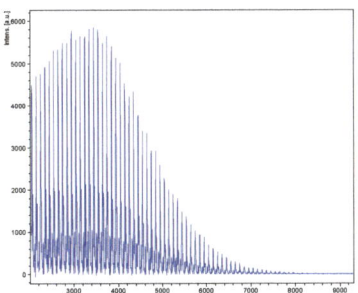 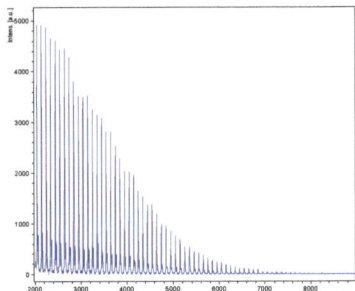

Figure 11. MALDI-ToF spectrum of PVL left using **1**: (Entry 1, Table 5); right using **2** (Entry 4, Table 5). For Entry 1, Table 5: The main families are (i) chain polymer (terminated by 2 OH) as sodium adducts [M = 17 (OH) + 1(H) + n × 100.12 (VL) + 22.99 (Na$^+$)], e.g., for n = 23, calc. 2343.8, obsv. 2345.8; n = 25

calc. 2544.0, obsv. 2545.9 with peaks offset by about 2 Da; (ii) cyclic polymers as the potassium adducts e.g., n = 50 calc. 5045.1, obsv. 5045.2; n = 60 calc 6046.3, obsv. 6045.2. For Entry 4, Table 5: The main families are (i) chain polymer (terminated by 2 OH) as sodium adducts [M = 17 (OH) + 1(H) + n × 100.12 (VL) + 22.99 (Na$^+$)] (e.g., for n = 23, calc. 2343.8, obsv. 2345.7; n = 25 calc. 2544.0, obsv. 2545.5 with peaks offset by 1.5–1.9 Da; (ii) cyclic polymers as the potassium adducts, e.g., n = 50 calc. 5045.1, obsv. 5045.2; n = 60 calc. 6046.3, obsv. 6045.5. Expansions of all the peaks highlighted above can be found in SI (Figures S41 and S42).

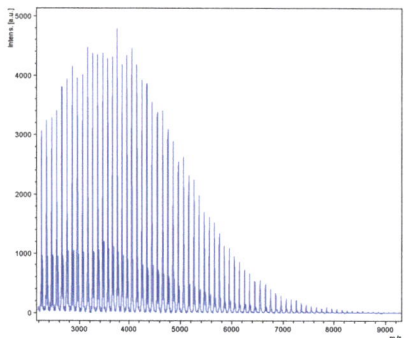 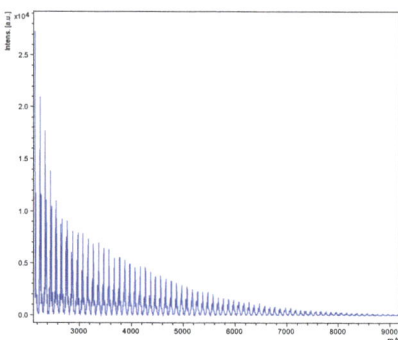

Figure 12. MALDI-ToF spectrum of PVL left using **3**: (Entry 5, Table 5); right using **II** (Entry 9, Table 5). For Entry 5, Table 5: The main families are (i) chain polymer (terminated by 2 OH) as sodium adducts [M = 17 (OH) + 1(H) + n × 100.12 (VL) + 22.99 (Na$^+$)], e.g., for n = 23, calc. 2343.8, obsv. 2346.4; n = 25 calc. 2544.0, obsv. 2546.1 with peaks offset by 1.6 to 2.1 Da; (ii) cyclic polymers as the potassium adducts, e.g., n = 50 calc. 5045.1, obsv. 5045.9; n = 60 calc. 6046.3, obsv. 6046.5. For Entry 9, Table 5: The main family is chain polymers (terminated by 2 OH) as sodium adducts [M = 17 (OH) + 1(H) + n × 100.12 (VL) + 22.99 (Na$^+$)], e.g., for n = 23, calc. 2343.8, obsv. 2345.0; n = 25 calc. 2544.0, obsv. 2544.8 with peaks offset by 0.8–1.2 Da. Expansions of all the peaks highlighted above can be found in SI (Figures S43 and S44).

Interestingly, in contrast to the ε-CL screening, when employing the dianilines for the ROP of δ-VL, high conversions were noted (see entries 11 to 13, Table 5). MALDI-ToF spectra (Figures S47–S49, SI) revealed the presence of families with OH end groups or cyclic polymers. This behavior contrasts with previous ROP studies where more robust conditions are usually needed for the ROP of δ-VL versus ε-Cl [46–50] and is inconsistent with the thermodynamic parameters for these lactones [48]. As for ε-CL, the use of 1,2-$(NH_2)(CO_2H)C_6H_4$ afforded a higher conversion than 2-$H_2NC_6H_3$-1,4-$(CO_2H)_2$ (99% vs. 13%), and in this case, anthranilic acid-derived end groups (Figure S50, SI) were assigned to a family of peaks offset by ca. 3Da; the ^{1}H NMR spectrum is given in Figure S51, SI.

TGA and DSC results (Figures S52–S55, SI) for the molybdenum catalysts employed herein indicate that at the temperature utilized for the ROP procedure, there is no degradation to other by-products.

4. Materials and Methods

All manipulations were carried out under an atmosphere of nitrogen using standard Schlenk line and cannula techniques or a conventional N_2-filled glovebox. Solvents were refluxed over the appropriate drying agents and distilled and degassed prior to use, i.e., dimethoxyethane was refluxed over Na-benzophenone/ketyl, and acetonitrile and triethylamine were refluxed over calcium hydride. Trimethylsilylchloride (TCI, Oxford, UK), 2-aminoterephthalic acid $H_2NC_6H_3$-1,4-$(CO_2H)_2$ (Thermo Fisher Scientific, Altrincham, UK), 2,2′-methylenedianiline [2,2′-$(NH_2)C_6H_4]_2CH_2$ (Enamine, Kyiv, Ukraine), 2,2′-ethylenedianiline [2,2′-$(NH_2)C_6H_4]_2CH_2CH_2$ (ChemCruz, Huissen, The Netherlands), and anthranilic acid 1,2-(NH$_2$)(CO$_2$H)C$_6$H$_4$ (Sigma Aldrich, UK) were purchased from commercial sources and used directly. 2,2′-Diaminobiphenyl was prepared via reduction of the dinitro precursors using NaBH$_4$ (Sigma Aldrich, Gillingham, UK) [51]. Elemental analyses were performed at the University of Hull. FTIR spectra (nujol mulls, KBr windows) were recorded on a Nicolet Avatar 360 FT-IR spectrometer (Thermo Fisher Scientific, Altrincham, UK). ^{1}H NMR spectra were recorded at 400.2 MHz on a JEOL ECZ 400S spectrometer (JEOL, Welwyn Garden City, UK), with TMS δ_H = 0 as the internal standard or residual protic solvent; chemical shifts are given in ppm (δ). Matrix Assisted Laser Desorption/Ionization-Time of Flight (MALDI-TOF) mass spectrometry was performed on a Bruker III smart beam in linear mode. MALDI-TOF mass spectra were acquired by averaging at least 100 laser shots. Molecular weight analyses were performed on a Viscotek gel permeation chromatograph (Malvern, Worcestershire, UK) equipped with two Agilent PL gel columns (5 µM mixedD, 300 × 7.5 mm; Polymer laboratories, Church Stratton, UK) and a Viscotek refractive index detector at 35 °C (Malvern Panalytical, Worcestershire, UK). Extra dry stabilized THF (Acros Organics, Loughborough, UK) was used as the eluent at a flow rate of 1.0 mL min^{-1} and polystyrene standards (Malvern Panalytical, Worcestershire, UK); 1.5 mg mL^{-1}, prepared and filtered (0.2.0 mmol L^{-1}) directly prior to injection were used for calibration. Obtained molecular weights were corrected by a Mark−Houwink factor of 0.56 (PCL) and 0.57 (PVL) [44,45]. Molecular weights were calculated from the experimental traces using the OmniSEC software (Malvern Panalytical Ltd., Malvern, UK, v11.35). For the TGA runs, data were collected on a PerkinElmer TGA 400 (Shelton, CT, USA) using Pyris^{TM} software and a rate of 10 °C per min over the 30 °C to 800 °C under N$_2$. Sample weights were typically between 3 and 5 mg. For DSC runs, data were collected on a PerkinElmer DSC 4000 (Shelton, CT, USA) using 5 mg of sample encapsulated with tin foil.

4.1. Synthesis of [(MoCl$_3$[2,2′-N(C$_6$H$_4$)]$_2$}{HNC$_6$H$_3$-1-(CO$_2$),4-(CO$_2$H)]·2[2,2′-NH$_2$(C$_6$H$_4$)]$_2$·3.5MeCN (1·3.5MeCN)

To Na$_2$MoO$_4$ (3.00 g, 14.6 mmol), H$_2$NC$_6$H$_3$-1,4-(CO$_2$H)$_2$ (2.64 g, 14.6 mmol), and [2,2′-NH$_2$(C$_6$H$_4$)]$_2$ (1.34 g, 7.30 mmol) in DME (150 mL) were added Et$_3$N (8.1 mL, 58 mmol) and Me$_3$SiCl (14.8 mL, 117 mmol), and the system was heated at reflux for 12 h. On cooling, the purple suspension was filtered, and the solvent was removed from the filtrate. Crystallization of the residue from MeCN (50 mL) afforded **1**·3.5MeCN as orange/red prisms. Yield 4.36 g, 41%. C$_{28}$H$_{18}$Cl$_6$Mo$_2$N$_4$O$_8$·2(C$_{12}$H$_{12}$N$_2$)·1.5(C$_2$H$_3$N) (sample dried in-vacuo for 2 h) requires C 48.11, H 3.41, N 9.69%. Found: C 47.78, H 3.36, N 9.43%. IR: 3444bw, 2622w, 2603w, 2497w, 1715w, 1592w, 1581m, 1557m, 1397m, 1301m, 1260s, 1208w, 1171m, 1092s, 1035s, 903w, 850m, 801s, 760m, 739w, 723m, 685w, 619w, 471w, 463w, 434w. M.S. (MALDI-ToF): 1188 (M$^+$−2MeCN−C$_{12}$H$_{12}$N$_2$), 1111 (M$^+$−3MeCN−Cl−C$_{12}$H$_{12}$N$_2$), 1004 (M$^+$−2MeCN−2C$_{12}$H$_{12}$N$_2$), 953 (M$^+$−1.5MeCN−2C$_{12}$H$_{12}$N$_2$).

4.2. Synthesis of [MoCl$_2$(O$_2$CC$_6$H$_3$NHCO$_2$SiMe$_3$)(NC$_6$H$_4$CH$_2$C$_6$H$_4$NH$_2$)]·3(C$_2$H$_3$N) (2·3MeCN)

As for **1**, but using Na$_2$MoO$_4$ (3.00 g, 14.6 mmol), H$_2$NC$_6$H$_3$-1,4-(CO$_2$H)$_2$ (2.64 g, 14.6 mmol), [2,2′-(NH$_2$)C$_6$H$_4$]$_2$CH$_2$ (1.45 g, 7.31 mmol), Et$_3$N (8.1 mL, 58 mmol), and Me$_3$SiCl (14.8 mL, 117 mmol) in DME (150 mL) affording **2**·3MeCN as red crystals. Yield 3.88 g, 36%. C$_{24}$H$_{25}$Cl$_2$MoN$_3$O$_4$Si·2(C$_2$H$_3$N) (sample dried in vacuo for 1 h) requires C 48.28, H 4.49, N 10.06%. Found: C 47.79, H 4.31, N 9.91%. IR: 3501w, 3446w, 3385w, 1692m, 1624m, 1591m, 1552m, 1309m, 1260m, 1239s, 1170m, 1155m, 1089m, 1020m, 918m, 891m, 850m, 796m, 752s, 722s, 679w, 657w, 623w. M.S. (MALDI-ToF): 499 (MH$^+$−MeCN−C$_{13}$H$_{14}$N$_2$).

4.3. Synthesis of [MoCl$_3${1,2-(NH)(CO$_2$)C$_6$H$_4$}{NC$_6$H$_4$CH$_2$CH$_2$C$_6$H$_4$NH$_3$}]·MeCN (3·MeCN)

To Na$_2$MoO$_4$ (3.00 g, 14.6 mmol), 1,2-(NH$_2$)(CO$_2$H)C$_6$H$_4$ (2.00 g, 14.6 mmol), and [2,2′-(NH$_2$)C$_6$H$_4$]$_2$CH$_2$CH$_2$ (1.55 g, 7.30 mmol) in DME (150 mL) were added Et$_3$N (8.1 mL, 58 mmol) and Me$_3$SiCl (14.8 mL, 117 mmol), and the system was heated at reflux for 12 h. On cooling, the purple suspension was filtered, and the solvent was removed from the filtrate. Crystallization of the residue from MeCN (50 mL) afforded 3·MeCN as purple crystals. Yield 2.32 g, 27%. $C_{21}H_{20}Cl_3MoN_3O_2$ (sample dried in vacuo for 1 h) requires C 45.97, H 3.67, N 7.66%. Found: C 45.33, H 3.52, N 7.37%. IR: 3452w, 3348w, 2358w, 2250w, 2139w, 1939w, 1914w, 1667m, 1614m, 1585m, 1562m, 1536m, 1397s, 1331m, 1260m, 1171s, 1072m, 1036s, 941m, 906m, 851m, 807s, 763m, 723m, 663w. M.S. (MALDI-ToF): 452 (M^+–anthranilic acid).

4.4. ROP of ε-Caprolactone (ε-CL)

The pre-catalyst (0.010 mmol) was added to a Schlenk tube in the glovebox at room temperature. The appropriate equivalent of BnOH (from a pre-prepared stock solution of 1 mmol BnOH in 100 mL toluene) was added, and the system was stirred for 5 min and then the solvent was removed in vacuo. The appropriate amount of ε-CL was added, and the reaction mixture was then placed into a sand bath pre-heated at 130 °C and heated for the prescribed time (24 h) under either N_2 or air. The polymerization mixture was quenched on addition of an excess of glacial acetic acid (0.2 mL) in methanol (50 mL). The resultant polymer was then collected on filter paper and was dried in vacuo. GPC (in THF) was used to determine molecular weights (M_n and PDI) of the polymer products.

4.5. X-ray Crystallography

In all cases, crystals suitable for an X-ray diffraction study were grown from a saturated MeCN solution at 0 °C. Diffraction data for 1·2[2,2′-$NH_2(C_6H_4)$]$_2$·3.5MeCN and 2·3MeCN were collected on pixel array detector-equipped Rigaku diffractometers using a rotating anode X-ray source, while that for 3·MeCN was collected on a Bruker SMART 1K CCD (Bruker AXS, Madison, WI, USA) diffractometer equipped with a sealed-tube X-ray source [52]. Data were corrected for absorption, polarization, and Lp effects. All of the structures were solved and refined routinely [53–56]. H atoms were included in a riding model except the N*H* hydrogens in 2·3MeCN and 3·MeCN, where the coordinates were refined with mild distance restraints. $U_{iso}(H)$ was set to 120% of that of the carrier atoms except for O*H*, N*H*$_3$, and C*H*$_3$ (150%). Further details are presented in Table 6. CCDC 2,286,024–2,286,026 contain the supplementary crystallographic data for this paper. These data can be obtained free of charge from The Cambridge Crystallographic Data Centre via www.ccdc.cam.ac.uk/structures (accessed 20 March 2024).

Table 6. Crystallographic data for 1·2$C_{12}H_{12}N_2$·3.5MeCN, 2·3MeCN, and 3·MeCN.

Compound	1·2$C_{12}H_{12}N_2$·3.5MeCN	2·3MeCN	3·MeCN
CCDC No.	2,286,024	2,286,025	2,286,026
Formula	$C_{28}H_{18}Cl_6Mo_2N_4O_8$· 2($C_{12}H_{12}N_2$)·3.5($C_2H_3N$)	$C_{24}H_{25}Cl_2MoN_3O_4Si$· 3($C_2H_3N$)	$C_{21}H_{20}Cl_3MoN_3O_2$· (C_2H_3N)
Formula weight	1455.20	737.56	589.74
Crystal system	Monoclinic	Monoclinic	Monoclinic
Space group	Cc	$P2_1/n$	$P2_1/c$
Unit cell dimensions			
a (Å)	12.2500(2)	15.57139(19)	17.9901(13)
b (Å)	31.4061(7)	13.33765(14)	16.5331(12)
c (Å)	16.2619(2)	17.8684(2)	18.5902(13)
β (°)	97.3910(13)	110.7234(14)	114.058(2)
V (Å^3)	6204.37(19)	3470.91(7)	5049.0(6)
Z	4	4	8
Temperature (K)	100(2)	100(2)	160(2)
Wavelength (Å)	1.54178	0.71073	0.71073
Calculated density (g.cm^{-3})	1.558	1.411	1.552
Absorption coefficient (mm^{-1})	6.22	0.61	0.87

Table 6. Cont.

Compound	1·2C$_{12}$H$_{12}$N$_2$·3.5MeCN	2·3MeCN	3·MeCN
Transmission factors (min./max.)	0.045 and 0.143	0.761 and 1.000	0.674 and 0.862
Crystal size (mm^3)	0.04 × 0.04 × 0.02	0.20 × 0.14 × 0.01	0.20 × 0.20 × 0.14
θ(max) (°)	75.7	33.6	28.5
Reflections measured	20,721	143,190	30,725
Unique reflections	8359	12,893	11,549
R_{int}	0.052	0.035	0.046
Reflections with $F^2 > 2\sigma(F^2)$	7358	11,128	7877
Number of parameters	798	412	621
R_1 [$F^2 > 2\sigma(F^2)$]	0.058	0.031	0.041
wR_2 (all data)	0.144	0.072	0.094
GOOF, S	1.03	1.05	0.99
Largest difference peak and hole (e Å^{-3})	1.67 and −0.36	0.87 and −0.80	0.51 and −0.96

5. Conclusions

In conclusion, the use of dianilines with (CH$_2$)$_n$ bridges where n = 0 to 2, when reacted with sodium molybdate in the presence of Et$_3$N/Me$_3$SiCL in 1,2-dimethoxyethane in combination with the acid-functionalized acids anthranilic acid or 2-aminoterephthalic acid, leads to the isolation of unusual H-bonded arrays. Relatively few molybdenum-based ROP catalysts have been reported in the literature. The systems reported herein are active as catalysts for the ROP of ε-caprolactone when employed as melts under either air or N$_2$. The products are of relatively high molecular weight (10,420–56,510 Da) and generally with broad Đ (1.12–3.08). A number of families were evident in the MALDI-ToF mass spectra with polymers present assigned to those terminated with OH, BnO (PCL only), as well as cyclic polymers. The parent dianilines were found to be capable of the ROP of δ-VL, but not ε-CL. The acids anthranilic acid or 2-aminoterephthalic acid were capable of the ROP of both cyclic esters; in terms of conversion, anthranilic acid outperformed 2-aminoterephthalic acid the best. All of these systems are thought to benefit from the presence of H-bonding, which has been shown in other studies to boost ROP performance [31–36].

Supplementary Materials: The following supporting information can be downloaded at: https://www.mdpi.com/article/10.3390/catal14030214/s1, Figure S1. IR spectrum of **1**; Figure S2. Different views of the molecular structures of **1**; Figure S3. MALDI-ToF spectrum of **1**; Figure S4. ^{1}H NMR spectrum of aromatic region of **1**; Figure S5. IR spectrum of **2**; Figure S6. Different views of the molecular structures of **2**; Figure S7. MALDI-ToF spectrum of **2**; Figure S8. IR spectrum of **3**; Figure S9. Different views of the molecular structures of **3**; Figure S10. MALDI-ToF spectrum of **3**; Figures S11–S14. Selected ^{1}H NMR spectra for %conversion measurements; Figures S15–S21. Selected gpc traces; Figures S22–S25. Expansions of MALDI-ToF spectra of PCL obtained from **1**–**3**; Figure S26. MALDI-ToF spectrum of PCL obtained using anthranilic acid; Figure S27. ^{1}H spectrum of PCL obtained using anthranilic acid; Figures S28–S35. Selected ^{1}H NMR spectra for %conversion measurements; Figures S36–S42. Selected gpc traces; Figures S43–S46. Expansions of MALDI-ToF spectra of PCL obtained from **1**–**3**; Figure S47. MALDI-ToF of PVL using [2,2′-NH$_2$(C$_6$H$_4$)]$_2$ (run 11, Table 5); Figure S48. MALDI-ToF of PVL using [2,2′-(NH$_2$)C$_6$H$_4$]$_2$CH$_2$ (run 12, Table 5); Figure S49. MALDI-ToF of PVL using [2,2′-(NH$_2$)C$_6$H$_4$]$_2$CH$_2$CH$_2$ (run 13, Table 5); Figure S50. MALDI-ToF spectrum of PVL obtained using anthranilic acid; Figure S51. ^{1}H NMR spectrum for PVL using anthranilic acid; Figure S52. TGA for **2**·3MeCN; Figure S53. DSC for **2**·3MeCN; Figure S54. TGA for **II**; Figure S55. DSC for **II**.

Author Contributions: W.C.: Investigation, Writing—review and editing. M.R.J.E.: Investigation, Writing—review and editing. C.R.: Conceptualization, Writing—original draft, Writing—review and editing. All authors have read and agreed to the published version of the manuscript.

Funding: This research was funded by UKRI Creative Circular Plastic grant (EP/S025537/1).

Data Availability Statement: Data is available upon requested.

Acknowledgments: The EPSRC National Crystallographic Service Centre at Southampton University is thanked for data collection of **1**·2[2,2′-NH$_2$(C$_6$H$_4$)]$_2$·3.5MeCN and **2**·3MeCN.

Conflicts of Interest: There are no conflicts of interest to declare.

References

1. Schrock, R.R.; Hoveyda, A.H. Molybdenum and Tungsten Imido Alkylidene Complexes as Efficient Olefin-Metathesis Catalysts. *Angew. Chem. Int. Ed.* **2003**, *42*, 4592–4633. [CrossRef] [PubMed]
2. Gibson, V.C.; Redshaw, C.; Clegg, W.; Elsegood, M.R.J. Synthesis and characterisation of molybdenum complexes bearing highly functionalised imido substituents. *Dalton Trans.* **1997**, 3207–3212. [CrossRef]
3. Redshaw, C.; Gibson, V.C.; Clegg, W.; Edwards, A.J.; Miles, B. Pentamethylcyclopentadienyl tungsten complexes containing imido, hydrazido and amino acid derived N-O chelate ligands. *Dalton Trans.* **1997**, 3343–3347. [CrossRef]
4. Gibson, V.C.; Redshaw, C.; Clegg, W.; Elsegood, M.R.J. Carboxylate versus imidobenzoate bonding for anthranilic acid derivatives. Crystals structures of [W(η-Cp*)Cl$_3$(η^2-O$_2$CC$_6$H$_4$NH$_2$-2)] and [ReCl(OEt)(PPh$_3$)$_2$(NC$_6$H$_4$CO$_2$)]. *Inorg. Chem. Commun.* **2001**, *4*, 95–99. [CrossRef]
5. Humphrey, S.M.; Redshaw, C.; Holmes, K.E.; Elsegood, M.R.J. Acid/amide bonding for anthranilic acid derivatives: Crystal structures of W(X)Cl$_3$(HO$_2$CC$_6$H$_4$NH-2) (X = O, NPh). *Inorg. Chim. Acta* **2005**, *358*, 222–226. [CrossRef]
6. Gibson, V.C.; Redshaw, C.; Clegg, W.; Elsegood, M.R.J.; Siemeling, U.; Turk, T. Synthesis and X-ray crystal structures of hydridotris(3,5-dimethylpyrazolyl)borate (Tp′) molybdenum(VI) bis(imido) complexes. *Polyhedron* **2004**, *23*, 189–194. [CrossRef]
7. Gibson, V.C.; Redshaw, C.; Clegg, W.; Elsegood, M.R.J. Supramolecular metallocalixarene chemistry: Linking metallocalixarenes through imido bridges. *Chem. Commun.* **1998**, 1969–1970. [CrossRef]
8. Gibson, V.C.; Redshaw, C.; Clegg, W.; Elsegood, M.R.J.; Siemeling, U.; Türk, T. Bridged bis(imido)molybdenum complexes: Isolobal analogues of *ansa*-zirconocenes and biziriconocenes. *J. Chem. Soc. Dalton Trans.* **1996**, 4513–4515. [CrossRef]
9. Siemeling, U.; Tűrk, T.; Schoeller, W.W.; Redshaw, C.; Gibson, V.C. Benefits of the chelate effect: Preparation of an unsymmetrical *ansa*-bis(imido)molybdenum complex containing a seven-membered chelate ring. *Inorg. Chem.* **1998**, *37*, 4738–4739. [CrossRef]
10. Redshaw, C.; Gibson, V.C.; Elsegood, M.R.J.; Clegg, W. New coordination modes at molybdenum for 2-diphenylphosphinoaniline derived ligands. *Chem. Commun.* **2007**, 1951–1953. [CrossRef]
11. Yang, W.; Zhao, Q.-K.; Redshaw, C.; Elsegood, M.R.J. Molybdenum complexes derived from the oxydianiline [(2-NH$_2$C$_6$H$_4$)$_2$O]: Synthesis, characterization and ε-caprolactone ROP capability. *Dalton Trans.* **2015**, *44*, 13133–13140. [CrossRef]
12. Kretzschmar, E.A.; Kipke, J.; Sundermeyer, J. The first chiral diimido chelate complexes of molybdenum and tungsten: Transition metal diimido complexes on the way to asymmetric catalysis. *Chem. Commun.* **1999**, 2381–2382. [CrossRef]
13. Singh, N.; Ogunseitan, O.A.; Wong, M.H.; Tang, Y. Sustainable materials alternative to petrochemical plastics pollution: A review analysis. *Sustain. Horiz.* **2022**, *2*, 100016. [CrossRef]
14. Keefe, B.J.; Hillmyer, M.A.; Tolman, W.B. Polymerization of lactide and related cyclic esters by discrete metal complexes. *J. Chem. Soc. Dalton Trans.* **2001**, 2215–2224. [CrossRef]
15. Dechy-Cabaret, O.; Martin-Vaca, B.; Bourissou, D. Controlled Ring-Opening Polymerization of Lactide and Glycolide. *Chem. Rev.* **2004**, *104*, 6147–6176. [CrossRef] [PubMed]
16. Labet, M.; Thielemans, W. Synthesis of polycaprolactone: A review. *Chem. Soc. Rev.* **2009**, *38*, 3484–3504. [CrossRef] [PubMed]
17. Thomas, C.M. Stereocontrolled ring-opening polymerization of cyclic esters: Synthesis of new polyester microstructures. *Chem. Soc. Rev.* **2010**, *39*, 165–173. [CrossRef] [PubMed]
18. Arbaoui, A.; Redshaw, C. Metal catalysts for ε-caprolactone polymerisation. *Polym. Chem.* **2010**, *1*, 801–826. [CrossRef]
19. Dove, A.P. Organic Catalysis for Ring Opening Polymerization. *ACS Macro Lett.* **2012**, *1*, 1409–1412. [CrossRef]
20. Sarazin, Y.; Carpentier, J.-F. Discrete Cationic Complexes for Ring-Opening Polymerization Catalysis of Cyclic Esters and Epoxides. *Chem. Rev.* **2015**, *115*, 3564–3614. [CrossRef]
21. Gao, J.; Zhu, D.; Zhang, W.; Solan, G.A.; Ma, Y.; Sun, W.-H. Recent progress in the application of group 1, 2 & 13 metal complexes as catalysts for the ring opening polymerization of cyclic esters. *Inorg. Chem. Front.* **2019**, *6*, 2619–2652.
22. Nifant'ev, I.; Ivchenko, P. Coordination Ring-Opening Polymerization of Cyclic Esters: A Critical Overview of DFT Modeling and Visualization of the Reaction Mechanisms. *Molecules* **2019**, *24*, 4117. [CrossRef]
23. Tazekas, E.; Lowy, P.A.; Rahman, M.A.; Lykkeberg, A.; Zhou, Y.; Chambenahalli, R.; Garden, J.A. Main group metal polymerisation catalysts. *Chem. Soc. Rev* **2022**, *51*, 8793–8814. [CrossRef]
24. Naz, F.; Abdur, R.M.; Mumtaz, F.; Elkadi, M.; Verpoort, F. Advances in cyclic ester ring-opening polymerization using heterogeneous catalysts. *Appl. Organomet. Chem.* **2023**, *37*, e7296. [CrossRef]
25. Wang, Y.; Wang, X.; Zhang, W.; Sun, W.-H. Progress of Ring Opening Polymerization of Cyclic Esters Catalyzed by Iron Compounds. *Organometallics* **2023**, *42*, 1680–1692. [CrossRef]
26. Báez, J.E.; Martínez-Richa, A. Synthesis and characterization of poly(ε-caprolactone) and copolyesters by catalysis with molybdenum compounds: Polymers with acid-functional asymmetric telechelic architecture. *Polymer* **2005**, *46*, 12118. [CrossRef]
27. Maruta, Y.; Abiko, A. Bis(salicylaldehydato)dioxomolybdenum complexes: Catalysis for ring-opening polymerization. *Polym. Bull.* **2014**, *71*, 1433. [CrossRef]

28. Al-Khafaji, Y.; Prior, T.J.; Elsegood, M.R.J.; Redshaw, C. Molybdenum(VI) imido complexes derived from chelating phenols: Synthesis, characterization and ε-caprolactone ROP capability. *Catalysts* **2015**, *5*, 1928–1947. [CrossRef]
29. Sun, Z.; Zhao, Y.; Prior, T.J.; Elsegood, M.R.J.; Wang, K.; Xing, T.; Redshaw, C. Mono-oxo molybdenum(VI) and tungsten(VI) complexes bearing chelating aryloxides: Synthesis, structure and ring opening polymerization of cyclic esters. *Dalton Trans.* **2019**, *48*, 1454–1466. [CrossRef] [PubMed]
30. Xing, T.; Elsegood, M.R.J.; Dale, S.H.; Redshaw, C. Pentamethylcyclopentadienyl molybdenum(V) complexes derived from iodoanilines: Synthesis, structure, and ROP of ε-caprolactone. *Catalysts* **2021**, *11*, 1554. [CrossRef]
31. Xu, S.; Zhu, H.; Li, Z.; Wei, F.; Gao, Y.; Xu, J.; Wang, H.; Liu, J.; Guo, T.; Guo, K. Tuning the H-bond donicity boosts carboxylic acid efficiency in ring-opening polymerization. *Eur. Polym. J.* **2019**, *112*, 799–808. [CrossRef]
32. Jehanno, C.; Mezzasalma, L.; Sardon, H.; Ruipérez, F.; Coulembier, O.; Taton, D. Benzoic Acid as an Efficient Organocatalyst for the Statistical Ring Opening Copolymerization of ε-Caprolactone and L-Lactide: A Computational Investigation. *Macromolecules* **2019**, *52*, 9238–9247. [CrossRef]
33. Liu, J.; Chen, C.; Li, Z.; Wu, W.; Zhi, X.; Zhang, Q.; Wu, H.; Wang, X.; Cui, S.; Guo, K. A squaramide and tertiary amine: An excellent hydrogen-bonding pair organocatalyst for living polymerization. *Polym. Chem.* **2015**, *6*, 3754–3757. [CrossRef]
34. Liu, J.; Xu, J.; Li, Z.; Xu, S.; Wang, X.; Wang, H.; Guo, T.; Gao, Y.; Zhang, L.; Guo, K. Squaramide and amine binary H-bond organocatalysis in polymerizations of cyclic carbonates, lactones, and lactides. *Polym. Chem.* **2017**, *8*, 7054–7068. [CrossRef]
35. Du, F.; Zheng, Y.; Yuan, W.; Shan, G.; Bao, Y.; Jie, S.; Pan, P. Solvent-Free Ring-Opening Polymerization of Lactones with Hydrogen Bonding Bisurea Catalyst. *J. Polym. Sci. Part A Polym. Chem.* **2019**, *57*, 90–100. [CrossRef]
36. Jain, I.; Malik, P. Advances in urea and thiourea catalyzed ring opening polymerization: A brief overview. *Eur. Polym. J.* **2020**, *133*, 109791. [CrossRef]
37. Printz, G.; Ryzhakov, D.; Gourlaouen, C.; Jacques, B.; Messaoudi, S.; Dumas, F.; Le Bideau, F.; Dagorne, S. First Use of Thiosquaramides as Polymerization Catalysts: Controlled ROP of Lactide Implicating Key Secondary Interactions for Optimal Performance. *ChemCatChem* **2024**, *16*, e202301207. [CrossRef]
38. Kazakov, O.I.; Datta, P.P.; Isajani, M.; Kiesewetter, E.T.; Kiesewetter, M.K. Cooperative Hydrogen-Bond Pairing in Organocatalytic Ring Opening Polymerization. *Macromolecules* **2014**, *47*, 7463–7468. [CrossRef] [PubMed]
39. Spink, S.S.; Kazakov, O.I.; Kiesewetter, E.T.; Kiesewetter, M.K. Rate Accelerated Organocatalytic Ring-Opening Polymerization of L-Lactide via the Application of a Bis(thiourea) H-bond Donating Cocatalyst. *Macromolecules* **2015**, *48*, 6127–6131. [CrossRef] [PubMed]
40. Etter, M.C.; MacDonald, J.C.; Bernstein, J. Graph-set analysis of hydrogen-bond patterns in organic crystals. *Acta Cryst.* **1990**, *B46*, 256–262. [CrossRef] [PubMed]
41. Bernstein, J.; Davis, R.E.; Shimoni, L.; Chang, N.-L. Patterns in Hydrogen Bonding: Functionality and Graph Set Analysis in Crystals. *Angew Chem. Int. Ed. Engl.* **1995**, *34*, 1555–1573. [CrossRef]
42. Groom, C.R.; Bruno, I.J.; Lightfoot, M.P.; Ward, S.C. The Cambridge Structural Database. *Acta Cryst.* **2016**, *B72*, 171–179. [CrossRef]
43. Basenko, S.V.; Soldatenko, A.S.; Vashchenko, A.V.; Smirnov, V.I. Silacyclophanones 3*. Cyclic organosilicon esters of *ortho*-phthalic acids. *Chem. Heterocycl. Compd.* **2019**, *55*, 1139. [CrossRef]
44. Shah, S.A.A.; Dorn, H.; Gindl, J.; Noltemeye, M.; Schmidt, H.-G.; Roesky, H.W. Synthesis and structural characterization of sulfonates, phosphinates and carboxylates of organometallic Group 4 metal fluorides. *J. Organomet. Chem.* **1998**, *550*, 1–6. [CrossRef]
45. Basenko, S.V.; Soldatenko, A.S.; Vashchenko, O.V.; Smirnov, V.I. Silacyclophanones 2*. Cyclic organosilicon esters of terephthalic acid. *Chem. Heterocycl. Compd.* **2018**, *54*, 826–828. [CrossRef]
46. Huang, C.-H.; Wang, F.-C.; Ko, B.-T.; Yu, T.-L.; Lin, C.-C. Ring-Opening Polymerization of ε-Caprolactone and L-Lactide Using Aluminum Thiolates as Initiator. *Macromolecules* **2001**, *34*, 356–361. [CrossRef]
47. Save, M.; Schappacher, M.; Soum, A. Controlled Ring-Opening Polymerization of Lactones and Lactides Initiated by Lanthanum Isopropoxide, 1. General Aspects and Kinetics. *Macromol. Chem. Phys.* **2002**, *203*, 889–899. [CrossRef]
48. Al-Khafaji, Y.F.; Elsegood, M.R.J.; Frese, J.W.A.; Redshaw, C. Ring opening polymerization of lactides and lactones by multimetallic alkyl zinc complexes derived from the acids $Ph_2C(X)CO_2H$ (X = OH, NH_2). *RSC Adv.* **2017**, *7*, 4510–4517. [CrossRef]
49. Wang, X.; Zhao, K.-Q.; Al-Khafaji, Y.; Mo, S.; Prior, T.J.; Elsegood, M.R.J.; Redshaw, C. Organoaluminium complexes derived from anilines or Schiff bases for the ring-opening polymerization of ε-caprolactone, δ-valerolactone and *rac*-lactide. *Eur. J. Inorg. Chem.* **2017**, *2017*, 1951–1965. [CrossRef]
50. *Handbook of Ring Opening Polymerization*; Dubois, P.; Coulembier, O.; Raquez, J.-M. (Eds.) Wiley-VCH: Weinheim, Germany, 2009. [CrossRef]
51. Aikawa, K.; Mikami, K. Dual chirality control of palladium(II) complexes bearing *Tropos* biphenyl diamine ligands. *Chem. Commun.* **2005**, 5799–5801. [CrossRef] [PubMed]
52. *APEX 2 & SAINT Software for CCD Diffractometers*; Bruker AXS Inc.: Madison, WI, USA, 2006.
53. *CrysAlisPRO Software, Oxford Rigaku Diffraction*; Rigaku Corporation: Wrocław, Poland, 2021 & 2023.
54. Sheldrick, G.M. Crystal structure refinement with SHELXL. *Acta Cryst.* **2015**, *C71*, 3–8. [CrossRef]

55. Sheldrick, G.M. *SHELXT*—Integrated space-group and crystal-structure determination. *Acta Cryst.* **2015**, *A71*, 3–8. [CrossRef] [PubMed]
56. Palatinus, L.; Chapuis, G. SUPERFLIP—A computer program for the solution of crystal structures by charge flipping in arbitrary dimensions. *J. Appl. Cryst.* **2007**, *40*, 786–790. [CrossRef]

Disclaimer/Publisher's Note: The statements, opinions and data contained in all publications are solely those of the individual author(s) and contributor(s) and not of MDPI and/or the editor(s). MDPI and/or the editor(s) disclaim responsibility for any injury to people or property resulting from any ideas, methods, instructions or products referred to in the content.

Molecular Catalysts for OER/ORR in Zn–Air Batteries

Evgeny V. Rebrov [1,2,*] **and Peng-Zhao Gao** [3]

[1] School of Engineering, University of Warwick, Coventry CV4 7AL, UK
[2] Department of Chemical Engineering and Chemistry, Eindhoven University of Technology, P.O. Box 513, 5600 MB Eindhoven, The Netherlands
[3] College of Materials Science and Engineering, Hunan University, Changsha 410082, China; gaopengzhao7602@hnu.edu.cn
[*] Correspondence: e.rebrov@warwick.ac.uk or e.rebrov@tue.nl

Abstract: Zn–air batteries are becoming the promising power source for small electronic devices and electric vehicles. They provide a relatively high specific energy density at relatively low cost. This review presents exciting advances and challenges related to the development of molecular catalysts for cathode reactions in Zn–air batteries. Bifunctional electrocatalysts for oxygen reduction reaction (ORR) and oxygen evolution reaction (OER) play the main role in improving performance of reversible fuel cell and metal–air batteries. The catalyst development strategies are reviewed, along with strategies to enhance catalyst performance by application of magnetic field. Proper design of bifunctional molecular ORR/OER catalysts allows the prolongment of the battery reversibility to a few thousand cycles and reach of energy efficiencies of over 70%.

Keywords: Zn–air battery; bifunctional electrocatalyst; ORR; OER; ferromagnetic electrocatalysts; AC magnetic field

1. Introduction

In many areas, technologies are changing from fossil fuel technology to electrified technologies that are driven by renewable electrons. Currently these transformations are taking place very rapidly, enabled by novel materials that reduce the cost of renewable electrons and their storage, and extend the range of these novel applications. The increasing role of electricity as an energy carrier in decarbonizing the economy results in a higher demand for electrical energy storage in the form of rechargeable batteries. It remains a great challenge to achieve next-generation rechargeable batteries with high energy density, excellent cycling stability, good rate capability, efficient active material utilization, and high Coulombic efficiency. Two battery applications driving demand growth are: (i) electric vehicles and (ii) stationary forms of energy storage. As renewable electrons become readily available and more cost competitive with conventional energy sources, the breadth of their potential application significantly increases. The solar industry has already achieved the US Department of Energy's original SunShot 2020 cost target of USD 0.06 per kWh for utility-scale photovoltaic solar power [1]. The most recent 700 MW PV facility in Portugal has a bid power cost of USD 0.013 per kWh, which makes this one of the lowest costs of electricity [2]. Yet, the amount of energy storage required in the future is significant. It requires 1.0 TWh of storage to mitigate the intermittency issue and to achieve 90% renewables by 2035 worldwide [3]. Europe is set to host around 280 gigafactories (6000 GWh) by 2030 and the growth is currently driven by automotive demand which is linked to the decarbonization of road transport.

Rechargeable batteries are considered as the primary energy source for future transportation and stationary applications. The growth observed in electric vehicle (EV) adoption is expected to continue due to government incentives for consumers in many markets of the world. Analysts project mobility storage demands in 2030 of 0.8 to 3.0 TWh, with the demand for light-duty EVs dominating near-term markets [4]. In today's EV market, lithium-ion batteries (LIBs) are extensively employed [5]. The specific energy of lithium-air batteries is 5928 Wh kg^{-1} and a nominal cell voltage is 2.96 V [6]. LIBs gain a competitive advantage in terms of stability and energy density and their rather high safety plays a crucial role in empowering the EV industry. However, they suffer from low ionic conductivity at room temperature and advanced functional electrolytes are still in the research stage. The leading manufacturers of LIBs are listed in Table 1.

Owing to its high reactivity, lithium is present only in compounds in nature: either in brines or hard rock minerals. The rarity of lithium in nature results in a rather high price of LIBs. Lithium demand will rise further from approximately 500,000 tons of lithium carbonate equivalent (LCE) in 2021 to some 3 to 4 million tons in 2030 [7]. However, satisfying the demand for lithium is not a trivial problem. Electric vehicle (EV) sales doubled to approximately 7 million units in 2021. At the same time, the Li price increased by 550% in a year [8]. In 2022, the lithium carbonate and lithium hydroxide price passed USD 75,000 and USD 65,000 per ton, respectively (compared with a five-year average of around USD 14,500 per ton) [8].

Table 1. Leading producers of rechargeable batteries [9,10].

Top 5 Countries/Regions, Share of Commissioned Capacity, 2020 (747 GWh)	Leading Producers	Main Countries in Which Leading Producers Own Battery Manufacturing Plants
China 76% (568 GWh) U.S.A. 8% (60 GWh) Europe 7% (52 GWh) Republic of Korea 5% (37 GWh) Japan 4% (30 GWh)	CATL (Ningde, China)	China (Shenyang) JV "BBA" with BMW Group, Germany (Erfurt)
	BYD (Shenzhen, China)	China (Pingshan in Shenzhen) France, Hungary
	Panasonic (Osaka, Japan)	Japan, China, USA (Nevada)
	LG (Seoul, Republic of Korea)	Republic of Korea (Ochang), China (Nanjing), USA (Holland, Michigan), Poland (Wroclaw)
	Samsung (Suwon, Republic of Korea)	Republic of Korea (Ulsan; Pohang), USA (Auburn Hill), China (Tianjin, Xi'an), Europe (Hungary, Austria), India, Malaysia, Vietnam
	SK Innovation (Seoul, Republic of Korea) Gotion High-Tech (Hefei, China), Envision (Shanghai, China), AESC (Yokohama, Japan)	Republic of Korea (Seosan), China (Changzhou, Jiangsu), USA (Commerce, Georgia; two locations in JV with Ford), Hungary (Komaron, Ivancsá) Japan (Kanagawa); USA (Tennessee), UK (Sunderland) and China (Jiangyin) China (Hefei), Germany (Salzgitter)

Therefore, much research is devoted to alternative batteries based on zinc (Zn) cathodes. They show a large potential for further development as energy storage devices. As competitive electrochemical technologies for energy storage and conversion, rechargeable zinc-based batteries have attracted increasing attention including the zinc-ion battery and zinc–air battery (ZAB) [11]. Compared with the zinc-ion battery, the rechargeable zinc–air battery (ZAB) shows high energy density because of the employment of O_2 as the active material, low negative potential (-0.762 V vs. SHE), abundance, low toxicity, and intrinsic safety advantages. Rechargeable ZABs are also considered as one of the most economically feasible battery solutions for grid-scale applications [12]. Currently, several companies have already started deploying ZABs for utility scale energy storage, including NantEnergy, who installed 3000 systems in nine countries at a manufacturing cost as low as USD 100 per kWh in 2019 [11]. Large ZABs with a total charge of 500–2000 Ah can be used in railway and maritime devices, while small scale batteries with a charge of 200–400 mAh are used for small hearing-aid devices [12].

The average Zn price in 2021 was USD 2950 per ton [13], indicating that this material is ca six times cheaper than lithium carbonate. The volumetric energy density of ZABs is relatively high (6136 Wh L^{-1}). Moreover, the currently achievable battery lifetime in practical conditions is around 150 cycles, and the energy efficiency is below 60%, which is still far from the operational requirements of EVs [14]. This is due to excessive dendrite formation on the electrode and electrolyte carbonation. Therefore, further improvements are required to increase energy density, safety, and to decrease costs.

A typical ZAB consists of a porous air cathode and a Zn metal anode, separated by an alkaline electrolyte. During discharge, O$_2$ from the air permeates the porous cathode and is reduced on the catalyst surface (Figure 1a).

$$O_2 + 2 H_2O + 4e^- \rightarrow 4 OH^- \ (E_0 = 0.40 \text{ V}) \qquad (1)$$

Here E_0 represents the standard electrode potential with respect to the standard hydrogen electrode. At the same time, the Zn anode reacts with hydroxide ions to produce soluble zincate ions as well as insoluble ZnO.

$$Zn + 4 OH^- \rightarrow Zn(OH)_4^{2-} + 2e^- \qquad (2)$$

$$Zn(OH)_4^{2-} \leftrightarrow ZnO + H_2O + 2 OH^- \qquad (3)$$

$$\text{Total: } Zn + 2 OH^- \rightarrow ZnO + H_2O + 2e^- \ (E_0 = -1.25 \text{ V}) \qquad (4)$$

$$\text{Overall reaction at the Zn anode is } 2 Zn + O_2 \rightarrow 2 ZnO \ (E_0 = 1.65 \text{ V}) \qquad (5)$$

However, the practical working voltages are usually lower than 1.20 V, much less than the theoretical potential (1.65 V) in consideration of the internal loss of battery, resulting from the overpotential over the catalysts, ohmic loss and concentration loss. In addition to these factors, Zn is thermodynamically unstable in alkaline solution as zinc has a higher negative reduction potential than hydrogen, resulting in hydrogen gas evolution at the solid/liquid interface gas:

$$Zn + H_2O \rightarrow ZnO + H_2 \qquad (6)$$

This reaction deteriorates the efficiency of the Zn anode and accelerates self-discharge. CO$_2$ from the air also reacts with the alkaline electrolyte to form carbonates (Equations (7) and (8)), that could block the gas diffusion layer to prevent the flow of oxygen.

$$CO_2 + OH^- \rightarrow HCO_3^- \qquad (7)$$

$$HCO_3^- + OH^- \rightarrow CO_3^{2-} + H_2O \qquad (8)$$

The mobility of carbonate/bicarbonate anions is lower than that of OH$^-$ [15], and hence, the carbonation process causes a decline in the ionic conductivity of the alkaline electrolyte.

The charge process is fulfilled by reversing the reactions (Figure 1b). The oxygen-evolution reaction (OER) at the air electrode generates oxygen. Zincate ions are reduced to the Zn(0) at the anode surface. Electrocatalysis plays an important role in nearly all rechargeable battery technologies. Electrocatalysis is one of the main technologies to directly use renewable electrical energy to drive reactions. A SciFinder search on the terms "electrocatalyst" and "rechargeable battery" returned >3700 items for the period 2020–2022, mainly related to the development of novel catalysts for oxygen evolution reaction (OER) and oxygen reduction reaction (ORR), with about 50% of all research papers coming from

China. Developing an exceptional catalyst capable of accelerating both oxygen reduction reaction (ORR) and OER remains challenging.

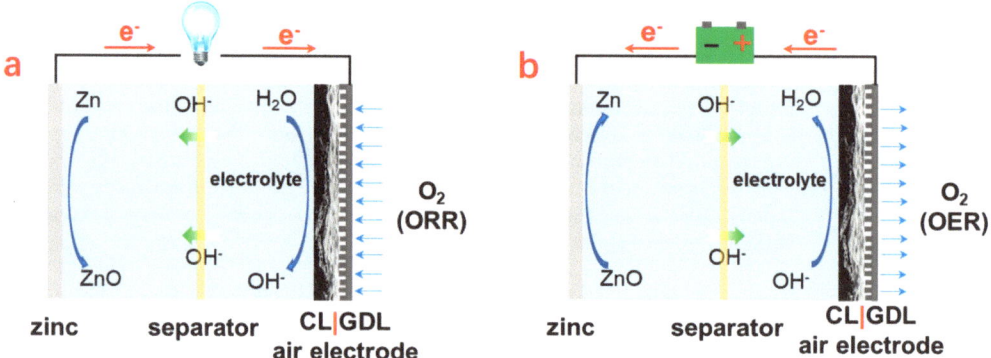

Figure 1. The schematic discharging (**a**) and charging (**b**) mechanisms of rechargeable zinc-air batteries [16]. During discharge, O_2 from the air permeates the porous catalyst layer (shown with blue arrows) and is reduced on the catalyst surface with water molecules (Equation (1)). The hydroxide ions diffuse via the separator to the zinc electrode (shown with green arrows). During charging, the OER at the air electrode generates oxygen (shown with blue arrows).

Among reported catalysts, single-atom catalysts (SACs) have attracted extensive attention due to their maximum atom utilization efficiency, homogenous active centers, and unique reaction mechanisms. The novel electrocatalysts also encompass metal clusters and bimetallic nanoparticles, and metal catalysts derived from metal organic frameworks (MOFs). Due to the environmental effects of metal usage, many research groups develop metal-free carbon-based catalysts [17]. The use of relatively cheap and electrically conductive carbon supports improves the catalyst dispersion and active area while decreasing the costs. However, despite the advanced methods such as defect engineering (heteroatom doping) to create favorable metal-free carbon-based catalysts, their catalytic activity is insufficient and not yet comparable to metal-based catalysts [17]. As a result, it is impractical to entirely abandon the use of metal materials in building bifunctional catalysts at the current state, highlighting the significance of integrating the advantages of different elements to construct high-performance catalysts to improve the ORR and OER. Several strategies, including structural engineering, defect engineering, and single atom engineering, have been used to develop effective catalysts to improve the electrochemical performance of ZABs. A great deal of research has been devoted to the development of novel air cathodes loaded with bifunctional catalysts, and they can be sorted by their constituents, such as metal oxides and carbon-based materials.

Electrocatalytic reactions may further be improved by employing magnetic heating of composite magnetic catalysts. Local heating of the electrodes was already employed to improve cell performance. However, conventional heating systems cannot avoid global heating of the whole electrochemical cell, and, therefore, inductive heating is often applied. It allows the localization of heating in the immediate vicinity of magnetic nanoparticles (MNPs) when these are positioned in an external AC magnetic field [18]. Magnetic hyperthermia requires the application of magnetic nanoparticles compatible with heterogeneous catalysts, when placed in an alternative magnetic field, to convert the energy into heat. The power generated by the magnetic nanoparticles is governed by their specific absorption rate (SAR). A SAR of above 1 kW g^{-1} is required for applications in catalysis, in the range of frequencies (100–500 kHz) and magnetic fields (20–100 mT) already used in industry. This is a relatively new research area and SciFinder search on the terms "magnetic catalyst" and "inductive heating" returns less than 50 items for the period 2019–2023, mainly related to the development of composite magnetic catalysts.

In this review, we discuss recent advances in the development of nanostructured catalysts, including single atom catalysts on carbon supports, mono and bi-metallic catalysts, mixed metal oxides and layered double hydroxides for ORR and/or OER electrocatalysis, with a focus towards improving their performance in ZABs. Structure–activity relationships in oxygen electrocatalysis are highlighted. Further enhancement of catalytic activity by application of an external magnetic field is also discussed. Finally, the remaining challenges for the application of nanostructured catalysts in ZABs are highlighted. The overall aim is to provide a snapshot over development for ORR and OER electrocatalysis.

2. Air Electrode Catalysts

The air cathode has a considerable effect on electrochemical behavior of rechargeable batteries. It is commonly composed of three components: (i) the current collector, (ii) the gas diffusion layer (GDL) and (iii) the catalyst. A binder-free air electrode was formed by deposition of cobalt oxide nanoparticles on a carbon nanofiber mat by thermally treating electrospun Co(II)-containing polyacrylonitrile fibers [19]. The battery exhibited better stability and cycling performance compared with a conventional ZAB assembled with a Pt/C catalyst. Aerosol printing is an emerging additive manufacturing technology which can deposit a catalytic coating on the electrode with high precision, so that the catalyst layer has excellent adhesion with a carbon cloth surface, and interface resistance has little influence on the electrochemical characteristics of the electrode and is suitable for a flexible battery. The technology for preparing the conformal catalytic layer based on the aerosol method was reported to form a catalytic coating on the cathode of the battery [20].

2.1. Activated Carbon Materials

Activated carbons (ACs), typically derived from biomass, have gained great interest due to their low price, reproducibility and environmentally friendly nature. Numerous biomass wastes (coconut shells, vegetable wastes, etc.) were used as precursors to prepare ACs with large effective specific surface area. Another considerable advantage of biomass precursors is that they contain rich sources of heteroatoms (e.g., N, P, O, S), which are natural dopants for porous carbon materials. Different oxygen-containing functional groups on the surface of ACs provide active sites for catalytic processes. These electron-withdrawal groups create a positive charge density on adjacent carbon atoms which can facilitate the adsorption of OH^- ions, promote transportation of electrons and assure easy and rapid recombination of two adsorbed oxygen atoms for OER. The oxygen-containing groups can favor the adsorption of water via hydrogen bonding [21]. The construction of hierarchically porous materials (HPAC) achieves fast ion transportation and provides a larger effective surface area for applications for the OER.

The use of templates makes it easy to create porous catalysts with different morphologies. Nanocasting involves the preparation of templates and the transfer of morphological information from the templates to the ACs. Montmorillonite, copper, halloysite, SiO_2, and block copolymers are often used as templates. For example, using a block polymer, poly(styrene)-block-poly(ethylene oxide), mesoporous carbon spheres with a large specific surface area of 356 $m^2\,g^{-1}$ were fabricated [22]. The activation methods of carbon materials are divided into chemical, physical, and combined (chemical–physical) methods. Among them, physical activation has the advantages of low cost and low environmental impact. Usually, the ACs activated by water vapor have a wide pore size distribution (PSD) due to the faster evaporation rate and the smaller diameter of water molecules. The ACs activated by CO_2 have a relatively narrow PSD due to the slower diffusion rate and large diameter of CO_2 molecules [23].

The heteroatom incorporated into ACs is advantageous to regulate their electron donor performance and, thus, adjust the electrical and chemical performance of their surface. Nitrogen (N), sulfur (S), oxygen (O), boron (B), and phosphorus (P) -doped ACs were investigated [24]. The synergistic effect of multiple-heteroatoms-doped ACs on the electrochemical properties was reported. Various N- and S-doped ACs with a low S content were obtained by pyrolysis of protic ionic liquids (PILs) of [R-NH$_x$][HSO$_4$] (R means carbon chain) [25].

The N-doped HPACs enhance both electronic conductivity and surface wettability and, therefore, they are considered as ideal electrode materials. The presence of nitrogen heteroatoms in hexagonal carbon rings raises the basicity, electrical conductivity, and oxidation resistance of ACs. The N-doped HPACs are typically produced by pyrolysis of polyacrylonitrile, polyaniline, phenolic resin, and biomass derivatives [20]. Nitrogen-doped carbon nanotube supported metal catalysts (M-N-C, M=Fe, Co) decrease the binding energies of the surface-adsorbed *OOH and *OH species, thus boosting the catalytic activity in the ORR reaction. Yet many carbon supports suffer from serious degradation under high potentials of OER process [26]. The potential above 0.207 V may already result in the oxidation of ACs to CO_2. This process is accelerated when a noticeable amount of oxygen is generated at a potential of 1.5 V [27]. The electrochemical stability of ACs is highly dependent on their morphology, structure, and dopants. For instance, amorphous carbon black is highly unstable, whereas highly graphitized carbon nanotubes can significantly improve their stability under OER conditions. However, N-doped graphene remains expensive, complex to set up, and difficult to scale up for practical applications. Chemical vapor deposition (CVD) for graphene synthesis requires metal substrates such as nickel or copper as the growth catalyst.

Extremely rapid heating for the synthesis of highly open structured N-doped graphene sheets (N-ex-G) was demonstrated [28]. The N-ex-G were produced by fast heating (>150 K s^{-1}) to 1100 °C under an Ar/NH$_3$ flow. The materials exhibited hexagonal symmetry with the existence of symmetrical three-fold sp^2-bonding of carbon atoms with pyridinic N, pyrrolic N, and quaternary N atoms. The N-ex-G nanosheets showed very comparable ORR activity to that of a commercial Pt/C catalyst. The respective ZAB with N-*ex*-G demonstrated a current density of 47.6 mA cm^{-2} at 0.8 V and a power density of 42.4 mW cm^{-2}.

The graphene/metal oxide composites combine the excellent conductivity of graphene and the high energy density of transition metal oxide [29]. Graphene oxide (GO) is mostly used as a precursor for preparing graphene/metal oxide composites, and it is mixed with a transition metal compound, then the transition metal oxide is generated in-situ on the surface of the graphene oxide by methods such as coprecipitation, hydrothermal or solvothermal synthesis.

Metal organic frameworks (MOFs), wherein transition metals are coordinated by organic ligands in a 3D network, are excellent precursors for the synthesis of porous carbon- or nitrogen-doped carbon architectures and modified porous carbon materials. MOFs are also an important platform material in the area of bifunctional ORR/OER electrocatalysts. For example, zeolitic imidazole framework-67 (ZIF-67) has been extensively employed for the synthesis of porous Co$_4$N/Co-Nx-C composites. Upon carbonization under N$_2$ atmosphere, Co-Nx-C nanocomposites that partially inherited the porous framework were formed (Figure 2).

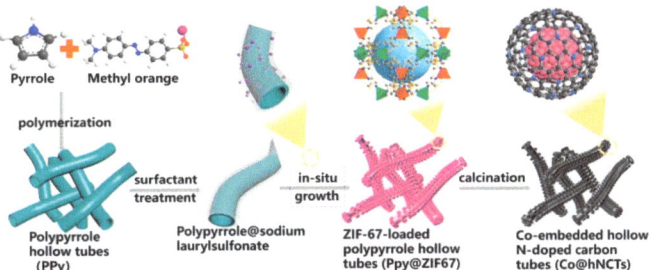

Figure 2. Schematic illustration of the synthesis of MOF-derived Co@hNCT catalysts. Adopted from [30].

A MOF-derived synthesis method was recently patented for an Fe-Eu-MOF/GO bimetallic cathode catalyst [31]. The prepared catalyst demonstrated a higher activity in both ORR and OER reactions and very high stability as compared with respective monometallic catalysts due to a large number of oxygen vacancies that enhance the interaction between metal nanoparticles and their support; in addition, they enhance the adsorption of oxygen on the surface of the catalyst and enable electrons to be rapidly transferred in an electrochemical reaction.

2.2. Monometallic Catalysts

The rates of ORR and OER in rechargeable batteries are relatively slow and they require different catalysts because of their different mechanisms onto the catalyst surface. Therefore, bifunctional electrocatalysts are needed for the best performances. The noble metal catalysts have high conductivity, high catalytic performance and good stability. The Pt-based nanocomposites are used for ORR [32]. However, the formation of irreversible Pt=O decreases its catalytic activity toward OER. On the contrary, IrO_2 and RuO_2 are effective for OER but not as active for ORR [33,34]. Yet, noble metal catalysts are costly, which limits their large-scale application. Therefore, the goal of catalyst development, including for Pt and other precious metal-based catalysts, is to increase the total number and/or the intrinsic activity of active sites [35]. This goal can be achieved by several strategies, which are summarized in Figure 3.

Nonprecious bifunctional electrocatalysts with low cost and efficient catalytic activities in ORR and OER are highly desired to promote the commercialization of ZABs. From the perspective of elemental composition, Fe and Ni have been paid much attention as the promising earth-abundant transition metals for the design of OER catalysts, especially Ni/Fe-based micro/nanostructures as a new generation of catalytical materials (such as metal hydroxide and mixed-metal oxides) which were developed over the last few years. They have attracted extensive attention because of their facile synthesis [36]. A gram of IrO_2 (99 wt %)) and RuO_2 (99.5 wt %) costs ca USD 43 and USD 95, respectively. The price of iron oxide and nickel oxide with a 99 wt % purity is 50–100 times lower. Therefore, these transition metal-based compounds are appealing candidates, especially for bifunctional catalysts [37,38]. In comparison with precious metals, they offer advantages of low prices and greater availability. Cobalt oxides, and Co-perovskites exhibit high intrinsic catalytic activity for OER in alkaline media (Table 2). However, their apparent activity is limited by either poor electronic conductivity or low surface area. Their durability is also limited by their tendency of sintering and/or aggregation. In alkaline solutions, corrosion limits the effective lifetime of Fe- and Ni-based OER electrocatalysts. Typically, monometallic Fe-based compounds on various substrates in a 1 M KOH aqueous solution with a constant overpotential were active for less than 180 h due to corrosion. Although the initial decay of a catalyst sometimes could increase the active surface area and catalytic activity, catalyst activity always reduces in the long-term.

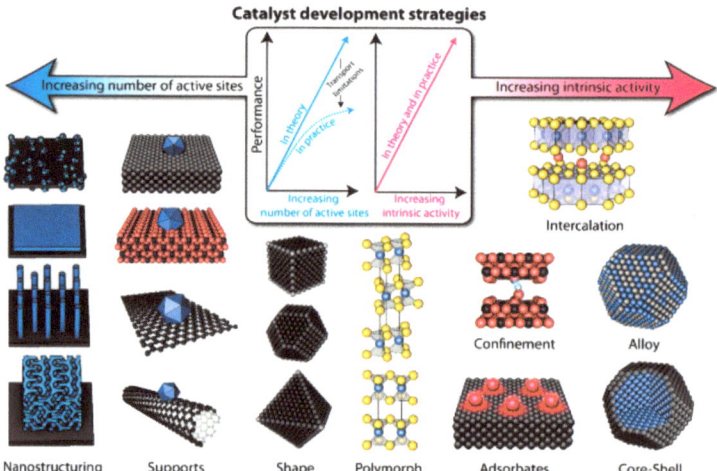

Figure 3. Different development strategies for metal catalyst preparation. Schematic of various catalyst development strategies, which aim to increase the number of active sites and/or increase the intrinsic activity of each active site. Adopted from [36].

Table 2. Recent advances in bifunctional MOF-derived catalysts.

Catalyst	OCV (V)	PPD (mW cm^{-2})	Discharge-Charge Voltage Gap (V) at Current Density	Cycling Stability	Ref
HCo@FeCo/N/C	1.45	125	0.84 (10 mA cm^{-2})	200 h (10 mA cm^{-2})	[39]
Co@N-CNT hollow nitrogen-doped carbon nanotubes	1.45	149	0.85 (5 mA cm^{-2})	500 h (5 mA cm^{-2})	[30]
Co/N-doped CNT/graphene hybrid	1.48	253	0.76 (5 mA cm^{-2})	9000 cycles, 3000 h (5 mA cm^{-2})	[40]
ZIF-derived Co$_9$S$_8$/GN graphene nanosheet		186	0.52 (2 mA cm^{-2})	2000 cycles, 147 h (2 mA cm^{-2})	[41]
CoP/(N,P) codoped porous carbon	1.4	186	1.00 (2 mA cm^{-2})	80 h (2 mA cm^{-2})	[42]
Ni–Co–S/(N,S) codoped porous carbon	1.43	137	0.73 (10 mA cm^{-2})	180 cycles (10 mA cm^{-2})	[43]
Co@SiO$_x$/N-doped carbon		138	0.82 (5 mA cm^{-2})	400 h (5 mA cm^{-2})	[44]
Co/Co$_9$S$_8$@CNT	1.44	185	0.75 (5 mA cm^{-2})	50 cycles, 2000 h (5 mA cm^{-2})	[45]
Ba$_{0.5}$Sr$_{0.5}$Co$_{0.8}$Fe$_{0.2}$O$_3$			0.83 (5 mA cm^{-2})	1800 cycles, 300 h (5 mA cm^{-2})	[46]
Co-MOF/LaCoO$_{3-\delta}$ hybrid	1.44	126	0.67 (5 mA cm^{-2})	120 h (5 mA cm^{-2})	[47]
Ni$_3$Fe/Co–N–C	1.39	68	0.75 (10 mA cm^{-2})	65 h (10 mA cm^{-2})	[48]
Co-NCS@nCNT	1.42	90	0.89 (5 mA cm^{-2})	480 cycles, 80 h (5 mA cm^{-2})	[49]
Co$_3$O$_4$/Co@NC	1.5	124	0.94 (10 mA cm^{-2})	3600 cycles, 600 h (10 mA cm^{-2})	[50]

These challenges are resolved by anchoring transition metal oxide/nitride nanoparticles into a porous substrate. Anchoring provides sufficient electronic conductivity to enhance charge transfer and greatly increase the dispersion of metal oxides in order to increase the number of active sites and to suppress the sintering of the individual nanoparticles. Furthermore, the doping of N heteroatoms in the carbon lattice enhances the ORR activity. The presence of transition metals (Fe, Co, Ni, Mn) or their alloys can also significantly improve the stability of the nitrogen-doped graphene nanotubes. Different bifunctional electrocatalysts with high activity and durability for both ORR and OER have been developed. Some of them are listed in Table 2.

Transition metals contain partially filled d-orbitals, conferring their compounds with a high electrochemical activity. However, most of them have good OER activity but weak ORR activity. Manganese or mixed cobalt oxides (Mn$_3$O$_4$, NiCo$_2$O$_4$) effectively catalyze the ORR because of the variable valence states and coordination structures [51–53]. Co spinel catalysts are the most used among those potential bifunctional electrocatalysts. NiCo$_2$O$_4$

was synthesized with various structures such as nanosphere, nanosheet and nanorods [54]. These hollow porous structures provide a larger specific surface area as compared with bulk samples of the same materials. The main synthetic methods for nanotube structures are template and electrospinning. The materials prepared by electrospinning have greater brittleness, while the catalysts prepared by the template method have surfactants residues, which reduce the rate of electrocatalytic reactions. In a relatively new approach, a Co_3O_4 catalyst was immobilized on a Ni foam by a spray pyrolysis method [55]. The respective ZAB showed an excellent cycling stability of more than 500 charge–discharge cycles at a current density of 5 mA cm^{-2}.

Manganese oxides are a specific example of the particularly active catalysts for use with the air electrode for commercialization of primary ZABs due to their rich oxidation states, chemical compositions and crystal structures. The electrochemical activity of MnO_2-based catalysts depends on their shape and crystallinity and crystallographic properties (phase composition and crystal size), which can be controlled via synthesis parameters [56]. Many theoretical investigations indicated the roles of oxygen vacancies on MnO_2 during the ORR in terms of electronic properties. The large variation in the structural disorder of MnO_2 and, therefore, in internal resistance at the electrode–electrolyte interface limits large-scale applications of this material. Various methods were studied for the preparation of MnO_2-based catalysts, such as sol-gel, rheological phase reaction, microemulsion, coprecipitation, solid-state template, microwave and hydrothermal synthesis. Often the density functional theory (DFT) was employed to investigate the effect of oxygen vacancies (OV) on the electrocatalytic performance of MnO_2. Defect-engineering of α-MnO_2 (211) and β-MnO_2 (110) by oxygen vacancy was studied to enhance its catalytic performance [57]. For example, Li et al. studied the electronic properties of OV β-MnO_2 (110), in which the oxygen vacancy in the sub-lattice of the surface structure was purposed [58]. Other DFT results involving the effects of OV on the electronic conductivity of MnO_2 were systematically examined in the bulk structure and discussed based on the density of states [59,60]. It was shown that the α-MnO_2 exhibits superior catalytic performance for OER (overpotential of 0.45 V at 50 mA cm^{-2}).

Doping of various quantities of metals into the MnO_2 lattice often improves the OER catalytic activity. The doping of 0.2 mol% of Ni to α-MnO_2 reduces overpotential in a 0.1 M KOH to 0.44 V at a 10 mA cm^{-2} current density [61]. The surface of the bare MnO_2 materials has inactive Mn^{4+} centers. After doping with Ni, a major part of surface Mn ions reduces to lower oxidation states. The presence of Ni ions results in an electron transfer to the Mn^{4+} ions, which creates a distortion in the Mn–O octahedral units and increases OER activity. The incorporation of Ag into α-MnO_2 improves electrical conductivity and increases the number of oxygen vacancies that also create a considerable distortion in the Mn–O octahedra [62]. As a result, the Ag/α-MnO_2 catalyst shows a threefold enhancement in current density in OER as compared with the undoped α-MnO_2 catalyst. The Ag/MnO_2 demonstrates a power density of 273 mW cm^{-2} and maintains a stable performance over 3200 cycles. It can be said that doping transition metal ions into less-active oxide surfaces could create a new avenue for OER mixed metal oxide research.

On the other side, isolated single atom catalysts (SACs) coordinated with nitrogen on a carbon support such as graphene, porous carbon (M-N-C) or om metal oxides have promising perspectives for replacement of precious metal catalysts. In particular, graphitic and diamond carbons show improved resistance to corrosion [63]. Their thermal stability, electronic properties, and catalytic activities can be controlled via interactions between the single atom center and neighboring heteroatoms such as nitrogen, oxygen, and sulfur [64]. The preparation of SACs requires a smaller amount of metal precursors, which relieves the metal availability issues. In general, SACs show higher specific catalytic activity (turnover frequency) as compared with conventional catalysts due to their high dispersion, which is close to 100%, as well as their increased stability. While SACs are still in the early stages of commercialization, their potential for improving catalytic activity and selectivity make them an exciting area of research and development. The market for SACs is fast developing, as the technology is relatively new and there are still many challenges that need to be addressed before these catalysts can be widely adopted. However, there is growing interest in SACs due to their high catalytic activity, selectivity, and stability. According to a recent report by Markets and Markets, the global market for SACs is expected to grow from USD 334.2 million in 2020 to USD 1.1 billion by 2025, at an annual growth rate of 26% [65].

Currently, Fe-N-C catalysts are considered to be among the most important ORR catalysts due to their hierarchical pores, oriented mesopores and high conductivity (Table 3) [66]. Their ORR catalytic activity is above that of non-noble-metal catalysts and is similar to that of a commercial Pt/C catalyst (Figure 4). In addition, Fe-N-C catalysts have a high stability. The intrinsic ORR activity of the Fe-N_4 sites relates to the electronic structure of carbon supports. However, the origin of ORR activity on Fe-N-C catalysts remains unclear, which hinders the further improvement of their catalytic properties. Several authors have reported that the electronic structure of carbon supports (electron-withdrawing or donating property) affects the d-band center or d-orbital level of the Fe site [67,68]. In their density functional theory (DFT) calculations, Liu et al. studied 42 types of divacancy defects around the Fe-N_4 active site by selectively removing C–C bonds or by rotating the different C–C bonds [69]. They established a relationship between the geometric, electronic structure, and ORR activity of the Fe-N_4 site. The first electron transfer step (*O_2 + H^+ + e → *OOH) was found to be the potential-determining step (PDS) on all Fe-N_4 configurations. The Fe-O bond length is an important descriptor, which influences the ORR activity. At larger distance between atoms in the Fe–O bond, hybridization between Fe $3d_{z^2}$ and O_2 $2p_z$ orbitals results in the activation of O_2. However, at shorter interatomic distance, the activation of O_2 occurs by the hybridization of Fe $3d_{yz}$ and $2p_y$ orbitals. Thus, the hybridization between Fe $3d_{z^2}$, $3d_{yz}$ and O_2 π^* orbitals is considered to be the origin of high activity of Fe-N_4 sites embedded in the graphene framework [69].

Further improvements are expected through the better understanding of the chemical origin of active sites on Fe-N-C catalysts [70].

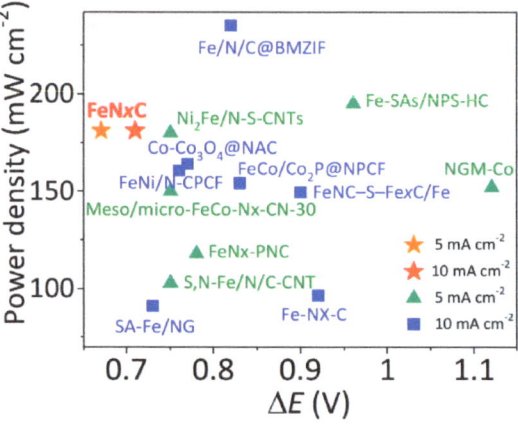

Figure 4. Comparison of ORR performance and ZAB performance (voltage gap, V) over different Fe-N-C electrocatalysts. Adopted from [71].

Despite the obvious advantages of Fe-N-C catalysts, their fabrication involves multi-step synthesis and lengthy post-processing procedures which make it difficult to scale up to industrial level. Moreover, the process requires the introduction of metal oxide nanoparticles which may also introduce some impurities decreasing the atom utilization efficiency and reducing catalyst performance. Therefore, developing cost-effective methods for preparation of high-purity Fe-N-C catalysts remains an important task to facilitate their large-scale application. Recently a one-pot synthesis was proposed, which provided a high density of the ORR active site and also facilitated the kinetics [71]. Fe-N-doped carbon (Fe-NC) was prepared via a pyrolysis of a mixture of glucose and dicyandiamide as carbon source and nitrogen source, respectively, with addition of $FeCl_2$ and $FeCl_3$. The pyrolysis was initially performed in a flow of Ar at 550 °C for 6 h and then at 900 °C for 3 h. The homogenous ink of as-prepared catalyst was used for preparing a ZAB cathode. The ZAB delivered a maximum power density of 181 mW cm^{-2} and a voltage gap between discharging and charging of 0.7 V at 10 mA cm^{-2}. The ZAB of Fe-N-C runs stably for more than 8 h at 1 mA cm^{-2}, and a 0.7 V voltage gap. No changes were observed in the structural, morphological and compositional properties after the reaction, suggesting that the Fe-N-C electrocatalyst is stable in alkaline solutions.

Table 3. Performance of ZABs using carbon-based single atom catalysts.

Air Electrode Catalyst	Active Material	Max Power Density (mW cm^{-2})	Specific Capacity (mAh g^{-1})	Duration of Tests (h)	Ref
Fe/N–C	FeN$_4$ embedded in N-doped carbon.	225	636	260	[72]
Fe-NCCs	Atomic Fe-N$_x$ dispersed in carbon.	66	705	67	[73]
FeN$_x$–PNC	FeN$_x$ on 2D N-doped carbon.	278	n/a	40	[74]
SA–Fe/NG	Fe-pyrrolic-N species on N-doped carbon.	91	n/a	20	[75]
Co/GO	Atomically dispersed Co on GO.	225	795	50	[76]
Zn/CoN–C	Zn and Co atoms coordinated via N on carbon.	230	n/a	28	[77]

n/a = not availble.

Long-term catalyst stability remains a key factor hindering SACs in ZAB applications. Single metal atom centers may be dissolved or they can aggregate to form larger nanoparticles in the course of electrocatalytic reactions, resulting in a significant activity loss due to the sharp reduction in the number of active sites. The undesirable Fenton reaction (Equation (9)) occurs over Fe–N–C catalysts during ORR resulting in hydroxyl radical formation that is detrimental to both the metal sites and the carbon support.

$$Fe^{2+} + H_2O_2 \rightarrow Fe^{3+} + OH^* + OH^- \tag{9}$$

The coordination of S, Se, or P metal atoms is a possible strategy to prevent unwanted side reactions, thus enhancing the stability of SACs without major loss of catalytic activity. Advanced support materials with high conductivity, corrosion resistance, and ability to support SACs at high metal loadings are highly desired. The scale-up process for the synthesis of SACs remains a challenge which limits their adoption by the market of low metal loading, leading to restrictions on a broader range of applications.

Overall, monometallic catalysts demonstrate unsatisfactory bifunctional performance. As discussed above, one of the well-known spinel materials, Co_3O_4, has also been studied for decades as a highly efficient and corrosion resistant OER catalyst, but its ORR activity is generally poor. Precise control of catalytic activity by varying the composition of these single metal catalysts remains a difficult task. Therefore, bi- and multi-metallic oxides attract a large interest. Many recent efforts have been made to design and prepare low-cost bifunctional catalysts to catalyze both the ORR and OER with high efficiency and long-term durability.

2.3. Mixed Metal Oxide Catalysts

Multicomponent spinel metal oxide nanocomposites demonstrate higher activity in OER as compared with their individual oxides [78]. The improved catalytic performance of spinel oxides is attributed to the presence of both tetrahedral (T_d) and octahedral (O_h) transition metal cation sites in these catalysts. Yet this complex structure with multiple sites also presents a significant challenge in identifying the precise OER active sites. For example, NiFe-, CoFe-, and NiCoFe-mixed metal oxides such as spinel $NiFe_2O_4$ and

CoFe$_2$O$_4$ are highly active for the OER. Incorporating other metals, such as W into the CoFe-mixed oxides, further enhances the OER activity [79]. It is also known that manganese, cobalt and nickel mixed oxides such as Mn$_3$O$_4$, Co$_3$O$_4$, NiCo$_2$O$_4$ and MnCo$_2$O$_4$ could effectively catalyze the ORR in alkaline solutions because of the variable valence states and coordination structures [80]. Liang et al. reported that the oxygen reduction strongly coupled with Co^{III}/Co^{II} redox reaction [81]. Other elements such as V, Cr, and Ni, were also studied to enhance the stabilization energy of high-entropy oxides and to tune the subtle relationship between surface electronic properties and catalytic activities. In general, the synthesis temperature, annealing time and temperature, as well as support type, have significant effects on the ORR performance.

Xu et al. obtained novel hybrid materials composed of La$_2$O$_3$ and Co$_3$O$_4$ supported on carbon-supported MnO$_2$ nanotubes (La$_2$O$_3$/Co$_3$O$_4$/MnO$_2$-CNTs) to assemble the air electrode for ZABs [82]. The hybrid catalyst was prepared by dissolving La nitrate and Co nitrate in an ammonia solution with an addition of MnO$_2$ nanotubes and CNTs. The mixture was sealed and maintained at 150 °C for 6 h in an autoclave, then washed and calcined in air at 300 °C for 1 h. The authors suggested that the high catalytic activity was due to formation of new active phases from inorganic materials and their intimacy with the underlying CNT network. The ZAB exhibits a small voltage gap, high reversibility and high stability, suggesting this battery can be used as a power source for portable electronics and in electrical vehicles. The voltage gap of charge/discharge increased by 0.1 V after 543 cycles at 10 mA cm^{-2}.

Morphological alterations and structural engineering can further boost catalytic performance by enriching the active sites and improving conductivity. A bifunctional catalyst with a narrow half-wave potential of 0.80 V for ORR and a potential of 1.58 V at 10 mA cm^{-2} for OER, was obtained by embedding of NiCo$_2$O$_4$ nanowhiskers (NCO) into a 3D honeycombed hierarchical porous N, O-codoped carbon (HHPC) [83] (Figure 5). The support was derived from gelatin biomass via a hard template-activator assisted pyrolysis. The large surface area provided enough space for dispersing NiCo$_2$O$_4$ evenly to expose abundant ORR and OER active sites. Grain-boundary strength between HHPC and NiCo$_2$O$_4$ nanowhiskers helped to stabilize the porous carbon network and protect it from corrosion.

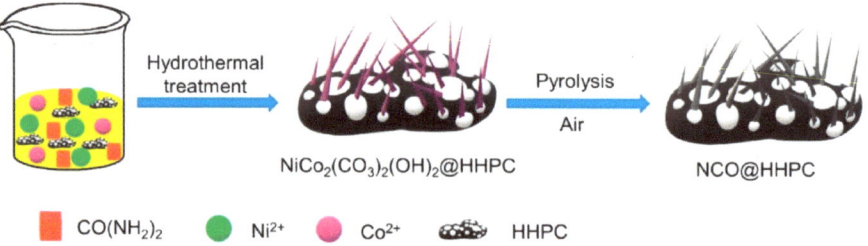

Figure 5. Schematic synthesis process of NiCo$_2$O$_4$ nanowhiskers (NCO) into biomass-derived honeycombed hierarchical porous N, O-codoped carbon. Adopted from [83].

CoFe/CoFe$_2$O$_4$ embedded in N-doped hollow carbon spheres (CoFe/CoFe$_2$O$_4$@NC) with a high degree of oxygen vacancies was synthesized through pyrolysis, carbonization and partial reduction [84]. The catalyst exhibited high ORR activity (half-wave potential: 0.86 V) and a low overpotential of 0.36 V at 10 mA cm^{-2} for the OER. The excellent performance was ascribed to the enhanced intrinsic activity of CoFe$_2$O$_4$ spinel and improved electrical conductivity due to the strong coupling effect CoFe and CoFe$_2$O$_4$.

2.4. Bi- and Multi-Metallic Catalysts

The design of bimetallic alloys with atomically precise noble metal loading attracts great interest in the field of ZABs. Unlike disordered perovskites, structurally ordered intermetallic phases provide much better control over electronic effects, resulting in compositional and positional order and, therefore, more uniform active sites.

Bimetallic CoFe clusters were dispersed onto hollow carbon nanospheres (designated as FeCo-SA/CS) to produce a bifunctional catalyst with a cyclability of over 100 h at a current density of 5 mA cm^{-2} [85]. The catalyst showed an ORR onset potential of 0.96 V. For OER, the catalyst attained a current density of 10 mA cm^{-2} at an overpotential of 0.36 V. Mesoporous FeCo-N-C nanofibers with embedding FeCo nanoparticles were prepared from electrospun FeCo-N coordination compounds with bicomponent polymers consisting of polyvinylpyrrolidone (PVP) and polyacrylonitrile (PAN). The hybrid nanofibers exhibited 1D mesoporous structure with a large surface area and uniformly distributed active sites. The co-existence of FeCo nanoparticles and FeCo-N active sites promoted the catalytic activity of ORR and OER simultaneously [86]. Ni$_3$Fe-N nanoparticles derived from ultrathin Ni$_3$Fe-LDHs were used by thermal ammonolysis and demonstrated a high OER activity [87]. The excellent performance of the Ni$_3$Fe-N catalyst derives from the intrinsic metallic character and unique electronic structure of this compound, which improves the electrical conductivity and adsorption of H$_2$O. In such intermetallic clusters, strong interaction between the two metals provides long-term stability and avoids deactivation.

Several Pt alloys with a transient metal possess superior ORR activities in which the origin of this higher activity is based on the modification of the electronic structure of Pt, which can affect the adsorption strength of oxygen-containing species on active sites. Here, a volcano relationship was demonstrated by Nørskov (Figure 6) [88]. For the best catalyst performance, the adsorption energy should be neither too small nor too big to balance reactant adsorption enthalpy and product desorption enthalpy. Similar relationships were established for PtM bimetallic catalysts with different metals and Pt/M molar ratios. For example, the TOF of the CuPt catalyst was increased by 60-fold over that of the Pt catalyst [89]. In addition, Pt-skin-like surfaces in bimetallic catalysts provide more active sites and higher ORR activities [90]. However, Pt alloyed with transition metals also tends to suffer from leaching and dissolution, which leads to insufficient stability and undermines the advantages of Pt alloy catalysts [91]. Therefore, core–shell bimetallic catalysts with Pt shell were developed to protect the transition metal core [92]. In these catalysts, the dissolution of core is mitigated and the interaction between Pt and the transition metal changes the electronic structure of Pt in a desirable direction. In addition, maximum Pt dispersion is achieved. In this way, a bifunctional Fe$_3$Pt nanoalloy supported onto Ni$_3$FeN was developed [93]. In this catalyst, the Ni$_3$FeN clusters considerably enhanced the activity of OER, whereas Fe$_3$Pt enhanced the activity of ORR as compared with commercial Pt/C and Ir/C catalysts at the same current density of 10 mA cm^{-2}.

Recent studies have indicated that the interaction between PtM catalysts and M-N-C supports demonstrates a synergistic effect to promote the ORR performance of these composite catalysts. For example, a highly active ternary PtZnCu catalyst within an N-doped carbon substrate was reported [94]. The catalyst was synthesized by pyrolysis of Pt, Zn and Cu precursors on the surface of the carbon support under a 5% H$_2$/Ar flow at 750 °C for 2 h. The synthesis method proposed by the authors provides a straightforward way for large-scale synthesis of bifunctional electrocatalysts reducing catalyst costs and improving atom efficiency. The half-wave potential was 0.93 V for the ORR in an acidic solution.

A magnetic field strategy was introduced in obtaining graphene-supported bimetallic Pt-Ni catalysts with predominant (111) facets as efficient ORR catalysts. In this way, a set of porous PtNi catalysts on graphene (G) with Pt/Ni molar ratios of 0.1, 0.5, 0.6, 1.8 and 2.4 was developed by Lyu et al. under high magnetic field in the range of 0.5–2 T [95]. The interaction between the nanoparticles produces microscopic Lorentz forces and improves interaction between Pt and Ni precursors, thus controlling the composition. The Pt$_{2.4}$Ni/G catalyst synthesized under 2 T showed a 5.1 times higher ORR rate than that of a commercial Pt/C catalyst.

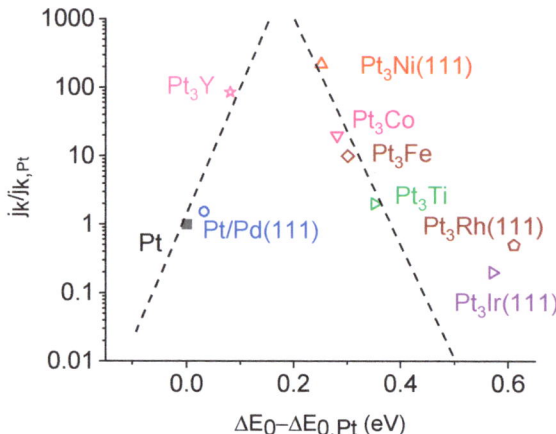

Figure 6. Volcano plot of ORR kinetic current density (j_k) as a function of calculated oxygen adsorption energy (ΔE_O) for Pt alloy electrocatalysts in acidic electrolytes. The dashed lines are the theoretical predictions from [96].

Ordered intermetallic compounds such as PdZn$_3$ were reported to possess excellent catalytic selectivity and stability for the selective semihydrogenation of acetylene alcohols [97]. However, so far, the studies on bimetallic PdM catalysts for ZABs remain very limited. This is mainly due to the lack of effective synthetic methods to control particle sizes during the high temperature annealing [98]. This requires answering two important questions: what annealing temperature is suitable for the formation of structurally ordered phases, and what factors control the ordering kinetics. Therefore, these factors should be studied and then a suitable synthetic route can be adopted. A successful synthesis of ordered Pd$_3$Pb intermetallic phase with small particle size was reported [99]. Potassium triethylborohydride (KEt$_3$BH) or lithium triethylborohydride (LiEt$_3$BH) were used as reducing agents in THF because they both have fast reduction kinetics and can provide the coreduction of Pd and Pb precursors. The obtained sample was annealed for 12 h. The rate of formation of the intermetallic phase was dominated by solid-state diffusion that increases exponentially with annealing temperature. Above 400 °C, there were superlattice peaks in XRD spectra (so-called "ordering peaks"), indicating that the intermetallic phases were formed. It was reported that a high annealing temperature promotes the formation of the ordered phase, but also leads to particle size growth of Pd$_3$Pb nanoparticles. For the samples annealed at 600 and 800 °C, the particle size increased from 4 nm to 7 and 12 nm, respectively. Therefore, a lower temperature of 400 °C was chosen to prepare the bimetallic Pd$_3$Pb/C catalyst. This catalyst demonstrated a significant enhancement in catalytic activity and durability for the ORR when compared with a Pd/C commercial catalyst.

Ag-based bimetallic formulations were found to have adequate ORR activity. The high reactivity for ORR over the PdAg$_9$ bimetallic catalysts was reported by the Stevenson group [100]. These authors produced bimetallic nanoalloys via coreduction of Ag and Pd carboxylic acid complex precursors in the presence of stabilizing ligands. In these catalysts, the Pd atoms facilitate the binding of the oxygen and subsequently the Ag atoms facilitate the desorption of the reaction products such as OH. The 9:1 Ag:Pd molar ratio exhibits a suitable geometric arrangement allowing to work by an ensemble effect. The full four-electron process takes place by combining the fast kinetics of Pd for the adsorption with rapid disproportionation on Ag for the desorption step. The data suggest a correlation between the bimetallic interaction developed at the nanoscale and the observed high catalytic activity.

Cu is a suitable sacrificial agent to be galvanically displaced by Pd for preparation of bimetallic alloys. In this way, Lüsi et al. showed the use of galvanic displacement on a Cu film to prepare PdCu catalysts for the ORR reaction in alkaline media [101]. A two-step method was employed by Betancourt et al. to obtain an active bimetallic catalyst. First, they reduced a Cu precursor onto a commercial Ag/C catalyst to form AgCu/C nanoparticles and then performed galvanic displacement of Cu to deposit an atomically dispersed loading of Pd [102]. The PdAg/C catalyst outperformed conventional Pd/C and Ag/C commercial catalysts in the ORR reaction. The synergistic effect of Pd and Ag plays a key role in electrode reactions. A similar method for preparation of a supported AuCu bimetallic catalyst for the air electrode was recently patented [103]. The method did not require the addition of a surfactant, is suitable for large-scale production, and the catalyst has large open porosity which provided a high diffusion rate to the catalytic sites.

2.5. Layered Double Hydroxides

The layered double hydroxides (LDH), $[M^{2+}_{1-x}M^{3+}_x(OH)_2][A^{n-}]_{x/n}\cdot zH_2O$, have unique structural properties due to incorporated anions and water between the large interlayer spaces. They belong to multi-metal clay materials which consist of charged balancing anions, A^{n-}, intercalated within brucite-like sheets of divalent, M^{2+}, and trivalent, M^{3+}, metal cations [104]. LDHs are able to combine electrochemically active elements such as Co, Ni, Fe, and many others. NiFe-LDH nanostructures are regarded as the most promising of OER catalysts due to their special electronic configuration as discussed in previous sections. The composition of LDHs can be easily tuned, resulting in the unique redox characteristics and notable catalytic activity during the OER [105]. Ni–O and Fe–O bond distances are comparable, thus, these two metals can replace each other and form edge-sharing FeO_6 and NiO_6 octahedra. Due to their semiconductor characteristics, a Schottky barrier may be formed when contacting the conductive substrate, which greatly hinders the charge transfer. Therefore, various strategies including size modulation, heteroatom doping, and surface and defect engineering were developed to improve the electrocatalytic performance of LDHs towards OER [106].

The physical and chemical properties of LDHs vary greatly when the dimensions of nanostructures are down to 1-nm scale, thus, the controllable and convenient fabrication of ultrathin nanocrystals is highly desirable for potential applications in ZABs. The exfoliation of LDH structure into a few layers and single sheets can expose more surface sites because of the large interlayer spacing between individual layers in the LDH structure. However, developing an effective strategy to exfoliate bulk NiFe-LDHs into stable single-layer nanosheets with more exposed active sites remains challenging. The synthesis requires a complex multistep process (i.e., hydrothermal process, anion exchange, and exfoliation) which takes several days to obtain the final product. Additionally, the refined morphology of as-synthesized LDH nanosheets obtained by exfoliation cannot be controlled precisely, as the high charge density of LDHs makes it harder to be exfoliated into monolayers.

In the bottom-up synthesis method, a surfactant (e.g., sodium dodecyl sulfonate) is introduced to form micelles in which LDH nanostructures are grown in a confined space. Recently two dimensional NiFe-LDHs nanosheets (1.3 nm) were obtained with a synthesis time of only 5 min [107]. They demonstrated enhanced OER performance compared with bulk NiFe-LDHs catalysts obtained via the coprecipitation method. In the bottom-up method modulating the crystallinity to fabricate a partially amorphous structure is a promising strategy in enhancing the electro-oxidation performances, which could achieve an optimal balance between active sites and conductivity. NiFe-LDH nanosheet arrays were in-situ grown on a NiFe alloy foam, which formed a self-supported anode with partially amorphous nanosheets providing reactive sites for the electrochemical generation of high-valence species. This also resulted in rich crystalline–amorphous interfaces to form and stabilize Ni^{3+} ions. Furthermore, the highly porous and conductive framework around the nanosheets further improved the electrocatalytic rate [108].

The higher intrinsic OER performance of NiFe-LDHs is further confirmed by other studies of NiFe-LDHs grown on different carbon materials (carbon quantum dots, graphitized carbon, 3D macro/mesoporous carbon, and graphene). A facile strategy to create ultrathin NiFe-LDHs nanosheets grown on carbon nanotubes (CNTs) was proposed [109]. Unlike bulk NiFe-LDH nanomaterials, the CNTs-supported catalysts demonstrated different 1D/3D layered structures, and electron transfer properties, thereby improving the OER performance.

The nature of active sites of NiFe-LDHs remain unclear, and thus, further mechanistic studies are helpful to understand the structure–activity correlation. Currently trial-and-error iterative design of bifunctional catalysts remains the main approach in this area. However, a more systematic approach is needed to achieve a major breakthrough. To this end, MOFs and single atom engineering are promising research directions that could pave the way for the systematic optimization of catalysts. Therefore, future studies should expand the limited understanding of the precise formation mechanisms behind MOFs to enable a more methodical tuning of catalyst materials. Upcoming research efforts should also focus on the stabilization of high-loading SACs, which is the main obstacle hindering their practical applications.

Other LDHs, such as CoMn, NiMn, and CoNi are still in the early stages of development of a bifunctional catalysts. They are typically anchored to porous conductive supports, such as N-doped graphene or carbon [110,111]. Such materials have large surface area and unique porous structure. Although the simple deposition method does not fully protect from corrosion in the alkaline electrolyte, excellent OER activity of NiMn LDH and $NiCo_2O_4$@NiMn composites was reported. It can be concluded that layered double hydroxides is one of the most effective OER catalysts due to their excellent activity and high stability in alkaline conditions.

3. Effect of an External Magnetic Field

Recent studies have shown that applying an external magnetic field is a simple and effective strategy to improve the electrocatalytic performance of non-precious metal catalysts. Electrocatalysis enhanced by the alternating magnetic field (AMF) may include the spin-polarization effect, magnetothermal effect, magnetohydrodynamic effect, and Néel relaxation. For example, magnetohydrodynamic effect occurs when a magnetic field interacts with charged ions, causing a macroscopic and microscopic convection which can change mass transfer rates of active species, reduce polarization resistance, and decrease bubble resistance caused by ohmic drop. The oxygen bubble escape rate increases significantly under AMF. The magnetothermal effect can increase the local temperature of magnetic catalysts resulting in enhancing the kinetics of OER [112]. In addition, composite magnetic catalysts can generate heat locally under AC magnetic field due to the internal flip of spins relative to the crystalline lattice (Néel relaxation) thereby increasing catalyst temperature [113–116]. Recent studies have revealed that structurally stable catalysts can undergo spin-state flipping changes by applying magneto-thermal perturbations using AMF, without altering their structure. Strain may also be the mechanism of magnetic field enhancement. Liu et al. [117] explored the change of the electrocatalytic performance of an FeCo catalyst in the OER under a magnetic field and proposed that strain would occur in the FeCo alloy, reducing OER overpotential. In general, magnetic field-enhanced electrocatalysis has multiple mechanisms acting together. In many cases, the stronger the magnetic properties of catalysts, the more significant the promoting effect is. However, most electrocatalysts undergo surface reconstruction in the alkaline OER process, and paramagnetic oxyhydroxide species can be formed on their surface reducing the magnitude of this effect.

3.1. Spin Polarization Effect

Research on magnetic spin-selective catalytic reactions and spin polarization dynamics is limited for OER in alkaline solutions. To quantify this effect, the exploration of promoting parallel spins to improve the OER activity was extensively studied. The formation of O–O bonds in the production of O_2 requires spin conservation of the paramagnetic triplet, making the catalyst's activity on the oxygen surface during spin polarization critical for achieving a parallel spin arrangement. The alternating magnetic field (AMF) was reported to reduce the energy barrier in transition metal oxide catalysts required for electron transition during the OER process [118]. Many transition metal oxide catalysts show that the magnetic susceptibility and AMF field enhances the spin energy states of electrons [119]. The formation of paramagnetic triplet oxygen molecules (↑O=O↑) during the OER is related to the spin of electrocatalysts [120–122].

Current research on the mechanism of electron spin interaction mainly involves catalysts with ferromagnetic properties. Ferromagnetic catalysts have an ordered magnetic moment structure and spin polarization, which makes them a good model system in understanding how spin polarization affects the OER mechanism and kinetics under AMF. For example, the OER overpotential over a $Co_{0.4}Ni_{0.6}$-MOF-74 catalyst reduced by 36% under a 150 kHz, 5.5 mT AMF at a current density of 10 mA cm^{-2}, as compared with that without magnetic field [123]. The promoting effect on OER often has a maximum at a specific intensity of AMF. For example, a maximum was observed at a magnetic field of 4.3 mT over ZIF-67-supported cobalt-based electrocatalysts with different magnetic properties (Co, CoO and Co_3O_4) (Figure 7a). At optimal conditions, the overpotential over ZIF-67-supported Co, CoO and Co_3O_4 catalysts was reduced by 140, 119, and 142 mV, respectively, relative to those in experiments without AMF. Further increase in the magnetic field intensity increased the overpotential due to electromagnetic induction [122]. The Co catalyst demonstrated the largest current density of 15.6 mA cm^{-2} (Figure 7b). Due to the thermal insulation of individual Co clusters, magnetic heating was strictly confined to magnetic ions. In the course of reaction, the adsorbed hydroxyl groups deprotonated on the surface to form O radicals, where a Co d_π-orbital electron and an p_π-orbital electron of oxygen showed the same spin configuration due to exchange interaction effects (Figure 7c) [124]. In the subsequent process of adsorption of OH-groups and deprotonation, electrons with specific spin directions were selectively removed through ferromagnetic exchange effects, thereby promoting the generation of triplet oxygen (Figure 7d) [125]. The formation of paramagnetic triplet oxygen occurs by similar mechanisms under DC and AC magnetic field. This can effectively reduce overpotential and increase the rate of electrochemical reaction. A similar behavior was reported over a $NiFe_2O_4$@MOF-74 catalyst. The overpotential in hydrogen evolution reaction was reduced by 30 mV at 10 mA cm^{-2} under an AMF of 2.3 mT [126]. It should be mentioned that this effect was not observed with CoO and other magnetic catalysts with weak exchange interaction.

A similar promoting effect was observed over core–shell $NiFe_2O_4$@(Ni, Fe)S/P catalysts when the overpotential in OER reduced from 323 to 176 mV at a 4.3 mT [127]. The authors noted hindering of the ion diffusion under AMF which could indicate that there is an optimum magnetic field at which the reaction rate has a maximum value. The OER process is a mass transfer limited reaction under higher reaction rates and, therefore, is influenced by the slow diffusion of reactants under AMF. In a catalytic reactor in which both electric and magnetic fields are present, moving ions are affected by the Lorentz forces of the electric and magnetic fields and change their original path, resulting in a slower mass transfer rate.

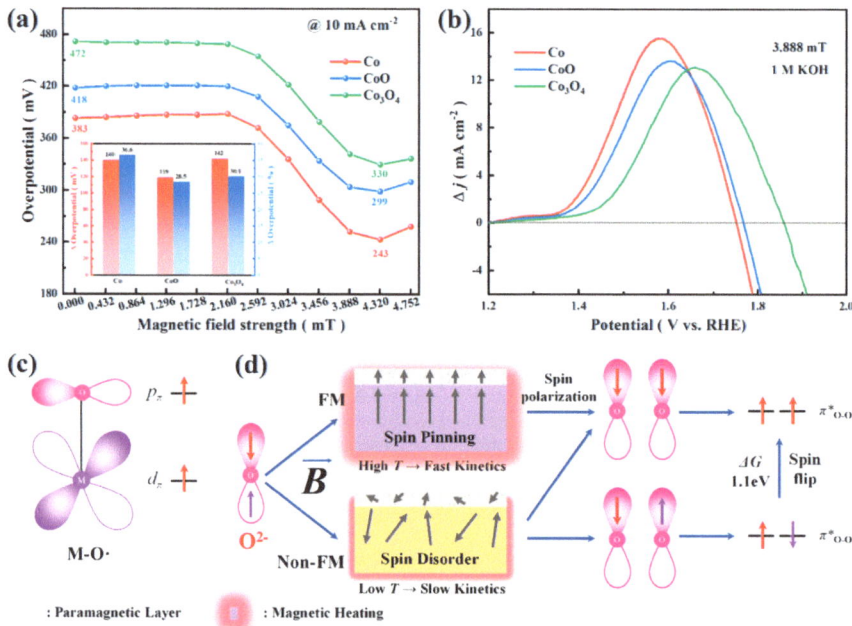

Figure 7. (**a**) Overpotential at a current density of 10 mA cm^{-2} under different AMF. (**b**) Maximum current density of Co, CoO, and Co$_3$O$_4$. (**c**) Spin configuration of M–O. (**d**) Schematic of the difference in the OER promotion mechanism of different magnetic electrocatalysts under AMF. Reprinted with permission from [122]. Copyright 2022, American Chemical Society.

As was discussed in Section 2.5, Ni, Co, and Fe metal layered double hydroxides (LDHs) are commonly used as catalysts for the OER in alkaline electrolytes. The iron group transition metals, including Fe^{3+} ($t_{2g}^3 e_g^2$), Co^{2+} ($t_{2g}^6 e_g^1$), and Ni^{2+} ($t_{2g}^6 e_g^2$), have a spin single electron, therefore, these catalysts are often seen as a model system for studying the spin-magnetic effect. The effective bonding p-orbitals of oxygen intermediates are crucial to enhancing OER reaction. Based on symmetry conservation and LDH planar structures, orbitals perpendicular to the plane orientation (out-of-plane orbitals including d_{z^2}, d_{xz}, and d_{yz}) favor reactant adsorption, while orbitals parallel to the LDH planes (in-plane orbitals including $d_{x^2-y^2}$ and d_{xy}) favor electron transport [128]. The introduction of an external magnetic field enhances the spin-electron exchange between LDH catalysts and oxygen-containing intermediates, thereby enhancing the OER rate.

In addition to direct application of ferromagnetic catalysts, doping specific ions into LDH is another strategy to create active OER catalysts. Sun et al. used this strategy to introduce Cu^{2+} ions into NiFe-LDH catalyst. The Cu^{2+} orbital can induce spin splitting at Fe(III) sites and excite electrons near the Fermi level, which facilitates adsorption of O$^-$ containing species [129]. The Cu$_1$Ni$_6$Fe$_2$-LDHs demonstrated a low overpotential of 210 mV at 10 mA cm^{-2}, which is one of the best OER activities reported. In general, the composite structure formed from two different magnetic materials can lead to magnetic inversion, thus triggering the catalytic activity. This strategy provides a new approach for catalyst design: doping magnetic materials to non-magnetic electrocatalysts and using magnetic field to enhance their performance [130]. Alternatively, composites of non-magnetic and magnetic catalysts can be designed using rational structural design by changing the electron configuration and breaking the symmetry, making the whole catalyst magnetic and, therefore, triggered by magnetic field.

3.2. Magnetothermal Effect

The spin polarization inversion of ferromagnetic materials under alternating magnetic fields induces a local magnetic heating effect to improve charge transfer efficiency, and the heated active site accelerates the adsorption/desorption process, resulting in faster catalytic reaction kinetics. This effect of local catalyst heating under AMF seems to be more important as compared with the spin-polarization effect. Gong et al. [131] showed that a Co@MoS$_2$ SAC demonstrated the appearance of in-plane ferromagnetic properties, that facilitated the parallel spin arrangement of oxygen atoms under 300 kHz, 4.55 mT AMF. As a result, the overpotential in OER reduced from 317 to 250 mV at a current density of 10 mA cm^{-2}, compared with when AMF was not applied. AMF is a promising methodology for further improving magnetic single-atom catalyst (SAC) activity. This approach is also applicable to other magnetic SACs, which provides a new avenue for improving the OER activity by application of a weak AMF.

A ferromagnetic Co/ZIF-67 catalyst showed the highest specific heating rate among ZIF-67-supported Co, CoO and Co$_3$O$_4$ catalysts and, therefore, the Co/ZIF-67 catalyst demonstrated the best performance in OER under AMF [122]. Magneto-thermal effect enhanced the catalytic performance of nickel-coated iron carbide nanoparticles (FeC–Ni) in the OER under AMF [132]. Local heating of the catalyst reduced the overpotential required for oxygen release in an alkaline electrolyte by 0.2 V. Using the same approach, Peng et al. developed a superparamagnetic NiFe@NiFeOOH core–shell catalyst [133]. The OER overpotential reduced to 209 mV at 10 mA cm^{-2} over this catalyst under AMF. The catalyst possesses a remarkable catalytic stability: no detachment of FeOOH or NiOOH was observed during the long-term operation.

Magnetic Cr$_2$Ge$_2$Te$_6$ nanosheet catalyst was demonstrated to improve the OER performance by the application of an external magnetic field [134]. The improvement was a combined effect of the Lorentz force generated by magnetic field and local magnetic heating. Furthermore, Cr$_2$Ge$_2$Te$_6$ with fewer layers has intrinsic spin and charge density dual bipolar tunable properties, which makes it a potential application in the field of spin electrocatalysis. Yuan synthesized 3D screw pyramid MoS$_2$ catalyst with abundant edge active sites [135]. Eddy current inside the nanosheets contributed to magnetic heating and the rate of HER over the screw pyramid MoS$_2$ catalysts are greatly improved compared with step pyramid MoS$_2$ under AMF. This example provides a promising guideline for eddy current assisted electrocatalytic enhancement of multilayered transition metal disulfides.

4. Conclusions and Outlook

Several categories of catalytic materials ranging from monometallic noble metal catalysts to alternative carbons, mixed metal oxides, spinels, perovskites and inorganic–organic composites have been discussed as catalysts for OER/ORR reactions as well as susceptor materials under alternating magnetic field with demonstrated influential factors to their performance and/or catalytic properties. Some basic synthetic strategies have also been presented. A few fundamental aspects including the OER/ORR reaction mechanisms over novel catalysts, should be studied to guide the design of high-efficiency air electrode catalysts. The large overpotential and slow kinetics at the cathode appear to be the major challenges that limit the practical electrochemical performances.

The single metal catalysts exhibit better ORR catalytic activity than the nanoparticles. The Pt-based noble metal family generally possesses virtues of both high activity and favorable stability but disadvantages of high cost. Making bimetallic catalysts with cheap transition metals and using Pt as a shell in core–shell catalysts are effective routes to lower the noble metal loading and further enhance the stability of the transition metals. Carbon-based materials, transition-metal oxides, and composite catalysts derived from polymers are promising abundant and cheap catalysts applicable for OER/ORR reactions. In particular, graphene and N-doped carbon nanotubes are attractive candidates for future research due to their high activity, high electrical conductivity and chemical stability. Layered double hydroxides (LDHs) are considered as cost-effective OER catalysts due to their excellent activity and stability in an OER in alkaline electrolytes. Their synthesis does not require lengthy, complicated steps, which opens good perspectives for scale-up. The combination of nitrogen-doped carbon materials with SAC or metal nanoparticles produces very active and stable bifunctional catalysts.

Proposed future research directions in these non-precious metal catalytic materials include: (i) exploiting the effect of physiochemical parameters (composition, morphology, oxidation state, defects, surface area, conductivity, etc.) on the intrinsic catalytic activity, (ii) increasing the density of active sites and their utilization through innovative design of materials and structures including single atom stabilization, and (iii) optimizing catalyst synthesis strategies (cheaper precursors, low temperature, shorter duration) to scale up their synthesis to industrial scale without decreasing their activity and stability.

The dissolution and irreversible oxidation of active metals are among the main reasons for ZAB failure and large overpotential in the charge–discharge process.

The application of an external magnetic field is an elegant strategy to enhance electrocatalytic rates and to reduce the overpotential. Studies performed help to understand the effect of the external magnetic field on the performance of ferromagnetic composite electrocatalysts employed in ZABs. Many studies were performed under a DC magnetic field, and there is still a lack of systematic understanding of the role of AC magnetic field in OER electrocatalysis. The capacity of inductive heating to drive heat directly to the catalytic sites turned out to be an enabling technology capable of increasing the rate of OER by an order of magnitude under AC magnetic field. The removal of mass transfer limitation over the reactor wall has reduced the concentration gradients in ZABs, largely contributing to the improvement of energy efficiency as well as the reduction in overpotential. Several studies showed that the magnetic field enhances the catalytic performance in a volcanic pattern, where high magnetic field intensity inhibits the reaction rate. This effect was explained as a counterplay between spin polarization and magnetohydrodynamic effect.

Although there have been several studies on the electron spin selectivity mechanism, related to changes in the electron spin structure of the catalyst, there is a need for in situ characterization of spintronic structures to better understand the theory and mechanism of spin catalysis.

Funding: This research was funded by the European Research Council Synergy Grant SCOPE No. 810182.

Data Availability Statement: The authors confirm that the data supporting the findings of this study are available within the article.

Conflicts of Interest: The authors declare no conflict of interest.

References

1. Cole, W.; Frew, B.; Gagnon, P.; Reimers, A.; Zuboy, J.; Margolis, R. Envisioning a low-cost solar future: Exploring the potential impact of achieving the SunShot 2030 targets for photovoltaics. *Energy* **2018**, *155*, 690–704. [CrossRef]
2. Bellini, E. Portuguese Government Confirms World Record Solar Price of $0.01316/kWh. *PV Mag. Int.* **2020**. Available online: https://www.pv-magazine.com/2020/08/27/portuguese-government-confirms-world-record-solar-price-of-0-01316-kwh (accessed on 24 July 2023).
3. He, W.; King, M.; Luo, X.; Dooner, M.; Li, D.; Wang, J. Technologies and economics of electric energy storages in power systems: Review and perspective. *Adv. Appl. Energy* **2021**, *4*, 100060. [CrossRef]
4. International Energy Agency Global EV Outlook 2020: Entering the Decade of Electric Drive? *IEA Paris* **2020**. Available online: https://www.iea.org/reports/global-ev-outlook-2020 (accessed on 24 July 2023).
5. Zeng, X.; Li, M.; Abd El-Hady, D.; Alshitari, W.; Al-Bogami, A.S.; Lu, J.; Amine, K. Commercialization of lithium battery technologies for electric vehicles. *Adv. Energy Mater.* **2019**, *9*, 1900161. [CrossRef]
6. Han, X.; Li, X.; White, J.; Zhong, C.; Deng, Y.; Hu, W.; Ma, T. Metal–air batteries: From static to flow system. *Adv. Energy Mater.* **2018**, *8*, 1801396. [CrossRef]
7. Shaulova, E.; Biagi, L. Lithium Industry Worldwide. Statistics Report on the Global Lithium Industry. 2023. Available online: https://www.statista.com/study/40094/lithium-statista-dossier (accessed on 24 July 2023).
8. Ellis, D. Lithium-Ion Batteries Demand to Grow 30% a Year—McKinsey. *Mining* **2022**. Available online: https://miningdigital.com/supply-chain-and-operations/lithium (accessed on 24 July 2023).

9. FCAB. Executive Summary. National Blueprint for Lithium Batteries 2021–2030. 2021. Available online: https://www.energy.gov (accessed on 24 July 2023).
10. BMI. BYD Becomes Benchmark's 7th Tier One EV Battery Manufacturer; 2nd China Cell Maker. 2021. Available online: https://www.benchmarkminerals.com/membership/byd-becomes-benchmarks-7th-tier-one-ev-battery-manufacturer-2nd-china-cell-maker-2/ (accessed on 19 May 2023).
11. Leong, K.W.; Wang, Y.; Ni, M.; Pan, W.; Luo, S.; Leung, D.Y.C. Rechargeable Zn–air batteries: Recent trends and future perspectives. *Renew. Sustain. Energy Rev.* **2022**, *154*, 111771. [CrossRef]
12. Khan, P.A.; Venkatesh, B. *Economic Analysis of Chemical Energy Storage Technologies*; Springer: Berlin/Heidelberg, Germany, 2016; pp. 277–291.
13. O'Neill, A. *World Bank Commodities Price Forecast*; World Bank: Washington, DC, USA, 2022.
14. Fu, J.; Liang, R.; Liu, G.; Yu, A.; Bai, Z.; Yang, L.; Chen, Z. Recent progress in electrically rechargeable zinc–air batteries. *Adv. Mater.* **2019**, *31*, 1805230. [CrossRef]
15. Kiss, A.M.; Myles, T.D.; Grew, K.N.; Peracchio, A.A.; Nelson, G.J.; Chiu, W.K.S. Carbonate and bicarbonate ion transport in alkaline anion exchange membranes. *J. Electrochem. Soc.* **2013**, *160*, F994–F999. [CrossRef]
16. Zhang, W.; Liu, Y.; Zhang, L.; Chen, J. Recent advances in isolated single-atom catalysts for zinc air batteries: A focus review. *Nanomaterials* **2019**, *9*, 1402. [CrossRef]
17. Hu, C.; Lin, Y.; Connell, J.W.; Cheng, H.; Gogotsi, Y.; Titirici, M.; Dai, L. Carbon-based metal-free catalysts for energy storage and Environmental Remediation. *Adv. Mater.* **2019**, *31*, 1806128. [CrossRef] [PubMed]
18. Meffre, A.; Mehdaoui, B.; Kelsen, V.; Fazzini, P.F.; Carrey, J.; Lachaize, S.; Respaud, M.; Chaudret, B. A Simple Chemical Route toward Monodisperse Iron Carbide Nanoparticles Displaying Tunable Magnetic and Unprecedented Hyperthermia Properties. *Nano Lett.* **2012**, *12*, 4722–4728. [CrossRef] [PubMed]
19. Li, B.; Ge, X.; Goh, F.W.T.; Hor, T.S.A.; Geng, D.; Du, G.; Liu, Z.; Zhang, J.; Liu, X.; Zong, Y. Co_3O_4 nanoparticles decorated carbon nanofiber mat as binder-free air-cathode for high performance rechargeable zinc-air batteries. *Nanoscale* **2015**, *7*, 1830–1838. [CrossRef] [PubMed]
20. Yuxin, Z.; Ying, Y.; Chunzhen, Z.; Meipeng, Z.; Yue, W.; Fengliang, Z.; Lili, R. Air cathode catalyst layer ink suitable for aerosol printing and preparation method thereof. Patent CN113871761A, 31 December 2021.
21. Pi, Y.-T.; Xing, X.-Y.; Lu, L.-M.; He, Z.-B.; Ren, T.-Z. Hierarchical porous activated carbon in OER with high efficiency. *RSC Adv.* **2016**, *6*, 102422–102427. [CrossRef]
22. Tang, J.; Liu, J.; Li, C.; Li, Y.; Tade, M.O.; Dai, S.; Yamauchi, Y. Synthesis of Nitrogen-doped mesoporous carbon spheres with extra-large pores through assembly of diblock copolymer micelles. *Angew. Chem. Int. Ed.* **2014**, *54*, 588–593. [CrossRef]
23. Fukuyama, H.; Terai, S. Preparing and characterizing the active carbon produced by steam and carbon dioxide as a heavy oil hydrocracking catalyst support. *Catal. Today* **2008**, *130*, 382–388. [CrossRef]
24. Zhang, X.; Yan, P.; Zhang, R.; Liu, K.; Liu, Y.; Liu, T.; Wang, X. A novel approach of binary doping sulfur and nitrogen into graphene layers for enhancing electrochemical performances of supercapacitors. *J. Mater. Chem. A* **2016**, *4*, 19053–19059. [CrossRef]
25. Sun, L.; Zhou, H.; Li, L.; Yao, Y.; Qu, H.; Zhang, C.; Liu, S.; Zhou, Y. Double soft-template synthesis of nitrogen/sulfur-codoped hierarchically porous carbon materials derived from protic ionic liquid for supercapacitor. *ACS Appl. Mater. Interfaces* **2017**, *9*, 26088–26095. [CrossRef]
26. Gupta, S.; Zhao, S.; Xu, H.; Wu, G. Highly stable nanocarbon catalysts for bifunctional oxygen reduction and evolution reactions in alkaline media. *ECS Meet. Abstr.* **2017**, *MA2017-01*, 1423. [CrossRef]
27. McCrory, C.C.L.; Jung, S.; Peters, J.C.; Jaramillo, T.F. Benchmarking Heterogeneous electrocatalysts for the oxygen evolution reaction. *J. Am. Chem. Soc.* **2013**, *135*, 16977–16987. [CrossRef]
28. Lee, D.U.; Park, H.W.; Higgins, D.; Nazar, L.; Chen, Z. Highly active graphene nanosheets prepared via extremely rapid heating as efficient zinc-air battery electrode material. *J. Electrochem. Soc.* **2013**, *160*, F910–F915. [CrossRef]
29. Chengen, G.H.; Chao, G.X. Carbon-coated graphene/metal oxide composite material and preparation method thereof. Patent CN113690429A, 23 August 2021.
30. Zhou, Q.; Zhang, Z.; Cai, J.; Liu, B.; Zhang, Y.; Gong, X.; Sui, X.; Yu, A.; Zhao, L.; Wang, Z.; et al. Template-guided synthesis of Co nanoparticles embedded in hollow nitrogen doped carbon tubes as a highly efficient catalyst for rechargeable Zn–air batteries. *Nano Energy* **2020**, *71*, 104592. [CrossRef]
31. Long, J.; Chen, C. Fe-based bimetallic zinc-air battery cathode catalyst based on layered MOF. Patent CN113097513A, 2 April 2021.
32. Jaksic, J.M.; Nan, F.; Papakonstantinou, G.D.; Botton, G.A.; Jaksic, M.M. Theory, substantiation, and properties of novel reversible electrocatalysts for oxygen electrode reactions. *J. Phys. Chem. C* **2015**, *119*, 11267–11285. [CrossRef]
33. Pan, J.; Xu, Y.Y.; Yang, H.; Dong, Z.; Liu, H.; Xia, B.Y. Advanced architectures and relatives of air electrodes in Zn–Air batteries. *Adv. Sci.* **2018**, *5*, 1700691. [CrossRef] [PubMed]
34. Li, Y.; Li, Q.; Wang, H.; Zhang, L.; Wilkinson, D.P.; Zhang, J. Recent Progresses in oxygen reduction reaction electrocatalysts for electrochemical energy applications. *Electrochem. Energy Rev.* **2019**, *2*, 518–538. [CrossRef]
35. Cheng, F.; Chen, J. Metal–air batteries: From oxygen reduction electrochemistry to cathode catalysts. *Chem. Soc. Rev.* **2012**, *41*, 2172. [CrossRef]

36. Sequeira, C.A.C. Carbon Anode in Carbon History. *Molecules* **2020**, *25*, 4996. [CrossRef] [PubMed]
37. Zamani-Meymian, M.-R.; Khanmohammadi Chenab, K.; Pourzolfaghar, H. Designing high-quality electrocatalysts based on CoO:MnO$_2$@C supported on carbon cloth fibers as bifunctional air cathodes for application in rechargeable Zn–Air battery. *ACS Appl. Mater. Interfaces* **2022**, *14*, 55594–55607. [CrossRef]
38. Song, Z.; Han, X.; Deng, Y.; Zhao, N.; Hu, W.; Zhong, C. Clarifying the Controversial Catalytic Performance of Co(OH)$_2$ and Co$_3$O$_4$ for oxygen reduction/evolution reactions toward efficient Zn–air batteries. *ACS Appl. Mater. Interfaces* **2017**, *9*, 22694–22703. [CrossRef] [PubMed]
39. Wu, Y.; Wu, X.; Tu, T.; Zhang, P.; Li, J.; Zhou, Y.; Huang, L.; Sun, S. Controlled synthesis of FeN$_x$-CoN$_x$ dual active sites interfaced with metallic Co nanoparticles as bifunctional oxygen electrocatalysts for rechargeable Zn–air batteries. *Appl. Catal. B Environ.* **2020**, *278*, 119259. [CrossRef]
40. Xu, Y.; Deng, P.; Chen, G.; Chen, J.; Yan, Y.; Qi, K.; Liu, H.; Xia, B.Y. 2D Nitrogen-doped carbon nanotubes/Graphene hybrid as bifunctional oxygen electrocatalyst for long-life rechargeable Zn–Air batteries. *Adv. Funct. Mater.* **2020**, *30*, 1906081. [CrossRef]
41. Sun, X.; Gong, Q.; Liang, Y.; Wu, M.; Xu, N.; Gong, P.; Sun, S.; Qiao, J. Exploiting a high-performance "double-carbon" structure Co$_9$S$_8$/GN bifunctional catalysts for rechargeable Zn–Air batteries. *ACS Appl. Mater. Interfaces* **2020**, *12*, 38202–38210. [CrossRef]
42. Wang, Y.; Wu, M.; Li, J.; Huang, H.; Qiao, J. In situ growth of CoP nanoparticles anchored on (N,P) co-doped porous carbon engineered by MOFs as advanced bifunctional oxygen catalyst for rechargeable Zn–air battery. *J. Mater. Chem. A* **2020**, *8*, 19043–19049. [CrossRef]
43. Wu, Z.; Wu, H.; Niu, T.; Wang, S.; Fu, G.; Jin, W.; Ma, T. Sulfurated Metal–Organic framework-derived nanocomposites for efficient bifunctional oxygen electrocatalysis and rechargeable Zn–air battery. *ACS Sustain. Chem. Eng.* **2020**, *8*, 9226–9234. [CrossRef]
44. Guo, X.; Zheng, S.; Luo, Y.; Pang, H. Synthesis of confining cobalt nanoparticles within SiO$_x$/nitrogen-doped carbon framework derived from sustainable bamboo leaves as oxygen electrocatalysts for rechargeable Zn–air batteries. *Chem. Eng. J.* **2020**, *401*, 126005. [CrossRef]
45. Zhang, H.-M.; Hu, C.; Ji, M.; Wang, M.; Yu, J.; Liu, H.; Zhu, C.; Xu, J. Co/Co$_9$S$_8$@carbon nanotubes on a carbon sheet: Facile controlled synthesis, and application to electrocatalysis in oxygen reduction/oxygen evolution reactions, and to a rechargeable Zn–air battery. *Inorg. Chem. Front.* **2021**, *8*, 368–375. [CrossRef]
46. Arafat, Y.; Azhar, M.R.; Zhong, Y.; Xu, X.; Tadé, M.O.; Shao, Z. A Porous nano-micro-composite as a high-performance bi-functional air electrode with remarkable stability for rechargeable Zinc–Air batteries. *Nano-Micro Lett.* **2020**, *12*, 130. [CrossRef]
47. Wang, X.; Ge, L.; Lu, Q.; Dai, J.; Guan, D.; Ran, R.; Weng, S.-C.; Hu, Z.; Zhou, W.; Shao, Z. High-performance metal-organic framework-perovskite hybrid as an important component of the air-electrode for rechargeable Zn-Air battery. *J. Power Sources* **2020**, *468*, 228377. [CrossRef]
48. Tan, J.; Thomas, T.; Liu, J.; Yang, L.; Pan, L.; Cao, R.; Shen, H.; Wang, J.; Liu, J.; Yang, M. Rapid microwave-assisted preparation of high-performance bifunctional Ni3Fe/Co-N-C for rechargeable Zn–air battery. *Chem. Eng. J.* **2020**, *395*, 125151. [CrossRef]
49. Chen, D.; Yu, J.; Cui, Z.; Zhang, Q.; Chen, X.; Sui, J.; Dong, H.; Yu, L.; Dong, L. Hierarchical architecture derived from two-dimensional zeolitic imidazolate frameworks as an efficient metal-based bifunctional oxygen electrocatalyst for rechargeable Zn–air batteries. *Electrochim. Acta* **2020**, *331*, 135394. [CrossRef]
50. Yu, N.-F.; Wu, C.; Huang, W.; Chen, Y.-H.; Ruan, D.-Q.; Bao, K.-L.; Chen, H.; Zhang, Y.; Zhu, Y.; Huang, Q.-H.; et al. Highly efficient Co$_3$O$_4$/Co@NCs bifunctional electrocatalysts for long life rechargeable Zn–air batteries. *Nano Energy* **2020**, *77*, 105200. [CrossRef]
51. Li, L.; Yang, J.; Yang, H.; Zhang, L.; Shao, J.; Huang, W.; Liu, B.; Dong, X. Anchoring Mn$_3$O$_4$ Nanoparticles on oxygen functionalized carbon nanotubes as bifunctional catalyst for rechargeable zinc-air battery. *ACS Appl. Energy Mater.* **2018**, *1*, 963–969. [CrossRef]
52. Huang, Z.; Qin, X.; Gu, X.; Li, G.; Mu, Y.; Wang, N.; Ithisuphalap, K.; Wang, H.; Guo, Z.; Shi, Z.; et al. Mn$_3$O$_4$ quantum dots supported on nitrogen-doped partially exfoliated multiwall carbon nanotubes as oxygen reduction electrocatalysts for high-performance Zn–Air batteries. *ACS Appl. Mater. Interfaces* **2018**, *10*, 23900–23909. [CrossRef]
53. Sidhureddy, B.; Prins, S.; Wen, J.; Thiruppathi, A.R.; Govindhan, M.; Chen, A. Synthesis and electrochemical study of mesoporous nickel-cobalt oxides for efficient oxygen reduction. *ACS Appl. Mater. Interfaces* **2019**, *11*, 18295–18304. [CrossRef] [PubMed]
54. Zhang, L.; Li, Y.; Peng, J.; Peng, K. Bifunctional NiCo$_2$O$_4$ porous nanotubes electrocatalyst for overall water-splitting. *Electrochim. Acta* **2019**, *318*, 762–769. [CrossRef]
55. Dilshad, K.A.J.; Rabinal, M.K. Rationally designed Zn-anode and Co$_3$O$_4$-cathode nanoelectrocatalysts for an efficient Zn–air battery. *Energy Fuels* **2021**, *35*, 12588–12598. [CrossRef]
56. Worku, A.K.; Ayele, D.W.; Habtu, N.G.; Teshager, M.A.; Workineh, Z.G. Recent progress in MnO$_2$-based oxygen electrocatalysts for rechargeable zinc-air batteries. *Mater. Today Sustain.* **2021**, *13*, 100072. [CrossRef]
57. Rittiruam, M.; Buapin, P.; Saelee, T.; Khajondetchairit, P.; Kheawhom, S.; Alling, B.; Praserthdam, S.; Ektarawong, A.; Praserthdam, P. First-principles calculation on effects of oxygen vacancy on α-MnO$_2$ and β-MnO$_2$ during oxygen reduction reaction for rechargeable metal-air batteries. *J. Alloys Compd.* **2022**, *926*, 166929. [CrossRef]
58. Li, L.; Feng, X.; Nie, Y.; Chen, S.; Shi, F.; Xiong, K.; Ding, W.; Qi, X.; Hu, J.; Wei, Z.; et al. Insight into the effect of oxygen vacancy concentration on the catalytic performance of MnO$_2$. *ACS Catal.* **2015**, *5*, 4825–4832. [CrossRef]

59. Gupta, P.K.; Bhandari, A.; Saha, S.; Bhattacharya, J.; Pala, R.G.S. Modulating oxygen evolution reactivity in MnO_2 through polymorphic engineering. *J. Phys. Chem. C* **2019**, *123*, 22345–22357. [CrossRef]
60. Zhou, Y.; Zhou, Z.; Hu, L.; Tian, R.; Wang, Y.; Arandiyan, H.; Chen, F.; Li, M.; Wan, T.; Han, Z.; et al. A facile approach to tailor electrocatalytic properties of MnO_2 through tuning phase transition, surface morphology and band structure. *Chem. Eng. J.* **2022**, *438*, 135561. [CrossRef]
61. Bera, K.; Karmakar, A.; Karthick, K.; Sankar, S.S.; Kumaravel, S.; Madhu, R.; Kundu, S. Enhancement of the OER kinetics of the less-explored α-MnO_2 via nickel doping approaches in alkaline medium. *Inorg. Chem.* **2021**, *60*, 19429–19439. [CrossRef]
62. Ni, S.; Zhang, H.; Zhao, Y.; Li, X.; Sun, Y.; Qian, J.; Xu, Q.; Gao, P.; Wu, D.; Kato, K.; et al. Single atomic Ag enhances the bifunctional activity and cycling stability of MnO_2. *Chem. Eng. J.* **2019**, *366*, 631–638. [CrossRef]
63. Zhang, J.-N. (Ed.) Carbon-Based Nanomaterials for Energy Conversion and Storage. In *Springer Series in Materials Science*; Springer Nature Singapore: Singapore, 2022; Volume 325, ISBN 978-981-19-4624-0.
64. Singh, B.; Gawande, M.B.; Kute, A.D.; Varma, R.S.; Fornasiero, P.; McNeice, P.; Jagadeesh, R.V.; Beller, M.; Zbořil, R. Single-atom (Iron-based) catalysts: Synthesis and applications. *Chem. Rev.* **2021**, *121*, 13620–13697. [CrossRef]
65. Global Industrial Catalyst Market Outlook to 2028. Available online: https://www.researchandmarkets.com/reports/5778274/global-industrial-catalyst-market-outlook (accessed on 11 July 2023).
66. Jiao, L.; Wan, G.; Zhang, R.; Zhou, H.; Yu, S.; Jiang, H. From Metal–organic frameworks to single-atom Fe implanted N-doped porous carbons: Efficient oxygen reduction in both alkaline and acidic media. *Angew. Chem. Int. Ed.* **2018**, *57*, 8525–8529. [CrossRef] [PubMed]
67. Yang, G.; Zhu, J.; Yuan, P.; Hu, Y.; Qu, G.; Lu, B.-A.; Xue, X.; Yin, H.; Cheng, W.; Cheng, J.; et al. Regulating Fe-spin state by atomically dispersed Mn-N in Fe-N-C catalysts with high oxygen reduction activity. *Nat. Commun.* **2021**, *12*, 1734. [CrossRef]
68. Mun, Y.; Lee, S.; Kim, K.; Kim, S.; Lee, S.; Han, J.W.; Lee, J. Versatile strategy for tuning ORR activity of a single Fe-N_4 site by controlling electron-withdrawing/donating properties of a carbon plane. *J. Am. Chem. Soc.* **2019**, *141*, 6254–6262. [CrossRef]
69. Liu, K.; Fu, J.; Lin, Y.; Luo, T.; Ni, G.; Li, H.; Lin, Z.; Liu, M. Insights into the activity of single-atom Fe-N-C catalysts for oxygen reduction reaction. *Nat. Commun.* **2022**, *13*, 2075. [CrossRef]
70. Shen, H.; Thomas, T.; Rasaki, S.A.; Saad, A.; Hu, C.; Wang, J.; Yang, M. Oxygen Reduction Reactions of Fe-N-C Catalysts: Current Status and the Way Forward. *Electrochem. Energy Rev.* **2019**, *2*, 252–276. [CrossRef]
71. Li, Y.; Xu, K.; Zhang, Q.; Zheng, Z.; Li, S.; Zhao, Q.; Li, C.; Dong, C.; Mei, Z.; Pan, F.; et al. One-pot synthesis of FeN_xC as efficient catalyst for high-performance zinc-air battery. *J. Energy Chem.* **2022**, *66*, 100–106. [CrossRef]
72. Chen, Y.; Ji, S.; Chen, C.; Peng, Q.; Wang, D.; Li, Y. Single-atom catalysts: Synthetic strategies and electrochemical applications. *Joule* **2018**, *2*, 1242–1264. [CrossRef]
73. Jia, N.; Xu, Q.; Zhao, F.; Gao, H.-X.; Song, J.; Chen, P.; An, Z.; Chen, X.; Chen, Y. Fe/N Codoped Carbon nanocages with single-atom feature as efficient oxygen reduction reaction electrocatalyst. *ACS Appl. Energy Mater.* **2018**, *1*, 4982–4990. [CrossRef]
74. Ma, L.; Chen, S.; Pei, Z.; Huang, Y.; Liang, G.; Mo, F.; Yang, Q.; Su, J.; Gao, Y.; Zapien, J.A.; et al. Single-Site Active Iron-Based Bifunctional Oxygen Catalyst for a Compressible and Rechargeable Zinc–Air Battery. *ACS Nano* **2018**, *12*, 1949–1958. [CrossRef]
75. Yang, L.; Cheng, D.; Xu, H.; Zeng, X.; Wan, X.; Shui, J.; Xiang, Z.; Cao, D. Unveiling the high-activity origin of single-atom iron catalysts for oxygen reduction reaction. *Proc. Natl. Acad. Sci. USA* **2018**, *115*, 6626–6631. [CrossRef]
76. Zhang, L.; Liu, T.; Chen, N.; Jia, Y.; Cai, R.; Theis, W.; Yang, X.; Xia, Y.; Yang, D.; Yao, X. Scalable and controllable synthesis of atomic metal electrocatalysts assisted by an egg-box in alginate. *J. Mater. Chem. A* **2018**, *6*, 18417–18425. [CrossRef]
77. Lu, Z.; Wang, B.; Hu, Y.; Liu, W.; Zhao, Y.; Yang, R.; Li, Z.; Luo, J.; Chi, B.; Jiang, Z.; et al. An isolated zinc–cobalt atomic pair for highly active and durable oxygen reduction. *Angew. Chem.* **2019**, *131*, 2648–2652. [CrossRef]
78. Li, S.; Zhou, X.; Fang, G.; Xie, G.; Liu, X.; Lin, X.; Qiu, H.-J. Multicomponent spinel metal oxide nanocomposites as high-performance bifunctional catalysts in Zn–Air batteries. *ACS Appl. Energy Mater.* **2020**, *3*, 7710–7718. [CrossRef]
79. Zhang, B.; Zheng, X.; Voznyy, O.; Comin, R.; Bajdich, M.; García-Melchor, M.; Han, L.; Xu, J.; Liu, M.; Zheng, L.; et al. Homogeneously dispersed multimetal oxygen-evolving catalysts. *Science* **2016**, *352*, 333–337. [CrossRef]
80. Song, W.; Ren, Z.; Chen, S.-Y.; Meng, Y.; Biswas, S.; Nandi, P.; Elsen, H.A.; Gao, P.-X.; Suib, S.L. Ni- and Mn-promoted mesoporous Co_3O_4: A stable bifunctional catalyst with surface-structure-dependent activity for oxygen reduction reaction and oxygen evolution reaction. *ACS Appl. Mater. Interfaces* **2016**, *8*, 20802–20813. [CrossRef]
81. Liang, Y.; Wang, H.; Zhou, J.; Li, Y.; Wang, J.; Regier, T.; Dai, H. Covalent hybrid of spinel manganese–cobalt oxide and Graphene as advanced oxygen reduction electrocatalysts. *J. Am. Chem. Soc.* **2012**, *134*, 3517–3523. [CrossRef] [PubMed]
82. Xu, N.; Qiao, J.; Zhang, X.; Ma, C.; Jian, S.; Liu, Y.; Pei, P. Morphology controlled $La_2O_3/Co_3O_4/MnO_2$–CNTs hybrid nanocomposites with durable bi-functional air electrode in high-performance zinc-air energy storage. *Appl. Energy* **2016**, *175*, 495–504. [CrossRef]
83. Xiao, X.; Hu, X.; Liang, Y.; Zhang, G.; Wang, X.; Yan, Y.; Li, X.; Yan, G.; Wang, J. Anchoring $NiCo_2O_4$ nanowhiskers in biomass-derived porous carbon as superior oxygen electrocatalyst for rechargeable Zn–air battery. *J. Power Sources* **2020**, *476*, 228684. [CrossRef]
84. Go, Y.; Min, K.; An, H.; Kim, K.; Eun Shim, S.; Baeck, S.-H. Oxygen-vacancy-rich $CoFe/CoFe_2O_4$ embedded in N-doped hollow carbon spheres as a highly efficient bifunctional electrocatalyst for Zn–air batteries. *Chem. Eng. J.* **2022**, *448*, 137665. [CrossRef]

85. Jose, V.; Hu, H.; Edison, E.; Manalastas, W.; Ren, H.; Kidkhunthod, P.; Sreejith, S.; Jayakumar, A.; Nsanzimana, J.M.V.; Srinivasan, M.; et al. Modulation of Single Atomic Co and Fe Sites on hollow carbon nanospheres as oxygen electrodes for rechargeable Zn–Air batteries. *Small Methods* **2021**, *5*, 2000751. [CrossRef]
86. Li, C.; Wu, M.; Liu, R. High-performance bifunctional oxygen electrocatalysts for zinc-air batteries over mesoporous Fe/Co-N-C nanofibers with embedding FeCo alloy nanoparticles. *Appl. Catal. B Environ.* **2019**, *244*, 150–158. [CrossRef]
87. Jia, X.; Zhao, Y.; Chen, G.; Shang, L.; Shi, R.; Kang, X.; Waterhouse, G.I.N.; Wu, L.-Z.; Tung, C.-H.; Zhang, T. Ni$_3$FeN Nanoparticles Derived from Ultrathin NiFe-Layered Double Hydroxide Nanosheets: An efficient overall water splitting electrocatalyst. *Adv. Energy Mater.* **2016**, *6*, 1502585. [CrossRef]
88. Nørskov, J.K.; Rossmeisl, J.; Logadottir, A.; Lindqvist, L.; Kitchin, J.R.; Bligaard, T.; Jónsson, H. Origin of the overpotential for oxygen reduction at a fuel-cell cathode. *J. Phys. Chem. B* **2004**, *108*, 17886–17892. [CrossRef]
89. Jensen, K.D.; Tymoczko, J.; Rossmeisl, J.; Bandarenka, A.S.; Chorkendorff, I.; Escudero-Escribano, M.; Stephens, I.E.L. Elucidation of the oxygen reduction volcano in alkaline media using a copper-platinum (111) alloy. *Angew. Chem. Int. Ed.* **2018**, *57*, 2800–2805. [CrossRef]
90. Shao, M.; Chang, Q.; Dodelet, J.-P.; Chenitz, R. Recent Advances in electrocatalysts for oxygen reduction reaction. *Chem. Rev.* **2016**, *116*, 3594–3657. [CrossRef]
91. Nie, Y.; Li, L.; Wei, Z. Recent advancements in Pt and Pt-free catalysts for oxygen reduction reaction. *Chem. Soc. Rev.* **2015**, *44*, 2168–2201. [CrossRef] [PubMed]
92. Xiong, Y.; Shan, H.; Zhou, Z.; Yan, Y.; Chen, W.; Yang, Y.; Liu, Y.; Tian, H.; Wu, J.; Zhang, H.; et al. Tuning surface structure and strain in Pd-Pt core-shell nanocrystals for enhanced electrocatalytic oxygen reduction. *Small* **2017**, *13*, 1603423. [CrossRef] [PubMed]
93. Cui, Z.; Fu, G.; Li, Y.; Goodenough, J.B. Ni$_3$FeN-Supported Fe$_3$Pt intermetallic nanoalloy as a high-performance bifunctional catalyst for Metal–Air Batteries. *Angew. Chem. Int. Ed.* **2017**, *56*, 9901–9905. [CrossRef] [PubMed]
94. Liu, T.; Sun, F.; Huang, M.; Guan, L. Ternary PtZnCu Intermetallic Nanoparticles as an efficient oxygen reduction electrocatalyst for fuel cells with ultralow Pt loading. *ACS Appl. Energy Mater.* **2022**, *5*, 12219–12226. [CrossRef]
95. Lyu, X.; Zhang, W.; Liu, S.; Wang, X.; Li, G.; Shi, B.; Wang, K.; Wang, X.; Wang, Q.; Jia, Y. A magnetic field strategy to porous Pt-Ni nanoparticles with predominant (111) facets for enhanced electrocatalytic oxygen reduction. *J. Energy Chem.* **2021**, *53*, 192–196. [CrossRef]
96. Rossmeisl, J.; Karlberg, G.S.; Jaramillo, T.; Nørskov, J.K. Steady state oxygen reduction and cyclic voltammetry. *Faraday Discuss.* **2009**, *140*, 337–346. [CrossRef]
97. Rebrov, E.V.; Klinger, E.A.; Berenguer-Murcia, A.; Sulman, E.M.; Schouten, J.C. Selective hydrogenation of 2-Methyl-3-butyne-2-ol in a wall-coated capillary microreactor with a Pd$_{25}$Zn$_{75}$/TiO$_2$ Catalyst. *Org. Process Res. Dev.* **2009**, *13*, 991–998. [CrossRef]
98. Cherkasov, N.; Ibhadon, A.O.; Rebrov, E.V. Novel synthesis of thick wall coatings of titania supported Bi poisoned Pd catalysts and application in selective hydrogenation of acetylene alcohols in capillary microreactors. *Lab Chip* **2015**, *15*, 1952–1960. [CrossRef]
99. Cui, Z.; Chen, H.; Zhao, M.; DiSalvo, F.J. High-Performance Pd$_3$Pb Intermetallic Catalyst for Electrochemical Oxygen Reduction. *Nano Lett.* **2016**, *16*, 2560–2566. [CrossRef] [PubMed]
100. Slanac, D.A.; Hardin, W.G.; Johnston, K.P.; Stevenson, K.J. Atomic ensemble and electronic effects in Ag-rich AgPd nanoalloy catalysts for oxygen reduction in alkaline media. *J. Am. Chem. Soc.* **2012**, *134*, 9812–9819. [CrossRef]
101. Lüsi, M.; Erikson, H.; Merisalu, M.; Kasikov, A.; Matisen, L.; Sammelselg, V.; Tammeveski, K. Oxygen electroreduction in alkaline solution on Pd coatings prepared by galvanic exchange of copper. *Electrocatalysis* **2018**, *9*, 400–408. [CrossRef]
102. Betancourt, L.E.; Rojas-Pérez, A.; Orozco, I.; Frenkel, A.I.; Li, Y.; Sasaki, K.; Senanayake, S.D.; Cabrera, C.R. Enhancing ORR performance of bimetallic PdAg electrocatalysts by designing interactions between Pd and Ag. *ACS Appl. Energy Mater.* **2020**, *3*, 2342–2349. [CrossRef]
103. Biao, K.; Dongwei, L.; Yanjun, H.; Zeng, W.M. Au/Cu$_2$O composite material, super-assembly preparation method and application. Patent No. CN113707890A, 17 August 2021.
104. Wang, Q.; O'Hare, D. Recent Advances in the synthesis and application of Layered Double Hydroxide (LDH) Nanosheets. *Chem. Rev.* **2012**, *112*, 4124–4155. [CrossRef] [PubMed]
105. Zhang, J.; Zhang, H.; Huang, Y. Electron-rich NiFe layered double hydroxides via interface engineering for boosting electrocatalytic oxygen evolution. *Appl. Catal. B Environ.* **2021**, *297*, 120453. [CrossRef]
106. Gao, R.; Yan, D. Recent Development of Ni/Fe-based micro/nanostructures toward Photo/Electrochemical water oxidation. *Adv. Energy Mater.* **2020**, *10*, 1900954. [CrossRef]
107. Zhang, H.; Li, H.; Akram, B.; Wang, X. Fabrication of NiFe layered double hydroxide with well-defined laminar superstructure as highly efficient oxygen evolution electrocatalysts. *Nano Res.* **2019**, *12*, 1327–1331. [CrossRef]
108. Xie, J.; Qu, H.; Lei, F.; Peng, X.; Liu, W.; Gao, L.; Hao, P.; Cui, G.; Tang, B. Partially amorphous nickel–iron layered double hydroxide nanosheet arrays for robust bifunctional electrocatalysis. *J. Mater. Chem. A* **2018**, *6*, 16121–16129. [CrossRef]
109. Zhao, D.; Jiang, P.; Pi, Y.; Huang, X. Superior Electrochemical Oxygen Evolution Enabled by Three-Dimensional Layered Double hydroxide nanosheet superstructures. *ChemCatChem* **2017**, *9*, 84–88. [CrossRef]
110. Li, K.; Guo, D.; Kang, J.; Wei, B.; Zhang, X.; Chen, Y. Hierarchical hollow spheres assembled with ultrathin CoMn double hydroxide nanosheets as trifunctional electrocatalyst for overall water splitting and Zn air battery. *ACS Sustain. Chem. Eng.* **2018**, *6*, 14641–14651. [CrossRef]

111. Guo, X.; Zheng, T.; Ji, G.; Hu, N.; Xu, C.; Zhang, Y. Core/shell design of efficient electrocatalysts based on $NiCo_2O_4$ nanowires and NiMn LDH nanosheets for rechargeable zinc–air batteries. *J. Mater. Chem. A* **2018**, *6*, 10243–10252. [CrossRef]
112. Loizou, K.; Mourdikoudis, S.; Sergides, A.; Besenhard, M.O.; Sarafidis, C.; Higashimine, K.; Kalogirou, O.; Maenosono, S.; Thanh, N.T.K.; Gavriilidis, A. rapid millifluidic synthesis of stable high magnetic moment Fe_xC_y nanoparticles for hyperthermia. *ACS Appl. Mater. Interfaces* **2020**, *12*, 28520–28531. [CrossRef] [PubMed]
113. Rebrov, E.V.; Gao, P.; Verhoeven, T.M.W.G.M.; Schouten, J.C.; Kleismit, R.; Turgut, Z.; Kozlowski, G. Structural and magnetic properties of sol–gel $Co_{2x}Ni_{0.5-x}Zn_{0.5-x}Fe_2O_4$ thin films. *J. Magn. Magn. Mater.* **2011**, *323*, 723–729. [CrossRef]
114. Houlding, T.K.; Gao, P.; Degirmenci, V.; Tchabanenko, K.; Rebrov, E.V. Mechanochemical synthesis of $TiO_2/NiFe_2O_4$ magnetic catalysts for operation under RF field. *Mater. Sci. Eng. B* **2015**, *193*, 175–180. [CrossRef]
115. Yan, B.; Gao, P.; Lu, Z.; Ma, R.; Rebrov, E.V.; Zheng, H.; Gao, Y. Effect of Pr^{3+} substitution on the microstructure, specific surface area, magnetic properties and specific heating rate of $Ni_{0.5}Zn_{0.5}Pr_xFe_{2-x}O_4$ nanoparticles synthesized via sol–gel method. *J. Alloys Compd.* **2015**, *639*, 626–634. [CrossRef]
116. Houlding, T.K.; Rebrov, E.V. Application of alternative energy forms in catalytic reactor engineering. *Green Process. Synth.* **2012**, *1*, 19–31. [CrossRef]
117. Liu, H.; Ren, Y.; Wang, K.; Mu, X.; Song, S.; Guo, J.; Yang, X.; Lu, Z. Magnetic-field-induced strain enhances electrocatalysis of FeCo alloys on anode catalysts for water splitting. *Metals* **2022**, *12*, 800. [CrossRef]
118. Zeng, Z.; Zhang, T.; Liu, Y.; Zhang, W.; Yin, Z.; Ji, Z.; Wei, J. Magnetic field-enhanced 4-electron pathway for well-aligned Co_3O_4/electrospun carbon nanofibers in the oxygen reduction reaction. *ChemSusChem* **2018**, *11*, 580–588. [CrossRef] [PubMed]
119. Li, Y.; Zhang, L.; Peng, J.; Zhang, W.; Peng, K. Magnetic field enhancing electrocatalysis of Co_3O_4/NF for oxygen evolution reaction. *J. Power Sources* **2019**, *433*, 226704. [CrossRef]
120. Li, X.; Cheng, Z.; Wang, X. Understanding the mechanism of the oxygen evolution reaction with consideration of spin. *Electrochem. Energy Rev.* **2021**, *4*, 136–145. [CrossRef]
121. Zhang, Y.; Guo, P.; Li, S.; Sun, J.; Wang, W.; Song, B.; Yang, X.; Wang, X.; Jiang, Z.; Wu, G.; et al. Magnetic field assisted electrocatalytic oxygen evolution reaction of nickel-based materials. *J. Mater. Chem. A* **2022**, *10*, 1760–1767. [CrossRef]
122. Zheng, H.; Wang, Y.; Xie, J.; Gao, P.; Li, D.; Rebrov, E.V.; Qin, H.; Liu, X.; Xiao, H. Enhanced alkaline oxygen evolution using spin polarization and magnetic heating effects under an AC magnetic field. *ACS Appl. Mater. Interfaces* **2022**, *14*, 34627–34636. [CrossRef]
123. Chen, H.; Zheng, H.; Yang, T.; Yue, S.; Gao, P.; Liu, X.; Xiao, H. AC magnetic field enhancement oxygen evolution reaction of bimetallic metal-organic framework. *Int. J. Hydrogen Energy* **2022**, *47*, 18675–18687. [CrossRef]
124. Wu, T.; Xu, Z.J. Oxygen evolution in spin-sensitive pathways. *Curr. Opin. Electrochem.* **2021**, *30*, 100804. [CrossRef]
125. Wu, T.; Ren, X.; Sun, Y.; Sun, S.; Xian, G.; Scherer, G.G.; Fisher, A.C.; Mandler, D.; Ager, J.W.; Grimaud, A.; et al. Spin pinning effect to reconstructed oxyhydroxide layer on ferromagnetic oxides for enhanced water oxidation. *Nat. Commun.* **2021**, *12*, 3634. [CrossRef]
126. Zheng, H.; Chen, H.; Wang, Y.; Gao, P.; Liu, X.; Rebrov, E.V. Fabrication of magnetic superstructure $NiFe_2O_4$@MOF-74 and its derivative for electrocatalytic hydrogen evolution with AC magnetic field. *ACS Appl. Mater. Interfaces* **2020**, *12*, 45987–45996. [CrossRef] [PubMed]
127. Wang, Y.; Yang, T.; Yue, S.; Zheng, H.; Liu, X.; Gao, P.; Qin, H.; Xiao, H. Effects of alternating magnetic fields on the OER of heterogeneous core–shell structured $NiFe_2O_4$@(Ni,Fe)S/P. *ACS Appl. Mater. Interfaces* **2023**, *15*, 11631–11641. [CrossRef]
128. Wang, K.; Yang, Q.; Zhang, H.; Zhang, M.; Jiang, H.; Zheng, C.; Li, J. Recent advances in catalyst design and activity enhancement induced by a magnetic field for electrocatalysis. *J. Mater. Chem. A* **2023**, *11*, 7802–7832. [CrossRef]
129. Sun, Z.; Lin, L.; He, J.; Ding, D.; Wang, T.; Li, J.; Li, M.; Liu, Y.; Li, Y.; Yuan, M.; et al. Regulating the Spin State of Fe III Enhances the Magnetic Effect of the Molecular Catalysis Mechanism. *J. Am. Chem. Soc.* **2022**, *144*, 8204–8213. [CrossRef] [PubMed]
130. Ge, J.; Chen, R.R.; Ren, X.; Liu, J.; Ong, S.J.H.; Xu, Z.J. Ferromagnetic–antiferromagnetic coupling core–shell nanoparticles with spin conservation for water oxidation. *Adv. Mater.* **2021**, *33*, 2101091. [CrossRef] [PubMed]
131. Gong, X.; Jiang, Z.; Zeng, W.; Hu, C.; Luo, X.; Lei, W.; Yuan, C. Alternating magnetic field induced magnetic heating in ferromagnetic cobalt Single-Atom Catalysts for efficient oxygen evolution reaction. *Nano Lett.* **2022**, *22*, 9411–9417. [CrossRef]
132. Niether, C.; Faure, S.; Bordet, A.; Deseure, J.; Chatenet, M.; Carrey, J.; Chaudret, B.; Rouet, A. Improved water electrolysis using magnetic heating of FeC–Ni core–shell nanoparticles. *Nat. Energy* **2018**, *3*, 476–483. [CrossRef]
133. Peng, D.; Hu, C.; Luo, X.; Huang, J.; Ding, Y.; Zhou, W.; Zhou, H.; Yang, Y.; Yu, T.; Lei, W.; et al. Electrochemical reconstruction of NiFe/NiFeOOH superparamagnetic core/catalytic shell heterostructure for magnetic heating enhancement of oxygen evolution reaction. *Small* **2023**, *19*, 2205665. [CrossRef] [PubMed]
134. Deng, J.; Qiao, H.; Li, C.; Huang, Z.; Luo, S.; Qi, X. Magnetic field enhanced surface activity of ferromagnetic $Cr_2Ge_2Te_6$ nanosheets for electrocatalytic oxygen evolution reaction. *Appl. Surf. Sci.* **2023**, *637*, 157899. [CrossRef]
135. Su, M.; Zhou, W.; Liu, L.; Chen, M.; Jiang, Z.; Luo, X.; Yang, Y.; Yu, T.; Lei, W.; Yuan, C. Micro eddy current facilitated by screwed MoS_2 structure for enhanced hydrogen evolution reaction. *Adv. Funct. Mater.* **2022**, *32*, 2111067. [CrossRef]

Disclaimer/Publisher's Note: The statements, opinions and data contained in all publications are solely those of the individual author(s) and contributor(s) and not of MDPI and/or the editor(s). MDPI and/or the editor(s) disclaim responsibility for any injury to people or property resulting from any ideas, methods, instructions or products referred to in the content.

Article

Vanadium Complexes Derived from *O,N,O*-tridentate 6-bis(*o*-hydroxyalkyl/aryl)pyridines: Structural Studies and Use in the Ring-Opening Polymerization of ε-Caprolactone and Ethylene Polymerization

Mark R. J. Elsegood [1], William Clegg [2] and Carl Redshaw [3,*]

1. Chemistry Department, Loughborough University, Loughborough LE11 3TU, UK; m.r.j.elsegood@lboro.ac.uk
2. Chemistry, School of Natural & Environmental Sciences, Newcastle University, Newcastle upon Tyne NE1 7RU, UK; bill.clegg@ncl.ac.uk
3. Plastics Collaboratory, Chemistry, School of Natural Sciences, University of Hull, Cottingham Road, Hull HU6 7RX, UK
* Correspondence: c.redshaw@hull.ac.uk

Citation: Elsegood, M.R.J.; Clegg, W.; Redshaw, C. Vanadium Complexes Derived from *O,N,O*-tridentate 6-bis(*o*-hydroxyalkyl/aryl)pyridines: Structural Studies and Use in the Ring-Opening Polymerization of ε-Caprolactone and Ethylene Polymerization. *Catalysts* **2023**, *13*, 988. https://doi.org/10.3390/catal13060988

Academic Editor: Moris S. Eisen

Received: 5 May 2023
Revised: 5 June 2023
Accepted: 6 June 2023
Published: 9 June 2023

Copyright: © 2023 by the authors. Licensee MDPI, Basel, Switzerland. This article is an open access article distributed under the terms and conditions of the Creative Commons Attribution (CC BY) license (https:// creativecommons.org/licenses/by/ 4.0/).

Abstract: Interaction of [VO(O*i*Pr)$_3$] with 6-bis(*o*-hydroxyaryl)pyridine, 2,6-{HOC(Ph)$_2$CH$_2$}$_2$(NC$_5$H$_3$), LH$_2$, afforded [VO(O*i*Pr)L] (**1**) in good yield. The reaction of LNa$_2$, generated in-situ from LH$_2$ and NaH, with [VCl$_3$(THF)$_3$] led to the isolation of [VL$_2$] (**2**) in which the pyridyl nitrogen atoms are *cis*; a regioisomer **3**·2THF, in which the pyridyl nitrogen atoms are *trans*, was isolated when using [VCl$_2$(TMEDA)$_2$]. The reaction of the 2,6-bis(*o*-hydroxyalkyl)pyridine {HOC(*i*Pr)$_2$CH$_2$}$_2$(NC$_5$H$_3$), L^{1}H$_2$, with [VO(OR)$_3$] (R = *n*Pr, *i*Pr) led, following work-up, to [VO(OR)L^1] (R = *n*Pr (**4**) and *i*Pr (**5**)). Use of the bis(methylpyridine)-substituted alcohol (*t*Bu)C(OH)[CH$_2$(C$_5$H$_3$Me-5)]$_2$, L^{2}H, with [VO(OR)$_3$] (R = Et, *i*Pr) led to the isolation of [VO(µ-O)(L^2)]$_2$ (**6**). Complexes **1** to **6** have been screened for their ability to act as pre-catalysts for the ring opening polymerization (ROP) of ε-caprolactone (ε-CL), δ-valerolactone (δ-VL), and *rac*-lactide (*r*-LA) and compared against the known catalyst [Ti(O*i*Pr)$_2$L] (**I**). Complexes **1**, **4**–**6** were also screened as catalysts for the polymerization of ethylene (in the presence of dimethylaluminium chloride/ethyltrichloroacetate). For the ROP of ε-CL, in toluene solution, conversions were low to moderate, affording low molecular weight products, whilst as melts, the systems were more active and afforded higher molecular weight polymers. For δ-VL, the systems run as melts afforded good conversions, but in the case of *r*-LA, all systems as melts exhibited low conversions (<10%) except for **6** (<54%) and **I** (<39%). In the case of ethylene polymerization, the highest activity (8600 Kg·mol·V^{-1}bar^{-1}h^{-1}) was exhibited by **1** in dichloromethane, affording high molecular weight, linear polyethylene at 70 °C. In the case of **4** and **5**, which contain the propyl-bearing chelates, the activities were somewhat lower (≤1500 Kg·mol·V^{-1}bar^{-1}h^{-1}), whilst **6** was found to be inactive.

Keywords: vanadium complexes; 6-bis(*o*-hydroxyaryl)pyridine; 2,6-bis(*o*-hydroxyalkyl)pyridine; molecular structures; ring-opening polymerization; ε-CL; *r*-LA; δ-VL; ethylene polymerization

1. Introduction

Plastics play a central role in society, and the recent COVID-19 pandemic has highlighted their continued importance [1]. Given this, there is still interest in the design of new post-metallocene systems for accessing new polyolefins, particularly as more than 60% of global plastic production is polyolefin-based [2], and partly driven by the need to generate new IP in developing countries [3]. However, petroleum-derived plastics and their subsequent disposal are problematic, and COVID-19 has only added to these problems [4]. There is, thus, also an urgent need to develop materials with greener characteristics, which can be applied to the needs of modern-day life. For example, biodegradable polymers have seen increased use in biomedical applications [5,6], although it should be noted, too, that polyolefins have medical applications [7]. To access such polymers via either olefin polymerization or ring opening polymerization (ROP), the choice of metal employed as the catalytic centre is important, and cheap and earth-abundant metals are favourable [8–10]. We note that organic catalysts can also be used for ROP [11]. For metal-catalyzed polymerizations, including both ethylene polymerization and ROP, the ligation at the metal not only plays a significant role in determining the catalytic activity of the system but also impacts the properties of the resultant products. This versatility stems from the ability to vary the electronic and/or steric properties of the ligands, as well as improving other properties such as solubility [12–16]. With this in mind, we have been exploring the use of bulky chelating phenolates/macrocycles as ancillary ligands in both α-olefin polymerization and for the ROP of cyclic esters [17,18]. Some promising results have been achieved by combining such chelates with the metal vanadium, which is the 20th most abundant metal in the Earth's crust. In particular, we and others have found that very high catalytic activities and thermally robust systems are accessible for α-olefin polymerization when conducted in the presence of co-catalysts such as dimethylaluminium chloride and the re-activator ethyl trichloroacetate [19–31], as well as ethylene copolymerization with the likes of propylene, 1-hexene or norbornene [32–41]. Moreover, a number of vanadyl-containing systems have shown promise in the ROP of cyclic esters [42,43], whilst other systems have been shown to be promising for both ROP and α-olefin polymerization [44–49]. We also note the favourable toxicity of vanadyl complexes; for example, vanadyl sulfate is not only used as a supplement but has shown promising behaviour against Type 2 diabetes [50]. We have also shown that vanadyl calixarenes exhibit low toxicity [51]. Given this, we were interested in reports concerning the use of *O,N,O*-tridentate 6-bis(*o*-hydroxyaryl)pyridine ligation at titanium, and the use of such complexes for ethylene polymerization [52], and more recently for the ROP of ε-caprolactone (ε-CL) and *L*-lactide (*L*-LA) [53]. These titanium species exhibited only low to modest catalytic activities for ethylene polymerization (using MAO as co-catalyst), whilst, for ROP, conversions of 93 and 98% were achievable at 60 °C for ε-CL and *L*-LA, respectively. We also note that this ligand set, bound to tungsten(VI), has been employed in the *cis*-specific polymerization of norbornene [54]. Herein, we report the use of such ligation extended to vanadium-based systems (see **1–5**, Scheme 1), as well as structural studies, and their ability to act as catalysts for both ethylene polymerization and the ring-opening polymerization (ROP) of ε-caprolactone (ε-CL), δ-valerolactone (δ-VL), and *rac*-lactide (*r*-LA). The bis(pyridine)alkoxide (*t*Bu)C(OH)[CH$_2$(C$_5$H$_3$Me-5)]$_2$, whose reported coordination chemistry is somewhat limited [55,56], is also investigated (see **6**, Scheme 1).

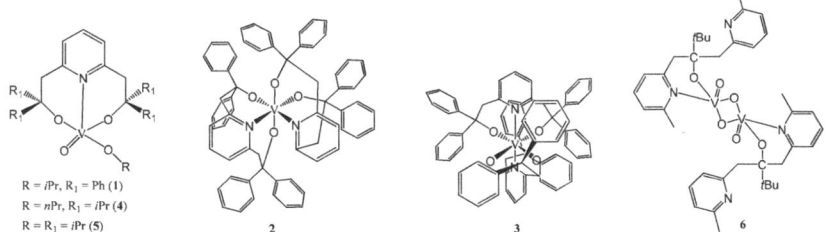

Scheme 1. Vanadium complexes **1–6**.

2. Results and Discussion

2.1. Synthesis and Characterization of V Complexes

2.1.1. Use of 2,6-bis(o-hydroxyaryl)pyridine, 2,6-{HOC(Ph)$_2$CH$_2$}$_2$(NC$_5$H$_3$), LH$_2$

The reaction of 6-bis(*o*-hydroxyaryl)pyridine, 2,6-{HOC(Ph)$_2$CH$_2$}$_2$(NC$_5$H$_3$), LH$_2$ [57], with one equivalent of [VO(O*i*Pr)$_3$] in refluxing toluene afforded, after workup (MeCN), yellow crystalline [VO(O*i*Pr)L] (**1**) in good yield. In the IR spectrum, the band at 1024 cm^{-1}

is assigned to vV=O, whilst the ^{51}V NMR spectrum (C$_6$D$_6$, 298 K) contains a single peak at δ −572.81 with Δω$_{1/2}$ 3.8 Hz. Single crystals of 1 were grown from a saturated solution in acetonitrile at ambient temperature. The molecular structure is shown in Figure 1, and selected bond lengths and angles are given in the caption. There is one molecule in the asymmetric unit. The trigonal bipyramidal vanadyl centre (τ = 0.97) [58] is chelated by the pyridine diolate ligand with the pyridyl N being axial, an oxo ligand equatorial, and an axial OiPr ligand. The two diolate oxygen atoms complete the equatorial donor set. There is an intramolecular C(3)–H(3)···π{C(27)} interaction at 2.83 Å, between two Ph groups at opposite ends of the pyridine diolate ligand. In the packing, there are some weak C–H···O/π interactions.

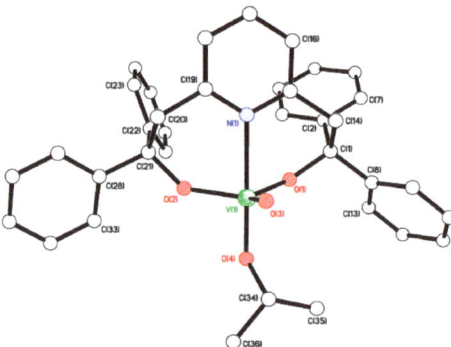

Figure 1. Molecular structure of [VO(OiPr)L] (**1**). Selected bond lengths (Å) and bond angles (°): V(1)–O(1) 1.7919(15), V(1)–O(2) 1.8019(15), V(1)–O(3) 1.6089(14), V(1)–O(4) 1.7841(15), V(1)–N(1) 2.3422(18); O(1)–V(1)–O(2) 119.61(7), N(1)–V(1)–O(4) 177.51(6).

Treatment of LH$_2$ with two equivalents of NaH in refluxing THF and subsequent reaction with 0.5 equiv. of [VCl$_3$(THF)$_3$] afforded the red complex [VL$_2$] (**2**) in moderate yield. The molecular structure is shown in Figure 2 (left), with selected bond lengths and angles given in the caption. There is one molecule in the asymmetric unit. The complex contains a central distorted octahedral vanadium(IV) centre, and the L ligands are bound such that the pyridyl N atoms are positioned *cis*. Interestingly, the use of the vanadium(II) complex [VCl$_2$(TMEDA)$_2$] as starting material led, after work-up, to red crystals of **3**·2THF which were found to be a regioisomer of **2**, with the V centre on a two-fold axis and in which the pyridyl N atoms are positioned *trans* (see Figure 2, right; for an alternative view, see Figure S3, ESI).

2.1.2. Use of 2,6-bis(o-hydroxy-i-propyl)pyridine L^{1}H$_2$

Use of the related potential *O,N,O*-donor 2,6-bis(*o*-hydroxy-*i*-propyl)pyridine L^{1}H$_2$ [52] with [VO(OnPr)$_3$] led, following work-up (MeCN), to the yellow complex [VO(OnPr)L^1] (**4**). In the IR spectrum, the band at 1018 cm^{-1} is assigned to vV=O, whilst the ^{51}V NMR spectrum (C$_6$D$_6$, 298 K) contains a single peak at δ −544.70 with Δω$_{1/2}$ 2.8 Hz (see Figure S4 left, ESI). Crystals suitable for an X-ray diffraction study were grown from acetonitrile at 0 °C. The molecular structure is shown in Figure 3, with selected bond lengths and angles given in the caption. There is one molecule in the asymmetric unit. The geometry of vanadium is best described as trigonal bipyramidal, with τ 0.94 [58]. The V(1) centre lies 0.2155(6) Å out of the trigonal plane O(1)/O(2)/O(3).

Similar use of [V(O)(OiPr)$_3$] with L^{1}H$_2$ afforded the complex [VO(OiPr)L^1] (**5**). In the IR spectrum, the band at 1017 cm^{-1} is assigned to vV=O, whilst the ^{51}V NMR spectrum (C$_6$D$_6$, 298 K) contains a single peak at δ −572.81 with Δω$_{1/2}$ 3.8 Hz. Again, the use of acetonitrile provided crystals suitable for an X-ray diffraction study (see Figure 4). There are two very similar molecules in the asymmetric unit. The geometry exhibited at both V atoms is trigonal bipyramidal. The V(1) centre lies 0.2086(11) Å away from the equatorial plane towards O(4), while V(2) lies 0.2228(11) Å away from the equatorial plane towards O(8). The τ value is close to 1 for both molecules [58]. Similar molecules pack in spirals along *a* with those involving V(1) in the central part of the unit cell and those involving V(2) along the *a*/*c* face (Figure 4, right).

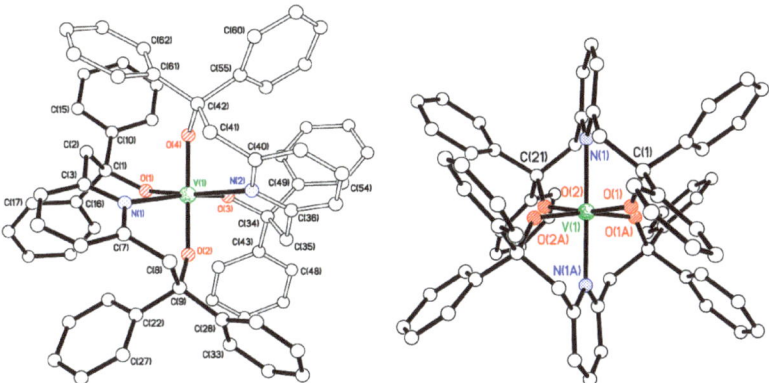

Figure 2. Left: Molecular structure of [V({OC(Ph)$_2$CH$_2$}$_2$(*cis*-NC$_5$H$_3$))$_2$] (**2**). Selected bond lengths (Å) and angles (°): V(1)–O(1) 1.8565(11), V(1)–O(2) 1.8340(10), V(1)–O(3) 1.8547(11), V(1)–O(4) 1.8230(10), V(1)–N(1) 2.2876(13), V(1)–N(2) 2.2692(13); O(2)–V(1)–O(4) 158.99(5), N(1)–V(1)–N(2) 105.87(5). **Right**: Molecular structure of [V({OC(Ph)$_2$CH$_2$}$_2$(*trans*-NC$_5$H$_3$))$_2$]·2THF (**3**·2THF). Selected bond lengths (Å) and angles (°): V(1)–O(1) 1.8819(13), V(1)–O(2) 1.8702(13), V(1)–N(1) 2.1720(16); O(1)–V(1)–O(2) 169.83(6), N(1)–V(1)–N(2) 179.68(9).

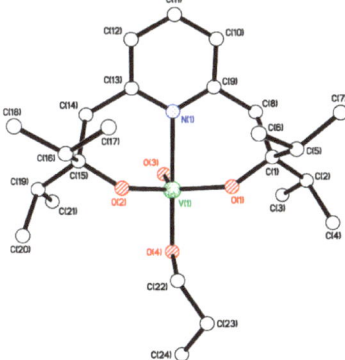

Figure 3. Molecular structure of [VO(O*n*Pr)L^1] (**4**). Selected bond lengths (Å) and angles (°): V(1)–O(1) 1.7985(9), V(1)–O(2) 1.7999(9), V(1)–O(3) 1.5984(9), V(1)–O(4) 1.8209(9), V(1)–N(1) 2.3045(11); O(1)–V(1)–O(2) 118.72(4), O(4)–V(1)–N(1) 175.13(4).

Note we have also re-determined the structure of L^{1}H$_2$ to higher precision with an *R*-factor of 0.052. It has previously been reported [52], though with a higher *R*-factor of 0.065 with CCDC refcode BAJSOX. The molecular structure is shown in Figures S1 and S2, ESI, together with details of the structure.

2.1.3. Use of the Bis(Methylpyridine)-Substituted Alcohol (*t*Bu)C(OH)[CH$_2$(C$_5$H$_3$Me-5)]$_2$

The reaction of (*t*Bu)C(OH)[CH$_2$(C$_5$H$_3$Me-5)]$_2$ [55] with one equivalent of either [VO(OEt)$_3$] or [VO(O*i*Pr)$_3$] in refluxing toluene afforded, after workup (MeCN), yellow crystalline [VO(μ-O)(L^2)]$_2$ (**6**) in moderate yield (*ca.* 40%). In the IR spectrum, the band at 1019 cm^{-1} is assigned to vV=O, and the ^{51}V NMR spectrum (C$_6$D$_6$, 298 K) contains a broad peak at δ −539.3 with Δω$_{1/2}$ 3520 Hz (see Figure S4 right, ESI). Small single crystals of **6** were grown from a saturated MeCN solution on standing for 24 h at 20 °C. The molecular structure is shown in Figure 5, with selected bond lengths and angles given in the caption. There is half a molecule in the asymmetric unit, and the molecule lies on an inversion centre. Only one of the two pyridyl nitrogen atoms binds to the metal ion from each ligand. There is approximate square-based pyramidal geometry at the V centre. The V$_2$O$_2$ diamond is fairly symmetrical, with very similar V–O bonds on each side. Pyridyl rings are not quite parallel, with a dihedral angle of 17.6° between them. This is illustrated by looking at a comparison of the distances C(5) . . . C(14) = 3.078 and C(2) . . . C(17) = 4.023 Å. The *anti*-arrangement of the Me groups on the pyridine rings is undoubtedly to minimize steric interaction. Molecules of **6** are arranged in layers with nothing unusual in terms of intermolecular interactions (see Figure S5, ESI).

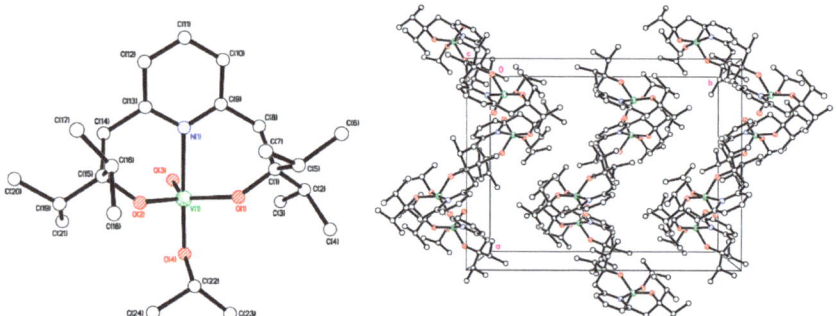

Figure 4. Left: Molecular structure of [VO(O*i*Pr)L^1] (**5**). Selected bond lengths (Å) and angles (°): V(1)–O(1) 1.8000(17), V(1)–O(2) 1.7902(16), V(1)–O(3) 1.6035(17), V(1)–O(4) 1.8097(18), V(1)–N(1) 2.285(2); O(1)–V(1)–O(2) 115.99(8), O(4)–V(1)–N(1) 177.06(7). **Right**: Packing found in **5**.

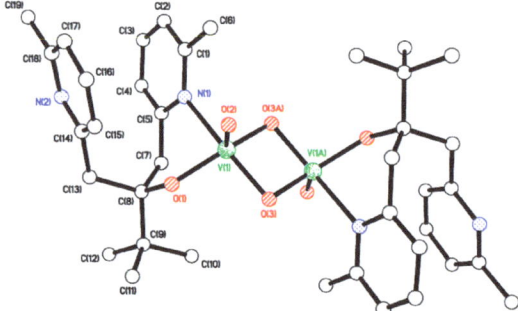

Figure 5. Molecular structure of [VO(μ-O)(L^2)]$_2$ (**6**). Selected bond lengths (Å) and angles (°): V(1)–O(1) 1.8067(9), V(1)–O(2) 1.5995(9), V(1)–O(3) 1.8214(9), V(1)–O(3A) 1.8350(9), V(1)–N(1) 2.2256(11); V(1)–O(3)–V(1A) 97.57(4), O(1)–V(1)–N(1) 83.25(4).

2.2. Ring Opening Polymerization (ROP) Studies

Complexes **1–6** were screened for their potential to act as catalysts for the ring-opening polymerization (ROP) of cyclic esters. Results are presented in Tables 1 and 2 (for ε-CL), Table 3 (for δ-VL), and Table 4 (for *r*-LA) and are compared against the known active catalyst [Ti(O*i*Pr)$_2$L] (**I**) [53].

Table 1. Results for ROP of ε-CL over 24 h using **1–6** and **I**.

Entry	Cat.	(CL):(Cat)	T/°C	Conv [a] (%)	M_n [b]	$M_{n,Cal}$ [c]	Đ [d]
1	1	500:1	15	0	-	-	-
2	1	500:1	60	20	270	11,430	1.63
3	1	500:1	90	23	380	26,270	2.10
4	1	500:1	130	>99	6760	56,520	2.00
5 [e]	1	500:1	130	>99	7610	56,520	1.67
6	1	100:1	130	50	550	5720	1.67
7	1	1000:1	130	7	250	8010	1.31
8	1	250:1	130	47	880	13,430	1.81
9	2	500:1	130	22	440	12,560	2.86
10	3	500:1	130	19	-	-	-
11	4	500:1	70	2	1275	1160	2.23
12	4	500:1	130	92	2510	52,520	1.35
13	4	100:1	130	64	270	7320	1.55
14	5	500:1	70	58	500	33,120	1.45
15	5	100:1	130	98	440	11,200	2.42
16 [e]	5	500:1	130	>99	4410	56,520	1.59
17	6	500:1	130	63	28,890/2260	35,970	1.22/1.17
18 [e]	6	500:1	90	49	620	27,980	1.62
19	I	500:1	130	>99	13,770	56,520	1.67
20 [e]	I	500:1	130	>99	2070	56,520	1.18

[a] Determined by ^{1}H NMR spectroscopy. [b] $M_{n/w}$, GPC values corrected considering Mark–Houwink method from polystyrene standards in THF, $M_{n/w}$ measured = $0.56 \times M_{n/w}$ GPC $\times 10^3$. [c] Calculated from ([monomer]$_0$/[cat]$_0$) × conv (%) × monomer molecular weight (M_{CL} = 114.14) + end groups. [d] From GPC. [e] Conducted in air.

Table 2. Synthesis of polycaprolactone using complexes **1–6** (and **I**) as melts (130 °C, 24 h).

Entry	Cat.	(CL):(Cat)	Conv [a] (%)	M_n [b]	$M_{n,Cal}$ [c]	Đ [d]
1	1	500:1	81	5280	46,240	2.14
2 [e]	1	500:1	85	3090	48,530	2.06
3	2	500:1	83	2990	47,390	1.24
4	3	500:1	84	2350	47,960	1.43
5	4	500:1	86	2460	49,100	2.02
6 [e]	4	500:1	98	2080	55,950	2.92
7	5	500:1	>99	5960	56,520	1.17
8 [e]	5	500:1	>99	3460	56,520	1.09
9	6	500:1	>99	2270/200	56,520	1.47
10 [e]	6	500:1	49	3580/280	27,980	1.04
11	I	500:1	74	6100	42,250	2.26
12 [e]	I	500:1	>99	8720	56,520	1.94

[a] Determined by ^{1}H NMR spectroscopy. [b] $M_{n/w}$, GPC values corrected considering Mark–Houwink method from polystyrene standards in THF, $M_{n/w}$ measured = $0.56 \times M_{n/w}$ GPC $\times 10^3$. [c] Calculated from ([monomer]$_0$/[cat]$_0$) × conv (%) × monomer molecular weight (M_{CL} = 114.14) + end groups. [d] From GPC. [e] Conducted in air.

Table 3. Synthesis of polyvalerolactone using complexes **1–6** (and **I**) as melts (130 °C, 24 h).

Entry	Cat.	[VL]:[Cat]	Conv [a] (%)	M_n [b]	$M_{n,Cal}$ [c]	Đ [d]
1	1	500:1	87	3100	43,570	1.35
2 [e]	1	500:1	91	920	45,570	2.00
3	2	500:1	77	500	38,580	1.49
4 [e]	2	500:1	83	330	41,570	1.27
5	3	500:1	75	500	49,580	1.55
6	4	500:1	97	3420	48,580	1.46
7 [e]	4	500:1	99	1550	49,580	1.87
8	5	500:1	89	2900	44,570	1.49

Table 3. *Cont.*

Entry	Cat.	[VL]:[Cat]	Conv ^{a} (%)	M_n ^{b}	$M_{n,Cal}$ ^{c}	Đ ^{d}
9 ^{e}	5	500:1	90	3990	45,070	1.18
10	6	500:1	75	380	37,560	1.60
11 ^{e}	6	500:1	54	390	27,050	2.16
12	I	500:1	72	500	36,060	1.76
13 ^{e}	I	500:1	46	360	23,050	1.85

^{a} Determined by ^{1}H NMR spectroscopy. ^{b} $M_{n/w}$, GPC values corrected considering Mark–Houwink method from polystyrene standards in THF, $M_{n/w}$ measured = 0.57 × $M_{n/w}$ GPC × 10^3. ^{c} Calculated from ([monomer]$_0$/[cat]$_0$) × conv (%) × monomer molecular weight (M_{VL} = 100.16). ^{d} From GPC. ^{e} Run in air.

Table 4. Synthesis of polylactide using complexes **1–6** (and **I**) as melts (130 °C, 24 h).

Entry	Cat.	(rLA):(Cat)	Conv ^{a} (%)
1	1	500:1	3
2 ^{b}	1	500:1	2
3	2	500:1	3
4	3	500:1	1
5	4	500:1	10
6 ^{b}	4	500:1	8
7	5	500:1	8
8 ^{b}	5	500:1	1
9	6	500:1	43
10 ^{b}	6	500:1	54
11	I	500:1	29
12 ^{b}	I	500:1	39

^{a} Determined by ^{1}H NMR spectroscopy. ^{b} Run in air.

2.2.1. ROP of ε-Caprolactone (ε-CL)

In the case of the ε-CL, for runs conducted in toluene, the products tended to be oils of low molecular weight terminated by OH groups (e.g., see Figure S6, ESI for MALDI-ToF spectra from entry 14, Table 1). Runs conducted at temperatures below 100 °C (entries **1–3**) resulted in poor conversion, whilst (ε-CL):(Cat) ratios either high or lower than 500:1 (entries **6–8**) also led to poor to moderate conversion. However, systems employing **1** under either N$_2$ or air at 130 °C employing the ratio 500:1 for (ε-CL:Cat) (entries **4** and **5**, respectively) and **I** under N$_2$ (entry **19**) afforded somewhat higher molecular weight products. A typical ^{1}H NMR spectrum of the obtained PCL is given in Figure S7, ESI. In all cases, the molecular weights of the products were still significantly lower than the theoretical values calculated via the NMR spectra. Such low molecular weights suggest that transesterification side reactions had occurred. In the case of **1** under air, the MALDI-ToF spectrum (Figure S8, ESI) revealed families of peaks terminated by either OH groups or by pyridine phenol/phenolate and OH. GPC results for dinuclear **6** under N$_2$ at 130 °C (entry **17**) indicated a bimodal distribution. Đ values range from 1.17/1.18 (for **I** and **6**), indicating good control to less controlled systems with larger Đ values. e.g., 2.86 (for **2**, entry **9**). Conversions were low to moderate for runs conducted at ≤90 °C or, in the case of **1**, when using a (ε-CL):(Cat) ratio other than 500:1; use of 100:1 for **5** (entry **15**) afforded a conversion of 98%. However, at 130 °C over 24 h with a ratio of 500:1, the systems employing **1**, **5**, or **I** under N$_2$ or air afforded > 98% conversion. Catalyst **4** also exhibited a high conversion (92%) at 130 °C under N$_2$ (entry **12**, Table 1). Results for coordinatively saturated **2** and **3** were similarly low, suggesting that either the *cis/trans* arrangement had little influence on behaviour or that, in the solution, the same regioisomer was present. Interestingly, the use of **1** (entry **5**) and **5** (entry **16**) under air led to improved performance in terms of %conversion and increased M_n, whereas the titanium catalyst **I**, whilst maintaining its %conversion, afforded a polymer of much lower molecular weight (M_n), albeit with better control (entry **20**). Kinetic studies were conducted on the structurally related complexes **1**, **4**, and **5** (see Figure 6). The results indicated the rate order was **5** > **4** > **1**, with the logarithmic dependence appearing nonlinear, which indicates deviation from the first order in ε-CL, particularly for **1** and, to a lesser extent, **4**. This is thought to be due to these systems exhibiting induction periods. The slower performance of **1** was thought to be due to restricted access to the metal centre caused by the increased steric bulk of the phenyl groups of the pyridinediolate ligand set. We note that the use of [Ti(OiPr)$_2$L] in the ROP of ε-CL in toluene at 60 °C over 20 h resulted in a conversion of 94% and a product with M_n 900 (by gpc). Use of the related complex [Ti(OiPr)$_2$L^{Me}] (where L^{Me}H$_2$ = 2,2′-(4-methylpyridine-2,6-diyl)bis(1,1-diphenylethan-1-ol) afforded a similar result in toluene, but when employed as a melt at 100 °C over 8 h achieved a conversion of 99% and afforded a higher molecular weight product with M_n 5800 (by gpc) [53]. In the next section, we also employ melt conditions using **1–6** (and **I**).

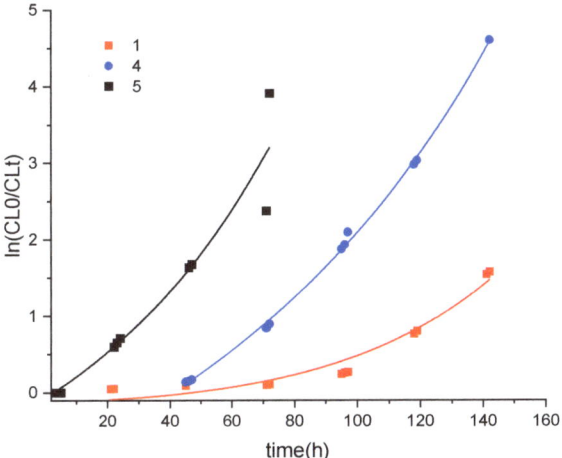

Figure 6. ln(CL$_0$/CL$_t$) versus time for PCL runs using 1, 4, and 5 (500:1, 130 °C, N$_2$).

Given the varied activity, control, and generally low molecular weights achieved in the above results, the runs were also conducted in the absence of solvent, i.e., as melts, using a 500:1 ratio for (ε-CL):(Cat), see Table 2. In general, the conversions were higher, with **4**, **5**, and **I** exhibiting the highest conversions. In all cases, the molecular weights of the products were higher versus solution studies. The majority of the systems were relatively well controlled (Ɖs < 1.47), though the Ɖs for **1**, **4**, and **I** were somewhat higher (1.94–2.92). Again, the use of **6** led to bimodal behaviour (entries **9** and **10**, Table 2). End-group analysis by ^{1}H NMR spectroscopy and MALDI-ToF mass spectrometry indicated the presence of multiple products. For example, in the case of **5** under N$_2$ (Table 2, entry **7**), the MALDI-ToF spectrum contained families of peaks associated with chain polymers terminated by 2 OH or terminated by OH/OiPr as well as cyclic products (Figure S9, ESI).

2.2.2. ROP of δ-Valerolactone (δ-VL)

In the case of δ-VL, runs were conducted as melts only (Table 3), and conversions were generally high; the lowest conversions were observed for **2** (≤83%), **3**, and **6** (≤75%), and the titanium complex **I** under air (46%, run **13**, Table 3). A typical ^{1}H NMR spectrum of the obtained PCL is given in Figure S10, ESI. As for PCL, molecular weights were far lower than calculated values, which was assumed to be due to transesterification; Ɖ values ranged from 1.18 to 2.16. The higher molecular weight products were afforded using complexes of the type [L/L^{1}VO(OR)], namely **1** under N$_2$, as well as **4** and **5** under either N$_2$ or air. MALDI-ToF spectra (e.g., Figures S11 and S12, ESI) of the products revealed the presence of a number of families of peaks assigned to chain polymers terminated by 2 OH groups and a smaller series assigned to cyclic polymers when using **5** as a melt under air (Table 3, entry **6**). In the case of **5** as a melt under N_2 (Table 3, entry **7**), there were chain polymers present, terminated by OH/O*i*Pr and cyclic polymers.

2.2.3. Ring Opening Polymerization Studies of Rac-Lactide (r-LA)

In the case of the ROP of *r*-LA, runs were, again, conducted in the absence of solvent, and the results (Table 4) revealed poor conversions (<40%), except for in the case of **6** (for entry **10**, M_n = 1260 (correction factor 0.58), D 2.94). From **6**, the resulting PLA was found to be atactic (Figures S13 and S14, ESI) [59]. Given these disappointing results, the ROP of *r*-LA was not further investigated. We note that the complexes [Ti(O*i*Pr)$_2$L] and [Ti(O*i*Pr)$_2$L^{Me}] in the ROP of *L*-LA in toluene at 60 °C over **6** or 3.5 h, respectively, resulted in a conversion of 96 or 98% and afforded a product with M_n 1100 (by gpc). The use of melts with the [Ti(O*i*Pr)$_2$L^{Me}] system led to increased M_n values (>8100) with good control [53].

2.3. Ethylene Polymerization Studies

The complexes **1**, **4**, **5**, and **6** were tested, in the presence of the co-catalyst dimethylaluminium chloride (DMAC) and the re-activator trichloroethylacetate (ETA), for their ability to act as pre-catalysts for the polymerization of ethylene. The results are presented in Table 5 and reveal that the observed activities are best described as moderate in terms of vanadium-based systems [19–41]. For example, under the same conditions, the imido complexes {[V(N*p*-RC$_6$H$_4$)]$_2$L^1} (L^1 = *p-tert*-butyltetrahomodioxacalix [6]arene; R = CF$_3$, Cl, F) afforded higher activities in the range 9.0–14.8 × 10^3 Kg·molV^{-1}bar^{-1}h^{-1} [48].

Table 5. Ethylene polymerization data for homogeneous catalysts (V complex + Me$_2$AlCl + ETA) [a].

Complex	PE Yield, g	Average Activity Kg PE/mol V Bar h	$M\eta$ × 10^{-3}	CH$_3$/1000C	Double Bonds/1000C		
					888 cm^{-1}	909 cm^{-1}	965 cm^{-1}
1 [b]	1.4	2800	450	4.9	-	0.16	-
1	4.3	8600	300	0.6	-	0.10	-
4	0.6	1200	380	1.8	0.09	0.90	0.04
5	0.74	1500	360	1.1	0.11	0.13	-
6	0	0	-	-	-	-	-

[a] Polymerization conditions: V-complex (0.5 μmol) was used as a solution in 0.5 mL of CH$_2$Cl$_2$; co-catalyst (Me$_2$AlCl 0.5 mmol+ ETA 0.5 mmol, molar ratio V: Me$_2$AlCl:ETA = 1:1000:1000), dissolved in 100 mL of toluene, 100 mL of heptane; T pol 70 °C, P C$_2$H$_4$ 2 bar, for 30 min. [b] **1** in 0.5 mL of toluene.

Given the poor yields, it was decided to measure M_v rather than M_w or M_n, whilst IR spectroscopy was employed to measure the methyl and C=C content; the high molecular weight and hence poor solubility of the polymers prevented the use of ^{13}C NMR spectroscopy. In all cases, the systems afforded high molecular weight linear polyethylene with a methyl branch content ≤4.9 per 1000 carbons.

In the case of complex **1**, the best results were obtained upon its introduction into the reactor as the solution in CH$_2$Cl$_2$ rather than in toluene, presumably as a result of increased solubility. If the groups bound to the carbon atoms adjacent to oxygen in the chelate are changed to a less bulky propyl group, then the activity is reduced from 8600 Kg·molV^{-1}bar^{-1}h^{-1} in **1** (bearing phenyl substituents) to 1200 Kg·molV^{-1}bar^{-1}h^{-1} for **4** (bearing *n*-propyl substituents) and to 1500 Kg·molV^{-1}bar^{-1}h^{-1} for **5** (bearing *n*-propyl substituents). Surprisingly, similar use of **6** led to inactivity, and this is tentatively assigned to the unfavourable flexibility of the free methylpyridine-containing arm, which has the potential to block access to the metal. The active species in these systems is thought to be a paramagnetic species formed on the reduction of the metal centre by the excess co-catalyst. We and others have investigated such species in related systems [60–63].

3. Materials and Methods

3.1. General

All reactions were conducted under an inert atmosphere using standard Schlenk techniques. Toluene was dried from sodium, acetonitrile was distilled from calcium hydride, and all solvents were degassed prior to use. IR spectra (nujol mulls, KBr windows) were recorded on a Nicolet Avatar 360 FTIR spectrometer, NMR spectrometer 400.2 MHz on a JEOL ECZ 400S spectrometer, with TMS $\delta H = 0$ as the internal standard or residual protic solvent [CD_3CN, $\delta H = 1.94$]. ^{51}V NMR spectra were recorded from C_6D_6 solutions at 105.1 MHz, ^{51}V chemical shifts were referenced to external $VOCl_3$ sample. Chemical shifts are given in ppm (δ), and coupling constants (J) are given in Hertz (Hz). Elemental analyses were performed by the elemental analysis service at the Department of Chemistry, the University of Hull, or London Metropolitan University. FTIR spectra (nujol mulls, KBr windows) were recorded on a Nicolet Avatar 360 FT-IR spectrometer. Matrix-Assisted Laser Desorption/Ionization Time of Flight (MALDI-TOF) mass spectrometry was performed in a Bruker autoflex III smart beam in linear mode, and the spectra were acquired by averaging at least 100 laser shots. 2,5-Dihydroxybenzoic acid was used as the matrix, and THF was used as a solvent. Sodium chloride was dissolved in methanol and used as the ionizing agent. Samples were prepared by mixing 20 µL of matrix solution in THF (2 mg·mL^{-1}) with 20 µL of matrix solution (10 mg·mL^{-1}) and 1 µL of a solution of ionizing agent (1 mg·mL^{-1}). Then, 1 mL of these mixtures was deposited on a target plate and allowed to dry in air at ambient temperature. The known compounds LH_2, L^1H_2, L^2H, [$VCl_3(THF)_3$] and [$VCl_2(TMEDA)_2$] were prepared by the literature methods [52,55,57,64,65]. The reagents [$VO(OR)_3$] (R = nPr, iPr) were purchased from Sigma Aldrich and were used as received. Molecular weights were calculated from the experimental traces using the OmniSEC software.

Crystal structures were determined from data collected at the UK National Crystallography Service (**1, 2, 4–6**) and in the home laboratory for **3**. Full details are given in the ESI. Crystal data are summarized in Tables S1 and S2.

3.2. Preparation of [VO(OiPr)L] (**1**)

To [$VO(OiPr)_3$] (0.50 mL, 2.12 mmol) and LH_2 (1.00 g, 2.12 mmol) was added toluene (30 mL), and then the system was refluxed for 12 h. On cooling, the volatiles were removed in a vacuo, and the residue was extracted into warm MeCN (30 mL). Yield: 1.09 g, 87%. $C_{36}H_{34}NO_4V$ requires C 72.59, H 5.75, and N 2.35%. Found: C 72.50, H 5.84, N 2.06%. IR: 1599 m, 1576 m, 1319 m, 1259 s, 1239 m, 1209 w, 1175 w, 1110 s, 1091 s, 1057 s, 1024 s, 956 s, 914 m, 892 m, 854 m, 813 m, 790 m, 765 w, 736 w, 704 s, 694 s, 669 w, 650 m, 620 w, 600 m, 532 w, 507 m, 486 m, 454 w. ^1H NMR (C_6D_6) δ: 7.69 (d, J = 7.6 Hz, 4 H, aryl$H$), 7.19 (m, 8 H, aryl$H$), 7.06 (m, 2 H, aryl$H$), 6.87 (overlapping m, 6 H, aryl$H$), 6.42 (t, J = 7.6 Hz,1 H, aryl$H$), 6.12 (d, J = 7.6 Hz, 2 H, aryl$H$), 6.02 (sept, J = 6.0 Hz, 1 H, CHMe_2), 4.22 (d, J = 14.4 Hz, 2 H, CH_2), 3.47 (d, J = 14.4 Hz, 2 H, CH_2), 1.36 (d, J = 6.0 Hz, 6 H, CHMe_2). ^{51}V (C_6D_6) δ: −572.09 ($\Delta\omega_{1/2}$ = 4 Hz).

3.3. Preparation of [V({OC(Ph)$_2$CH$_2$}$_2$(cis-NC$_5H_3$))$_2$] (**2**)

To LH_2 (1.00 g, 2.12 mmol) in THF (30 mL) was added NaH (0.10 g, 4.17 mmol), and the system was refluxed for 12 h. On cooling (to −78 °C), [$VCl_3(THF)_3$] (0.40 g, 1.07 mmol) was added, and the system was stirred for 12 h. The system was filtered, concentrated to about 20 mL, and left to stand at 0 °C to afford red crystals of **2**. Yield: 0.58 g, 55%. $C_{66}H_{54}N_2O_4V$ requires C 80.06, H 5.50, and N 2.83%. Found: C 80.51, H 5.55, N 2.80%. IR: 1975 w, 1886 w, 1815 w, 1667, 1601 m, 1579 w, 1317 m, 1261 s, 1151 m, 1097 s, 1021 s, 940 m, 918 w, 800 s, 722 m, 700 s, 639 w, 594 w, 547 w, 522 w, 492 w, 467 w. M.S. (MALDI): 836 (M—2Ph).

3.4. Preparation of [V({OC(Ph)₂CH₂}₂(trans-NC₅H₃))₂] (3·2THF)

As for **2**, but using [VCl₂(TMEDA)₂] (0.38 g, 1.07 mmol), LH₂ (1.00 g, 2.12 mmol) and NaH (0.10 g, 4.17 mmol) afforded **3**·2THF as red prisms. Yield: 0.45 g, 37%. $C_{74}H_{70}N_2O_6V$—THF (sample dried in vacuo for 1 h) requires C 79.15, H 5.88, N 2.63%. Found: C 79.46, H 6.02, N 2.33%. IR: 1667 w, 1631 w, 1601 w, 1551 w, 1413 m, 1304 m, 1261 s, 1094 bs, 1020 bs, 940 w, 918 w, 865 w, 800 s, 722 m, 700 m, 639 w, 598 w, 466 w.

3.5. Preparation of [VO(OnPr)L¹] (4)

To [VO(OnPr)₃] (0.68 mL, 3.00 mmol) and L¹H₂ (1.00 g, 2.98 mmol) was added toluene (30 mL), and then the system was refluxed for 12 h. On cooling, the volatiles were removed in vacuo, and the residue was extracted into warm MeCN (30 mL). On cooling to −20 °C, golden brown crystals of **4** formed. Yield: 0.89 g, 65%. $C_{24}H_{42}NO_4V$ requires C 62.73, H 9.21, and N 3.05%. Found: C 62.37, H 9.61, N 3.29%. IR: 2008 w, 1599 m, 1575 m, 1331 m, 1261 s, 1234 w, 1177 m, 1138 m, 1094 s, 1063 s, 1018 s, 981 s, 964 s, 913 w, 901 m, 863 w, 800 s, 772 m, 723 w, 708 m, 684 w, 658 m, 640 s, 628 s, 559 s, 494 w, 450 s, 410 s. ¹H NMR (C₆D₆) δ: 7.10 (1 H, aryl*H*), 6.82 (1 H, aryl*H*), 6.42 (1 H, aryl*H*), 5.20 (t, *J* = 6.4 Hz, 2 H, OCH₂), 3.62 (d, *J* = 14.0 Hz, 2 H, CH₂), 2.61 (d, *J* = 14.4 Hz, 2 H, CH₂), 2.05 (sept, *J* = 6.8 Hz, C*H*Me₂), 1.83 (m, *J* = 7.2 Hz, 2 H, OCH₂C*H*₂CH₃), 1.71 (sept, *J* = 6.8 Hz, C*H*Me₂), 1.25 (d, *J* = 6.8 Hz, 6 H, CH*Me*₂), 1.16 (d, *J* = 6.8 Hz, 6 H, CH*Me*₂), 1.05 (t, *J* = 7.2 Hz, 3 H, OCH₂CH₂C*H*₃), 0.76 (d, *J* = 6.8 Hz, 6 H, CH*Me*₂), 0.62 (d, *J* = 6.8 Hz, 6 H, CH*Me*₂). ⁵¹V (C₆D₆) δ: −544.70 ($\Delta\omega_{1/2}$ = 2.8 Hz).

3.6. Preparation of [VO(OiPr)L¹] (5)

To [VO(OiPr)₃] (0.71 mL, 3.01 mmol) and L¹H₂ (1.00 g, 2.98 mmol) was added toluene (30 mL), and then the system was refluxed for 12 h. On cooling, the volatiles were removed in vacuo, and the residue was extracted into warm MeCN (30 mL). On cooling to −20 °C, yellow crystals of **5** formed. Yield: 1.17 g, 85%. $C_{24}H_{42}NO_4V$ requires C 62.73, H 9.21, and N 3.05%. Found: C 61.92, H 9.55, N 3.16%.* IR: 1993 w, 1599 m, 1574 m, 1317 m, 1261 m, 1231 w, 1203 w, 1165 m, 1137 m, 1105 s, 1017 s, 972 s, 943 s, 905 m, 866 w, 838 m, 815 m, 770 m, 721 m, 705 m, 645 s. ¹H NMR (C₆D₆) δ: 6.93 (bm, 1 H, aryl*H*), 6.51 (bd, 2 H, aryl*H*), 5.73 (sept, *J* = 6.8 Hz, 1 H, C*H*Me₂), 3.61 (d, *J* = 14.4 Hz, 2 H, CH₂), 2.62 (d, *J* = 14.4 Hz, 2 H, CH₂), 2.06 (br sept, *J* = 6.0 Hz, 2 H, C*H*Me₂), 1.68 (sept, *J* = 6.8 Hz, 2 H, C*H*Me₂), 1.44 (d, *J* = 6.8 Hz, 6 H, CH*Me*₂), 1.26 (d, *J* = 6.8 Hz, 6 H, CH*Me*₂), 1.13 (d, *J* = 6.8 Hz, 6 H, CH*Me*₂), 0.74 (d, *J* = 6.8 Hz, 6 H, CH*Me*₂), 0.62 (d, *J* = 6.8 Hz, 6 H, CH*Me*₂). ⁵¹V (C₆D₆) δ: −572.81 ($\Delta\omega_{1/2}$ = 3.8 Hz). *Despite repeated attempts, this was the best fit for elemental analysis.

3.7. Preparation of [VO(μ-O)(L²)]₂ (6)

To [VO(OiPr)₃] (0.79 mL, 3.35 mmol) and L²H (1.00 g, 3.35 mmol) was added toluene (30 mL), and then the system was refluxed for 12 h. On cooling, the volatiles were removed in vacuo, and the residue was extracted into warm MeCN (30 mL). On cooling to −20 °C, small yellow crystals of **6** formed. Yield: 0.52 g, 41%. $C_{38}H_{50}N_4O_6V_2$ requires C 59.99, H 6.63, and N 7.37%. Found: C 60.70, H 7.07, N 7.45%. IR: 1639 w, 1593 s, 1578 s, 1304 m, 1261 s, 1203 w, 1155 m, 1097 s, 1019 s, 862 w, 794 s, 765 m, 666 w, 614 w, 581 w, 504 w. ¹H NMR (CD₃CN) δ: 7.47 (t, 4 H, *J* = 7.6 Hz, aryl*H*), 7.03 (d, 4 H, *J* = 7.6 Hz, aryl*H*), 6.91 (d, 4 H, *J* = 7.4 Hz, aryl*H*), 3.00 (s, 8 H, CH₂), 2.33 (s, 12 H, CH₃), 0.97 (s, 18 H, C(CH₃)₃). ⁵¹V (C₆D₆) δ: −539.3 ($\Delta\omega_{1/2}$ = 3520 Hz). Complex **5** is also available via the use of [VO(OEt)₃].

3.8. Procedure for ROP of ε-Caprolactone

A toluene solution of pre-catalyst (0.010 mmol, 1.0 mL toluene) was introduced into a Schlenk tube in the glove box at room temperature. The solution was stirred for 2 min, and then the appropriate equivalent of BnOH (from a pre-prepared stock solution of 1 mmol BnOH in 100 mL toluene) and the appropriate amount of ε-CL along with 1.5 mL toluene were added to the solution. For example, for Table 2, entry 1, a toluene solution of precatalyst **1** (0.010 mmol, 1.0 mL toluene) was introduced into a Schlenk tube, then 2 mL BnOH solution (1 mmol BnOH/100 mL toluene) and 20 mmol ε-CL along with 1.5 mL toluene was added to the solution. The reaction mixture was then placed into an oil/sand bath pre-heated at 130 °C, and the solution was stirred for the prescribed time (8 or 24 h). The polymerization mixture was quenched with the addition of an excess of glacial acetic acid (0.2 mL) to the solution, and the resultant solution was then poured into methanol (200 mL). The resultant polymer was then collected on filter paper and dried in vacuo.

3.9. Kinetic Studies

The polymerizations were carried out at 130 °C in toluene (2 mL) using 0.010 mmol of the complex. The molar ratio of monomer to initiator was fixed at 500:1, and at appropriate time intervals, 0.5 μL aliquots were removed (under N_2) and were quenched with wet $CDCl_3$. The percentage conversion of monomer to polymer was determined using ^{1}H NMR spectroscopy.

3.10. Procedure for Ethylene Polymerization

Ethylene polymerization experiments were performed in a steel 500 mL autoclave. The reactor was evacuated at 80 °C, cooled to 20 °C, and then charged with the freshly prepared solution of the co-catalyst in heptane/toluene. Pre-catalysts were introduced into the reactor in sealed glass ampoules containing 0.5 or 1.0 μmol of appropriate V-complex in 0.5 mL of solvent. After setting up the desired temperature and ethylene pressure, the reaction was started by breaking the ampoule with the pre-catalyst. During the polymerization, ethylene pressure (2 bar), temperature (70 °C), and stirring speed (2000 rpm) were maintained constant. After 30 min (during which time the ethylene consumption rate declined to nearly zero level), the reactor was opened to the atmosphere, and the polymeric product was dried in a fume-hood to a constant weight.

Polymerization conditions. For entry 1 of Table 5: V loading 1.0 μmol (dissolved in CH_2Cl_2), co-catalyst Et_2AlCl + ETA (molar ratio V: Et_2AlCl:ETA = 1:1000:500) in 50 mL of toluene + 100 mL of heptane, T pol 70 °C, P C_2H_4 = 2 bar, for 30 min. For entries 2–8 of Table 5: V complex was dissolved in toluene, co-catalyst Me_2AlCl + ETA (molar ratio V:Me_2AlCl:ETA = 1:1000:1000) in 100 mL of toluene + 100 mL of heptane; T pol 70 °C, P C_2H_4 = 2 bar, for 30 min.

4. Conclusions

The products resulting from the interaction of the potentially O,N,O-tridentate 6-bis(o-hydroxyaryl)pyridines 2,6-{HOC(X)$_2$CH$_2$}$_2$(NC$_5$H$_3$) (X = Ph, LH$_2$; iPr, L^{1}H$_2$) with the vanadyl trisalkoxides [VO(OR)$_3$] (R = n-Pr, i-Pr) have been isolated in good yield and characterized as [VO(Oi/n-Pr)L/L^1]. The use of LNa$_2$ with [VCl$_3$(THF)$_3$] afforded [VL$_2$] in which the pyridyl N atoms are positioned *cis*, whereas the use of [VCl$_2$(TMEDA)$_2$] afforded the regioisomer with *trans* pyridine rings. The related potentially N,O,N-tridentate bis(methylpyridine)-substituted alcohol (tBu)C(OH)[CH$_2$(C$_5$H$_3$Me-5)]$_2$, L^{2}H, on treatment with [VO(OR)$_3$] (R = Et, i-Pr) afforded [VO(μ-O)(L^2)]$_2$, in which L^2 binds in N,O-bidentate fashion. These products were screened for their ability to act as catalysts for the ring-opening polymerization of ε-caprolactone (ε-CL), δ-valerolactone (δ-VL), and *rac*-lactide (*r*-LA). In the case of ε-CL, products formed in solution were both linear and cyclic and of low molecular weight. Higher molecular weight products were formed when melt conditions were employed. In the case of δ-VL, linear and cyclic products were also formed, with the higher molecular weight products accessed when using systems employing [L/L^{1}VO(OR)], i.e., **1**, **4**, and **5**. For both PCL and PVL, the low molecular weights (versus the theoretical values) suggested transesterifaction side reactions had occurred. For *r*-LA, only the bis(methylpyridine)-substituted alcohol-derived complex **6** exhibited any appreciable activity. Complexes **1**, **4**, and **5** displayed moderate activities in ethylene polymerization in the presence of dimethylaluminium chloride (DMAC) activator and the re-activator trichloroethylacetate (ETA), affording linear high-MW polyethylene, whilst **6** was inactive under the conditions.

Supplementary Materials: The following supporting information can be downloaded at: https://www.mdpi.com/article/10.3390/catal13060988/s1, Figure S1: Molecular structure of L^1H_2; Figure S2: Intermolecular interaction and packing of L^1H_2; Figure S3: Alternative view of the molecular structure of V({OC(Ph)$_2$CH$_2$}$_2$(*trans*-NC$_5$H$_3$))$_2$]·2THF (**3**·2THF); Packing of **6**; Table S1: Crystallographic data for **1**, **2** and **3**·2THF; Table S2: Crystallographic data for **4–6** and L^1H_2; Figure S4: ^{51}V NMR spectra of **4** and **6**; Figure S5: Packing of **6**; Figure S6: MALDI-ToF of PCL using **5** under N$_2$ at 70 °C (entry 14, Table 1); Figure S7: ^{1}H NMR spectrum of PCL (using **1** in air, entry 5, Table 1); Figure S8: MALDI-ToF of PCL using **1** under air (entry 5, Table 1); Figure S9: MALDI-ToF of PCL using **5** as a melt under N$_2$ (entry 7, Table 2); Figure S10: ^{1}H NMR spectrum of PVL (using **4** under N$_2$, entry 6, Table 3); Figure S11: MALDI-ToF of PVL using **5** as a melt under air (entry 8, Table 3); Figure S12: MALDI-ToF of PVL using **5** as a melt under N$_2$ (entry 9,Table 3); Figure S13: ^{13}C NMR spectrum of the methine region of the PLA using **6** under N$_2$ (entry 9, Table 4); Figure S14: ^{13}C NMR spectrum of the methine region of the PLA using **6** under air (entry 10, Table 4) [66,67].

Author Contributions: M.R.J.E. formal analysis; W.C. formal analysis; C.R. conceptualization, funding acquisition, writing—review and editing. All authors have read and agreed to the published version of the manuscript.

Funding: The EPSRC National X-ray Crystallography Service (Southampton) is thanked for data collection. CR also thanks the British Council for funding a workshop, and the EPSRC (grant number EP/S025537/1) for funding.

Data Availability Statement: The supplementary crystallographic data can be found free of charge at the Cambridge Crystallographic Data Centre. Deposition numbers 2239107-2239113.

Acknowledgments: We thank the EPSRC National Crystallographic Service at Southampton for data collection. We thank the British Council Newton Fund for support. We thank V.C. Gibson (Imperial College) for the use of the synthesis of **3**.

Conflicts of Interest: The authors declare no conflict of interest. The funders had no role in the design of the study; in the collection, analyses, or interpretation of data; in the writing of the manuscript, or in the decision to publish the results.

References

1. Czigany, T.; Ronkay, F. The coronavirus and plastics. *Express Polym. Lett.* **2020**, *14*, 510–511. [CrossRef]
2. Jehanno, C.; Alty, J.W.; Roosen, M.; De Meester, S.; Dove, A.P.; Chen, E.Y.-X.; Leibfarth, F.A.; Sardon, H. Critical advances and future opportunities in upcycling commodity polymers. *Nature* **2022**, *603*, 803–814. [CrossRef] [PubMed]
3. Reichman, J.H. Intellectual Property in the Twenty-First Century: Will the Developing Countries Lead or Follow? In *Intellectual Property Rights: Legal and Economic Challenges for Development*; Cimoli, M., Dosi, G., Maskus, K.E., Okediji, R.L., Reichman, J.H., Eds.; Oxford University Press: Oxford, UK, 2014. [CrossRef]
4. Silva, A.L.P.; Prata, J.C.; Duarte, A.C.; Barceló, D.; Rocha-Santos, T. An urgent call to think globally and act locally on landfill disposable plastics under and after COVID-19 pandemic: Pollution prevention and technological (Bio) remediation solutions. *Chem. Eng. J.* **2021**, *426*, 131201. [CrossRef] [PubMed]
5. Albertsson, A.-C.; Varma, I.K. Recent Developments in Ring Opening Polymerization of Lactones for Biomedical Applications. *Biomacromolecules* **2003**, *4*, 1466–1486. [CrossRef]
6. Tian, H.; Tang, Z.; Zhuang, X.; Chen, X.; Jing, X. Biodegradable Synthetic Polymers: Preparation, functionalization and biomedical application. *Prog. Polym. Sci.* **2012**, *37*, 237–280. [CrossRef]
7. Kim, Y.K. The use of polyolefins in industrial and medical applications. In *The Textile Institute Book Series, Polyolefin Fibres*, 2nd ed.; Woodhead Publishing: Sawston, UK, 2017; pp. 135–155. ISBN 9780081011324. [CrossRef]
8. Arbaoui, A.; Redshaw, C. Metal Catalysts for ε-caprolactone polymerisation. *Polym. Chem.* **2010**, *1*, 801–826. [CrossRef]
9. Zhang, X.; Fevre, M.; Jones, G.O.; Waymouth, R.M. Catalysis as an Enabling Science for Sustainable Polymers. *Chem. Rev.* **2018**, *118*, 839–885. [CrossRef]
10. Chen, C. Designing catalysts for olefin polymerization and copolymerization: Beyond electronic and steric tuning. *Nat. Rev. Chem.* **2018**, *2*, 6–14. [CrossRef]
11. Dove, A.P. Organic Catalysis for Ring-Opening Polymerization. *ACS Macro Lett.* **2012**, *1*, 1409–1412. [CrossRef]
12. Kamber, N.E.; Jeong, W.; Waymouth, R.M.; Pratt, R.C.; Lohmeijer, B.G.G.; Hedrick, J.L. Organocatalytic Ring-Opening Polymerization. *Chem. Rev.* **2007**, *107*, 5813–5840. [CrossRef]

13. Williams, C.K.; Hillmyer, M.A. Polymers from Renewable Resources: A Perspective for a Special Issue of Polymer Reviews. *Polym. Rev.* **2008**, *48*, 1–10. [CrossRef]
14. Place, E.S.; George, J.H.; Williams, C.K.; Stevens, M.M. Synthetic Polymer Scaffolds for Tissue Engineering. *Chem. Soc. Rev.* **2009**, *38*, 1139–1151. [CrossRef]
15. Labet, M.; Thielemans, W. Synthesis of Polycaprolactone: A review. *Chem. Soc. Rev.* **2009**, *38*, 3484–3504. [CrossRef]
16. Lecomte, P.; Jér, C. *Synthetic Biodegradable Polymers*; Rieger, B., Künkel, A., Coates, G., Reichardt, R., Dinjus, E., Zevaco, T., Eds.; Springer: Berlin/Heidelberg, Germany, 2011; Volume 245, pp. 173–211. ISBN 978-3642271533.
17. Redshaw, C. Metallocalixarene catalysts: α-olefin polymerization and ROP of cyclic esters. *Dalton Trans.* **2016**, *45*, 9018–9030. [CrossRef]
18. Redshaw, C. Use of Metal Catalysts Bearing Schiff Base Macrocycles for the Ring Opening Polymerization (ROP) of Cyclic Esters. *Catalysts* **2017**, *7*, 165. [CrossRef]
19. Redshaw, C.; Warford, L.; Dale, S.H.; Elsegood, M.R.J. Vanadyl complexes bearing bi- and triphenolate chelate ligands: Highly active ethylene polymerisation procatalysts. *Chem. Commun.* **2004**, 1954–1955. [CrossRef]
20. Redshaw, C.; Rowan, M.A.; Homden, D.M.; Dale, S.H.; Elsegood, M.R.J.; Matsui, S.; Matsuura, S. Vanadyl C and N-capped tris(phenolate) complexes: Influence of pro-catalyst geometry on catalytic activity. *Chem. Commun.* **2006**, 3329–3331. [CrossRef]
21. Redshaw, C.; Rowan, M.A.; Warford, L.; Homden, D.M.; Arbaoui, A.; Elsegood, M.R.J.; Dale, S.H.; Yamato, T.; Casas, C.P.; Matsui, S.; et al. Oxo- and Imidovanadium Complexes Incorporating Methylene- and Dimethyleneoxa-Bridged Calix[3]- and -[4]arenes: Synthesis, Structures and Ethylene Polymerisation Catalysis. *Chem.-Eur. J.* **2007**, *13*, 1090–1107. [CrossRef]
22. Homden, D.; Redshaw, C.; Warford, L.; Hughes, D.L.; Wright, J.A.; Dale, S.H.; Elsegood, M.R.J. Synthesis, structure and ethylene polymerisation behaviour of vanadium(iv and v) complexes bearing chelating aryloxides. *Dalton Trans.* **2009**, 8900–8910. [CrossRef]
23. Lorber, C.; Despagnet-Ayoub, E.; Vendier, L.; Arbaoui, A.; Redshaw, C. Amine influence in vanadium-based ethylene polymerisation pro-catalysts bearing bis(phenolate) ligands with 'pendant' arms. *Catal. Sci. Technol.* **2011**, *1*, 489–494. [CrossRef]
24. Nomura, K.; Zhang, S. Design of Vanadium Complex Catalysts for Precise Olefin Polymerization. *Chem. Rev.* **2011**, *111*, 2342–2362. [CrossRef] [PubMed]
25. Wu, J.-Q.; Li, Y.-S. Well-defined vanadium complexes as the catalysts for olefin polymerization. *Coord. Chem. Rev.* **2011**, *255*, 2303–2314. [CrossRef]
26. Ma, J.; Zhao, K.-Q.; Walton, M.J.; Wright, J.A.; Frese, J.W.A.; Elsegood, M.R.J.; Xing, Q.; Sun, W.-H.; Redshaw, C. Vanadyl complexes bearing bi-dentate phenoxyimine ligands: Synthesis, structural studies and ethylene polymerization capability. *Dalton Trans.* **2014**, *43*, 8300–8310. [CrossRef]
27. Phillips, A.M.F.; Suo, H.; Silva, M.D.F.C.G.D.; Pombeiro, A.J.; Sun, W.-H. Recent developments in vanadium-catalyzed olefin coordination polymerization. *Coord. Chem. Rev.* **2020**, *416*, 213332. [CrossRef]
28. Qian, J.; Comito, R.J. A Robust Vanadium(V) Tris(2-pyridyl)borate Catalyst for Long-Lived High-Temperature Ethylene Polymerization. *Organometallics* **2021**, *40*, 1817–1821. [CrossRef]
29. Wu, R.; Niu, Z.; Huang, L.; Xia, Z.; Feng, Z.; Qi, Y.; Dai, Q.; Cui, L.; He, J.; Bai, C. Vanadium complexes bearing the bulky bis(imino)pyridine ligands: Good thermal stability toward ethylene polymerization. *Eur. Polym. J.* **2022**, *180*, 111569. [CrossRef]
30. Suo, H.; Phillips, A.M.F.; Satrudhar, M.; Martins, L.M.D.R.S.; Silva, M.d.F.G.d.; Pombeiro, A.J.L.; Han, M.; Sun, W. Achieving ultra-high molecular weight polyethylenes by vanadium aroylhydrazine-arylolates. *J. Polym. Sci.* **2023**, *61*, 482–490. [CrossRef]
31. Qian, J.; Comito, R.J. Ethylene Polymerization with Thermally Robust Vanadium(III) Tris(2-pyridyl)borate Complexes. *Organometallics* **2023**. [CrossRef]
32. Homden, D.M.; Redshaw, C.; Hughes, D.L. Vanadium Complexes Possessing $N_2O_2S_2$-Based Ligands: Highly Active Procatalysts for the Homopolymerization of Ethylene and Copolymerization of Ethylene/1-hexene. *Inorg. Chem.* **2007**, *46*, 10827–10839. [CrossRef]
33. Redshaw, C.; Walton, M.; Michiue, K.; Chao, Y.; Walton, A.; Elo, P.; Sumerin, V.; Jiang, C.; Elsegood, M.R.J. Vanadyl calix[6]arene complexes: Synthesis, structural studies and ethylene homo-(co-)polymerization capability. *Dalton Trans.* **2015**, *44*, 12292–12303. [CrossRef]
34. Redshaw, C.; Walton, M.J.; Elsegood, M.R.J.; Prior, T.J.; Michiue, K. Vanadium(v) tetra-phenolate complexes: Synthesis, structural studies and ethylene homo-(co-)polymerization capability. *RSC Adv.* **2015**, *5*, 89783–89796. [CrossRef]
35. Redshaw, C.; Walton, M.J.; Lee, D.S.; Jiang, C.; Elsegood, M.R.J.; Michiue, K. Vanadium(V) Oxo and Imido Calix[8]arene Complexes: Synthesis, Structural Studies, and Ethylene Homo/Copolymerisation Capability. *Chem.-Eur. J.* **2015**, *21*, 5199–5210. [CrossRef]
36. Diteepeng, N.; Tang, X.; Hou, X.; Li, Y.-S.; Phomphrai, K.; Nomura, K. Ethylene polymerisation and ethylene/norbornene copolymerisation by using aryloxo-modified vanadium(v) complexes containing 2,6-difluoro-, dichloro-phenylimido complexes. *Dalton Trans.* **2015**, *44*, 12273–12281. [CrossRef]
37. Zanchin, G.; Bertini, F.; Vendier, L.; Ricci, G.; Lorber, C.; Leone, G. Copolymerization of ethylene with propylene and higher α-olefins catalyzed by (imido)vanadium(iv) dichloride complexes. *Polym. Chem.* **2019**, *10*, 6200–6216. [CrossRef]
38. Ochędzan-Siodłak, W.; Siodłak, D.; Banaś, K.; Halikowska, K.; Wierzba, S.; Doležal, K. Naturally Occurring Oxazole Structural Units as Ligands of Vanadium Catalysts for Ethylene-Norbornene (Co)polymerization. *Catalysts* **2021**, *11*, 923. [CrossRef]

39. Wu, R.; Niu, Z.; Huang, L.; Yang, Y.; Xia, Z.; Fan, W.; Dai, Q.; Cui, L.; He, J.; Bai, C. Thermally stable vanadium complexes supported by the iminophenyl oxazolinylphenylamine ligands: Synthesis, characterization and application for ethylene (co-)polymerization. *Dalton Trans.* **2021**, *50*, 16067–16075. [CrossRef]
40. Białek, M.; Fryga, J.; Spaleniak, G.; Matsko, M.A.; Hajdasz, N. Ethylene homo- and copolymerization catalyzed by vanadium, zirconium, and titanium complexes having potentially tridentate Schiff base ligands. *J. Catal.* **2021**, *400*, 184–194. [CrossRef]
41. Nie, J.; Ren, F.; Li, Z.; Tian, K.; Zou, H.; Hou, X. Design and synthesis of binuclear vanadium catalysts for copolymerization of ethylene and polar monomers. *Polym. Chem.* **2022**, *13*, 3876–3881. [CrossRef]
42. Ge, F.; Dan, Y.; Al-Khafaji, Y.; Prior, T.J.; Jiang, L.; Elsegood, M.R.J.; Redshaw, C. Vanadium(v) phenolate complexes for ring opening homo- and co-polymerisation of ε-caprolactone, l-lactide and rac-lactide. *RSC Adv.* **2016**, *6*, 4792–4802. [CrossRef]
43. Białek, M.; Fryga, J.; Spaleniak, G.; Dziuk, B. Ring opening polymerization of ε-caprolactone initiated by titanium and vanadium complexes of ONO-type schiff base ligand. *J. Polym. Res.* **2021**, *28*, 79. [CrossRef]
44. Arbaoui, A.; Redshaw, C.; Homden, D.M.; Wright, J.A.; Elsegood, M.R.J. Vanadium-based imido-alkoxide pro-catalysts bearing bisphenolate ligands for ethylene and ε-caprolactone polymerisation. *Dalton Trans.* **2009**, 8911–8922. [CrossRef] [PubMed]
45. Clowes, L.; Redshaw, C.; Hughes, D.L. Vanadium-Based Pro-Catalysts Bearing Depleted 1,3-Calix[4]arenes for Ethylene or ε-Caprolactone Polymerization. *Inorg. Chem.* **2011**, *50*, 7838–7845. [CrossRef] [PubMed]
46. Clowes, L.; Walton, M.; Redshaw, C.; Chao, Y.; Walton, A.; Elo, P.; Sumerin, V.; Hughes, D.L. Vanadium(iii) phenoxyimine complexes for ethylene or ε-caprolactone polymerization: Mononuclear versus binuclear pre-catalysts. *Catal. Sci. Technol.* **2013**, *3*, 152–160. [CrossRef]
47. Ma, J.; Zhao, K.-Q.; Walton, M.; Wright, J.A.; Hughes, D.L.; Elsegood, M.R.J.; Michiue, K.; Sun, X.; Redshaw, C. Tri- and tetra-dentate imine vanadyl complexes: Synthesis, structure and ethylene polymerization/ring opening polymerization capability. *Dalton Trans.* **2014**, *43*, 16698–16706. [CrossRef] [PubMed]
48. Xing, T.; Prior, T.J.; Elsegood, M.R.J.; Semikolenova, N.V.; Soshnikov, I.E.; Bryliakov, K.P.; Chen, K.; Redshaw, C. Vanadium complexes derived from oxacalix[6]arenes: Structural studies and use in the ring opening homo-/co-polymerization of ε-caprolactone/δ-valerolactone and ethylene polymerization. *Catal. Sci. Technol.* **2021**, *11*, 624–636. [CrossRef]
49. Ishikura, H.; Neven, R.; Lange, T.; Galetová, A.; Blom, B.; Romano, D. Developments in vanadium-catalysed polymerisation reactions: A review. *Inorg. Chim. Acta* **2021**, *515*, 120047. [CrossRef]
50. Cusi, K.; Cukier, S.; DeFronzo, R.A.; Torres, M.; Puchulu, F.M.; Redondo, J.C.P. Vanadyl Sulfate Improves Hepatic and Muscle Insulin Sensitivity in Type 2 Diabetes. *J. Clin. Endocrinol. Metab.* **2001**, *86*, 1410–1417. [CrossRef]
51. Redshaw, C.; Elsegood, M.R.J.; Wright, J.A.; Baillie-Johnson, H.; Yamato, T.; De Giovanni, S.; Mueller, A. Cellular uptake of a fluorescent vanadyl sulfonylcalix[4]arene. *Chem. Commun.* **2012**, *48*, 1129–1131. [CrossRef]
52. Suzuki, N.; Kobayashi, G.; Hasegawa, T.; Masuyama, Y. Syntheses and structures of titanium complexes having O,N,O-tridentate ligands and their catalytic ability for ethylene polymerization. *J. Organomet. Chem.* **2012**, *717*, 23–28. [CrossRef]
53. Lai, F.-J.; Huang, T.-W.; Chang, Y.-L.; Chang, H.-Y.; Lu, W.-Y.; Ding, S.; Chen, H.-Y.; Chiu, C.-C.; Wu, K.-H. Titanium complexes bearing 2,6-Bis(o-hydroxyalkyl)pyridine ligands in the ring-opening polymerization of L-Lactide and ε-caprolactone. *Polymer* **2020**, *204*, 122860. [CrossRef]
54. Nakayama, Y.; Ikushima, N.; Nakamura, A. Cis-Specific Polymerization of Norbornene Catalyzed by Tungsten Based Complex Catalysts Bearing an O–N–O Tridentate Ligand. *Chem. Lett.* **1997**, *26*, 861–862. [CrossRef]
55. Evans, W.J.; Anwander, R.; Berlekamp, U.H.; Ziller, J.W. Synthesis and Structure of Lanthanide Complexes Derived from the O,N-Chelating, Bis(methylpyridine)-Substituted Alcohol HOC(CMe3)(2-CH2NC5H3Me-6)2. *Inorg. Chem.* **1995**, *34*, 3583–3588. [CrossRef]
56. Gibson, V.C.; Newton, C.; Redshaw, C.; Solan, G.A.; White, A.J.P.; Williams, D.J. Low valent chromium complexes bearing N,O-chelating pyridyl-enolate ligands [OC(Bu^t)(=2-CHN$_5$H$_3$Me-x)]$^−$ (x = 3–6). *Dalton Trans.* **2003**, 4612–4617. [CrossRef]
57. Berg, J.M.; Holm, R.H. Model for the active site of oxo-transfer molybdoenzymes: Synthesis, structure, and properties. *J. Am. Chem. Soc.* **1985**, *107*, 917–925. [CrossRef]
58. Addison, A.W.; Rao, T.N.; Reedijk, J.; van Rijn, J.; Verschoor, G.C. Synthesis, structure, and spectroscopic properties of copper(II) compounds containing nitrogen–sulphur donor ligands; the crystal and molecular structure of aqua[1,7-bis(N-methylbenzimidazol-2′-yl)-2,6-dithiaheptane]copper(II) perchlorate. *J. Chem. Soc. Dalton Trans.* **1984**, *7*, 1349–1356. [CrossRef]
59. Li, H.; Wang, C.; Bai, F.; Yue, J.; Woo, H.-G. Living Ring-Opening Polymerization of l-Lactide Catalyzed by Red-Al. *Organometallics* **2004**, *23*, 1411–1415. [CrossRef]
60. Soshnikov, I.E.; Semikolenova, N.V.; Shubin, A.A.; Bryliakov, K.P.; Zakharov, V.A.; Redshaw, C.; Talsi, E.P. EPR Monitoring of Vanadium(IV) Species Formed upon Activation of Vanadium(V) Polyphenolate Precatalysts with AlR$_2$Cl and AlR$_2$Cl/Ethyltrichloroacetate (R = Me, Et). *Organometallics* **2009**, *28*, 6714–6720. [CrossRef]
61. Soshnikov, I.E.; Semikolenova, N.V.; Bryliakov, K.P.; Shubin, A.A.; Zakharov, V.A.; Redshaw, C.; Talsi, E.P. An EPR Study of the V(IV) Species Formed Upon Activation of a Vanadyl Phenoxyimine Polymerization Catalyst with AlR$_3$ and AlR$_2$Cl (R = Me, Et). *Macromol. Chem. Phys.* **2009**, *210*, 542–548. [CrossRef]
62. Soshnikov, I.E.; Semikolenova, N.V.; Bryliakov, K.; Zakharov, V.A.; Redshaw, C.; Talsi, E.P. An EPR study of the vanadium species formed upon interaction of vanadyl N and C-capped tris(phenolate) complexes with AlEt$_3$ and AlEt$_2$Cl. *J. Mol. Catal. A Chem.* **2009**, *303*, 23–29. [CrossRef]

63. Deng, S.; Liu, Z.; Liu, B.; Jin, Y. Unravelling the Role of Al-alkyl Cocatalyst for the VO_x/SiO_2 Ethylene Polymerization Catalyst: Diethylaluminum Chloride vs. Triethylaluminum. *ChemCatChem* **2021**, *13*, 2278–2292. [CrossRef]
64. Manzer, L. Tetrahydrofuran Complexes of Selected Early Transition-Metals. In *Inorganic Syntheses*; Fackler, J.P., Ed.; John Wiley & Sons: Hoboken, NJ, USA, 1982; Volume 21, pp. 135–140. ISBN 978-0-470-13287-6.
65. Edema, J.J.H.; Stauthamer, W.; Van Bolhuis, F.; Gambarotta, S.; Smeets, W.J.J.; Spek, A.L. Novel vanadium(II) amine complexes: A facile entry in the chemistry of divalent vanadium. Synthesis and characterization of mononuclear L4VCl2 [L = amine, pyridine]: X-ray structures of trans-(TMEDA)2VCl2 [TMEDA = N,N,N′,N′-tetramethylethylenediamine] and trans-Mz2V(py)2 [Mz = o-C6H4CH2N(CH3)2, py = pyridine]. *Inorg. Chem.* **1990**, *29*, 1302–1306. [CrossRef]
66. Sheldrick, G.M. SHELXT-Integrated Space-Group and Crystal-Structure Determination. *Acta Crystallogr. Sect. A* **2015**, *A71*, 3–8. [CrossRef]
67. Sheldrick, G.M. Crystal Structure Refinement with SHELXL. *Acta Crystallogr. C* **2015**, *C71*, 3–8. [CrossRef]

Disclaimer/Publisher's Note: The statements, opinions and data contained in all publications are solely those of the individual author(s) and contributor(s) and not of MDPI and/or the editor(s). MDPI and/or the editor(s) disclaim responsibility for any injury to people or property resulting from any ideas, methods, instructions or products referred to in the content.

Article

Overview of Catalysts with MIRA21 Model in Heterogeneous Catalytic Hydrogenation of 2,4-Dinitrotoluene

Alexandra Jakab-Nácsa [1,2], Viktória Hajdu [2,3], László Vanyorek [2], László Farkas [1,2] and Béla Viskolcz [2,3,*]

1. BorsodChem Ltd., Bolyai tér 1, H-3700 Kazincbarcika, Hungary
2. Institute of Chemistry, Faculty of Materials and Chemical Engineering, University of Miskolc, H-3515 Miskolc, Hungary
3. Higher Education and Industrial Cooperation Centre, University of Miskolc, H-3515 Miskolc, Hungary
* Correspondence: bela.viskolcz@uni-miskolc.hu

Abstract: Although 2,4-dinitrotoluene (DNT) hydrogenation to 2,4-toluenediamine (TDA) has become less significant in basic and applied research, its industrial importance in polyurethane production is indisputable. The aim of this work is to characterize, rank, and compare the catalysts of 2,4-dinitrotoluene catalytic hydrogenation to 2,4-toluenediamine by applying the Miskolc Ranking 21 (MIRA21) model. This ranking model enables the characterization and comparison of catalysts with a mathematical model that is based on 15 essential parameters, such as catalyst performance, reaction conditions, catalyst conditions, and sustainability parameters. This systematic overview provides a comprehensive picture of the reaction, technological process, and the previous and new research results. In total, 58 catalysts from 15 research articles were selected and studied with the MIRA21 model, which covers the entire scope of DNT hydrogenation catalysts. Eight catalysts achieved the highest ranking (D1), whereas the transition metal oxide-supported platinum or palladium catalysts led the MIRA21 catalyst ranking list.

Keywords: hydrogenation; catalyst ranking; catalyst comparison; 2,4-dinitrotoluene; 2,4-toluenediamine

Citation: Jakab-Nácsa, A.; Hajdu, V.; Vanyorek, L.; Farkas, L.; Viskolcz, B. Overview of Catalysts with MIRA21 Model in Heterogeneous Catalytic Hydrogenation of 2,4-Dinitrotoluene. *Catalysts* **2023**, *13*, 387. https://doi.org/10.3390/catal13020387

Academic Editor: Carl Redshaw

Received: 27 January 2023
Revised: 2 February 2023
Accepted: 6 February 2023
Published: 10 February 2023

Copyright: © 2023 by the authors. Licensee MDPI, Basel, Switzerland. This article is an open access article distributed under the terms and conditions of the Creative Commons Attribution (CC BY) license (https://creativecommons.org/licenses/by/4.0/).

1. Introduction

Polyurethanes, referred to as urethanes, PUs, or PUR, are characterized by the urethane linking –NH–C (= O)–O–, which is established by the reaction of the organic isocyanate (NCO) groups and hydroxyl (OH) groups [1]. Due to their versatility and excellent mechanical, chemical, physical, and biological properties, they have a wide range of applications and a variety of uses, such as in appliances, automotive, construction, furniture, clothing, and the wood industries. Although the impact of COVID-19 has been startling, the global polyurethane market size was USD 56.45 billion in 2020 and it is projected to grow [2]. The rising demand for foams in furniture and in the construction industry has been driving the toluene diisocyanate (TDI) market growth.

TDI is one of the main materials of polyurethane production. TDI is produced in three different steps: the nitration of toluene, the dinitrotoluene hydrogenation to toluenediamine (TDA), and in the phosgenation of diaminotoluene. The general industrial process of TDA formation is the catalytic hydrogenation of dinitrotoluene in the liquid phase in the presence of a catalyst. Six isomers of TDA can be generated, but the major intermediate of TDI production is 2,4-toluenediamine (2,4-TDA).

In addition to the production volume and the versatility of its application, the industrial importance of TDA production is also shown by its patent history. A search on the Google Patents website using the keywords 'dinitrotoluene', 'hydrogenation', and 'toluenediamine' yields 400 patents that have been published since 1953 [3]. While the first patents in the 1950s described some general reaction conditions and some catalyst components, the newest patents provide much more detailed descriptions of multicomponent catalysts, their composition, and their preparation [4–10]. Although the latest patents describe the high performance catalysts comprising activated metal, one or more auxiliary metals, and a special support material such as oxide [11], the most commonly used catalyst in the industry is the nickel catalyst [12]. Despite the fewer number of published scientific research papers [13], a high conversion and selectivity were achieved with the catalysts of many different formulations of nickel, platinum, or palladium on carbon, oxide, or zeolite supports [14–19]. However, in addition to the catalytic performance, sustainability parameters also play an increasingly important role in the chemical industry, such as reversibility and stability [20–22]. There are many steps between the fundamental research on catalysts to their industrial application. Nevertheless, new scientific findings are essential for the development of applied technological innovations if the new knowledge is to be used effectively [23].

The Miskolc Ranking 21 (MIRA21) model is a new, multi-step, functional mathematical method to extract the knowledge from the heterogeneous catalyst data through the catalyst characterization, comparison, and ranking of a series of catalysts [24]. In our previous work we discussed the method and application possibilities through the reaction of nitrobenzene hydrogenation to aniline. The ranking model applies a fifteen-parameter descriptor system to facilitate the comparison of the experimental and scientific publication results of a determined reaction to support catalyst development. The parameters of the descriptor system can be divided into four groups: catalyst performance, reaction conditions, catalyst conditions, and sustainability parameters. The model qualifies and ranks the catalysts based on these parameters.

This overview summarized the advances in the selective hydrogenation of dinitrotoluene to form toluenediamine, based on the catalysts used to carry out this process in the last 50 years. As the focus point of this work, we characterized and ranked 58 catalysts from 15 articles according to the MIRA21 model to make the systematic comparison of them.

Figure 1 describes the technological process:

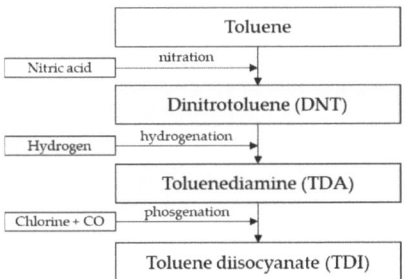

Figure 1. TDI production process from toluene [25].

The production of TDI is carried out in a three-step continuous process (Figure 1). Dinitrotoluene is produced in the first step by the nitration of toluene. The second and key step is the catalytic hydrogenation of dinitrotoluene to toluene diamine. In the last step, toluene diamine is phosgenated to form TDI.

The formation of DNT by the mixed acid nitration of toluene occurred at atmospheric pressure and between 40 °C and 70 °C. The main product of the process is a mixture of the 2,4- and 2,6-dinitrotoluene isomer mixture (Figure 2) [26]. These are the starting reagents for hydrogenation. The side products of the reaction are 2,3- and 3,4-DNT isomers, whereas the 2,5- and 3,5- isomers and other byproducts can also be found in smaller quantities.

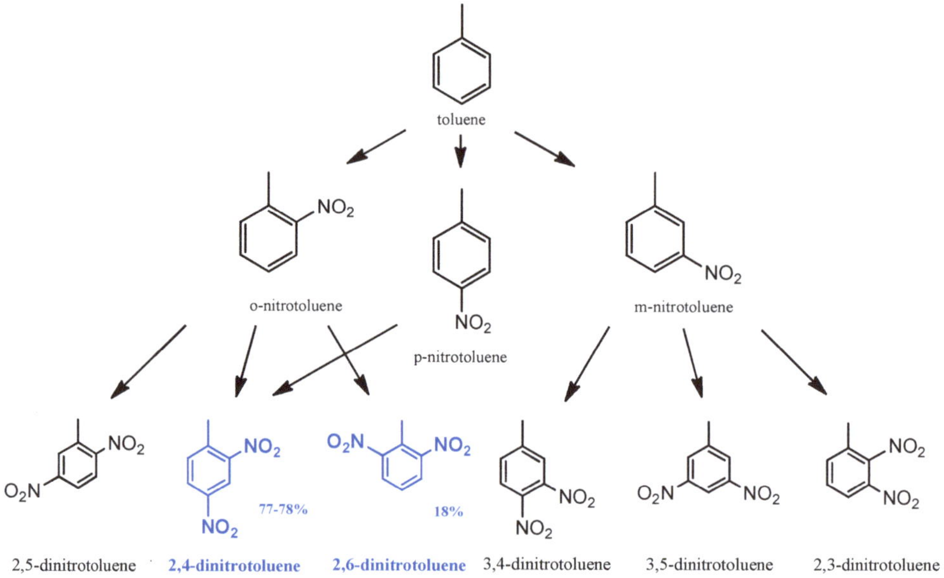

Figure 2. Raw material of hydrogenation process [26].

The second step of the industrial process is the catalytic hydrogenation to dinitrotoluene to toluenediamine using a solid catalyst at a high pressure and high temperature (100–150 °C, 5–8 bar). This step was previously carried out in the presence of iron filings and aqueous hydrochloric acid [27], but today it is hydrogenated using a Ra-Ni or Pd/C catalyst. In strong industrial conditions (high pressure and temperature), an extremely high-quality product is produced with a high yield. Furthermore, Figure 3 shows the general reaction equation with the main product of dinitrotoluene hydrogenation.

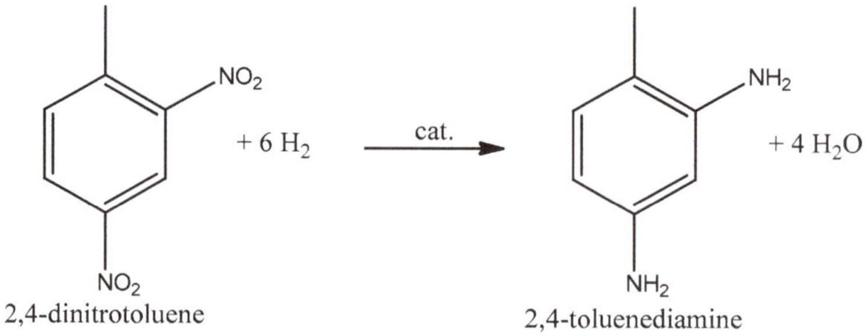

Figure 3. General reaction equation with the main 2,4 isomer [27].

The process occurred in a continuously stirred tank reactor where the DNT isomer mixture usually reacts with hydrogen gas in the presence of the supported precious metal catalyst in a TDA/water medium. In order to achieve a high conversion, the correct catalyst composition and reaction conditions (temperature, pressure, etc.) are crucial [28]. The spent catalyst is removed from the system through a catalyst filter and the new catalyst is added. It is important that the catalyst can be easily removed and regenerated. Westerwerp et al. made a pilot installation of a 2,4-DNT synthesis plant and studied the reactor design and operation process [29–31]. The experiments took place in a continuously stirred, three-phase slurry reactor with an evaporating solvent. They mentioned that, in addition to a good catalyst, it is important to choose the ideal hydrogenation reactor unit and optimal reaction parameters, and to solve the deactivation problem of the catalyst.

2. Reaction Mechanism and Kinetics

The kinetics and reaction mechanism of the catalytic hydrogenation of 2,4-dinitrotoluene to 2,4-toluenediamine was investigated by several research groups [15,16,32–34]. In the 1990s, Janssen et al. studied the reaction scheme and modelled the reaction rates and catalyst activity to evaluate the performance of a batch slurry reactor at 308–357 K and over the pressure range of 0–4 MPa [35,36]. The reaction rates are described by the Langmuir-Hinshelwood model. They found that 2,4-dinitrotoluene can be converted to 2,4-toluenediamine through two parallel pathways with consecutive reaction steps. They found that 4-hydroxyamino-2-nitrotoluene, 4-amino-2-nitrotoluene, and 2-amino-4-nitrotoluene are the most stable intermediates, but the presence of 2-amino-4-hydroxyaminotoluene and another azoxy compound were also observed. One of the reaction pathways is the direct conversion of an o-nitro group to an amino group. The other one is the conversion of the p-nitro group to an amino group in a two-step reaction.

While Janssen et al. was the first to describe the two reaction pathways, Neri et al. wrote a more complex reaction mechanism [37,38]. Neri et al. investigated this hydrogenation reaction over a supported Pd/C catalyst and found that 4-hydroxyamino-2-nitrotoluene, 2-amino-4-nitrotoluene, and 4-amino-2-nitrotoluene can form directly from 2,4-dinitrotoluene. Figure 4 compares the Janssen et al. and Neri et al. reaction schemes. The latter found that the hydrogenation of the hydroxylamine intermediate occurred via a triangular reaction pathway. Their further studies focused on 2-hydroxyamino-4-nitrotoluene as a reaction intermediate, which accumulates in the reaction mixture instead of 2-hydroxiamino-4-nitrotoluene [37,39,40]. It was shown that the formation of the nitro group depends on the presence of electron-donating substituents and steric effects [41].

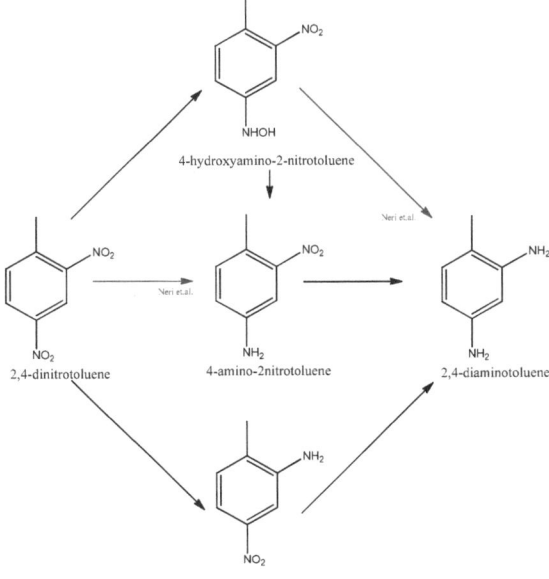

Figure 4. Dinitrotoluene hydrogenation pathways according to Janssen et al. and Neri et al. (Blue lines shows the new pathways comparable to Janssen et al.'s results.) [37].

Electronic structure computational studies can be a great help in studying the reaction mechanism of three-phase catalytic hydrogenation reactions. In such a study, Barone et al. applied the Monte Carlo algorithm to simulate the batch hydrogenation of 2,4-dinitrotoluene on a carbon-supported palladium catalyst [31,42,43]. They investigated the influence of the molecular adsorption modes, the steric hindrance, and the metal dispersion on the reaction mechanism. They found that the steric hindrance of the different surface species had the largest influence on the mechanism.

Hajdu et al. worked on a new catalyst that contains precious metal on chromium-oxide nanowires for 2,4-toluenediamine synthesis [44]. In our previous work, we examined and described a possible reaction mechanism based on the GC-MS results. Our study confirms the mechanism by Neri et.al., as we detected the presence of nitroso and hydroxylamine compounds (Figure 5).

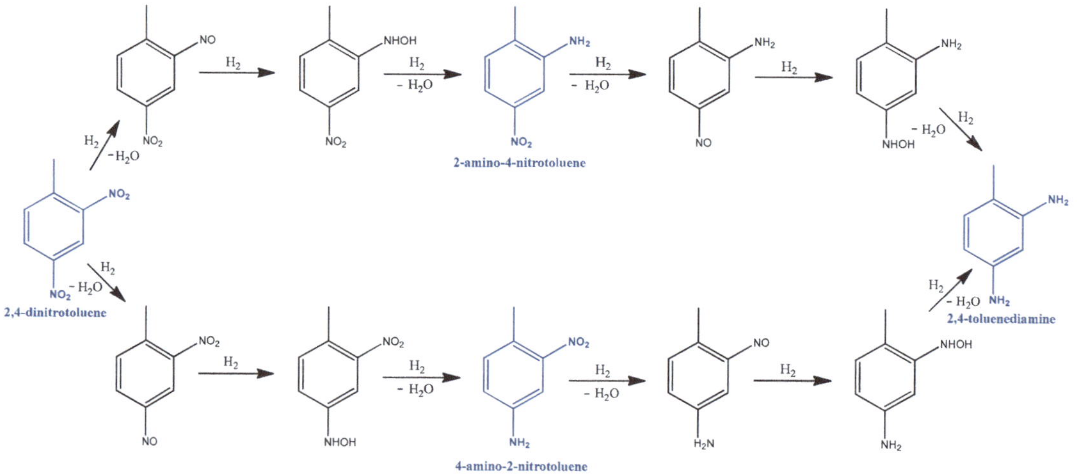

Figure 5. Reaction mechanism according to Hajdu et al. [44].

According to our results, 2,2-dinitro-4,4-azoxytoluene was found in the system (Figure 6), which could form through the reaction between the nitroso and hydroxylamine functional groups. In addition to the two semi-hydrogenated intermediates (4-amino-2-nitrotoluene, 2-amino-4-nitrotoluene), we detected other side products, which further supports the reaction mechanism of Neri et al. Figure 7 shows the detected and assumed side products obtained during the formation of TDA.

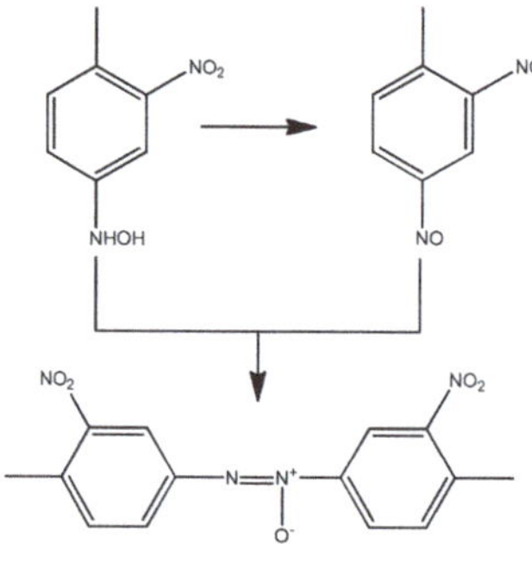

Figure 6. Formation of 2,2-dinitro-4,4-azoxytoluene [44].

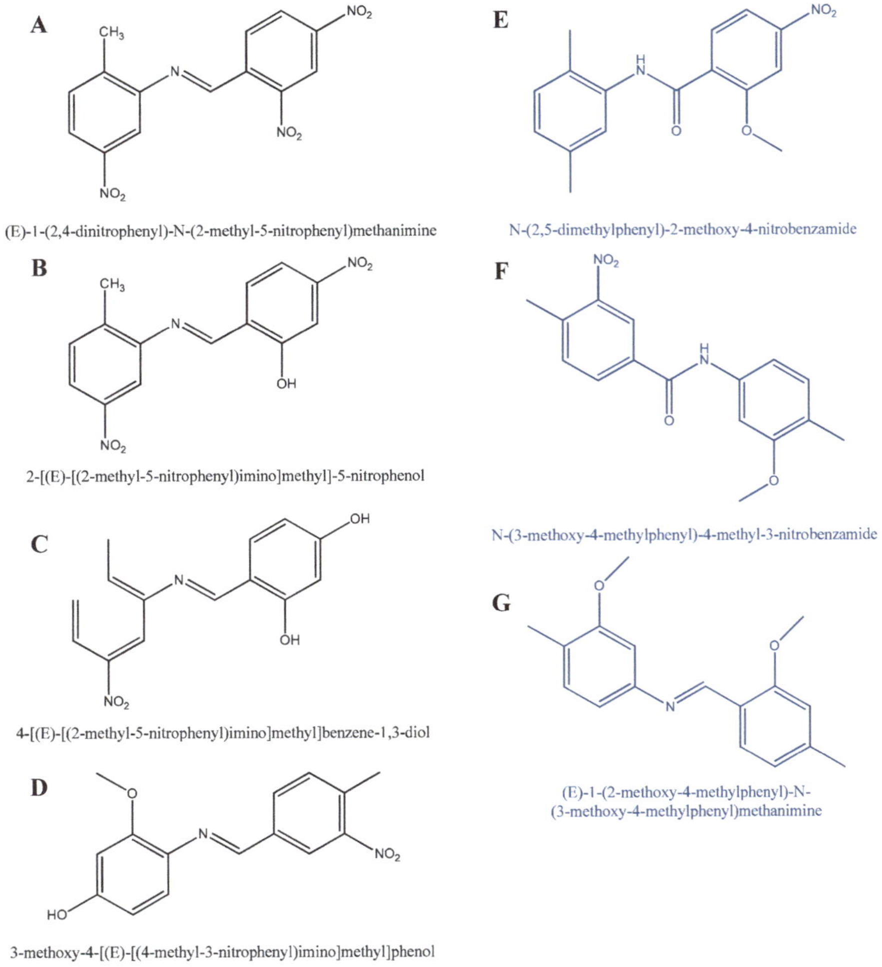

A (E)-1-(2,4-dinitrophenyl)-N-(2-methyl-5-nitrophenyl)methanimine

B 2-[(E)-[(2-methyl-5-nitrophenyl)imino]methyl]-5-nitrophenol

C 4-[(E)-[(2-methyl-5-nitrophenyl)imino]methyl]benzene-1,3-diol

D 3-methoxy-4-[(E)-[(4-methyl-3-nitrophenyl)imino]methyl]phenol

E N-(2,5-dimethylphenyl)-2-methoxy-4-nitrobenzamide

F N-(3-methoxy-4-methylphenyl)-4-methyl-3-nitrobenzamide

G (E)-1-(2-methoxy-4-methylphenyl)-N-(3-methoxy-4-methylphenyl)methanimine

Figure 7. Possible side products of the TDA synthesis according to Hajdu et al.'s research (black line—detected molecules, blue line—assumed molecules) [44].

We also demonstrated that E-1-(2,4-dinitrophenyl)-N-(2-methyl-5-nitrophenyl) methanimine and 2-[(E)-[(2-methyl-5-nitrophenyl)imino]methyl]-5-nitrophenol was formed by the water loss in the condensation reaction (**A** and **B**). Molecule **C** was formed by the reaction between 4-methylbenzene-1,3-diol and 2-nitroso-4-nitrotoluene. As shown in Figure 7, 2-nitro-4-nitrosotoluene reacted with 2-methoxy-4-methylphenol to yield compound **D**. Isomers **E** and **F** were formed by the reaction between 2-methoxy-1-methyl-4-nitrobenzene and dimethyl-2-nitrobenzene. Compound **G** was formed by the reaction between 2-methoxy-1,4-dimethylbenzene and 2-methoxy-4-nitrosotoluene.

3. Results and Discussion of TDA Synthesis Catalysts

The hydrogenation of 2,4-dinitrotoluene to 2,4-toluenediamine is an essential technological step in the polyurethane industry. Although the technological process, the reaction mechanism, and the reaction kinetics have been investigated and have come to be generally accepted, there is still much to learn about the catalysis of this process. That is why the mapping of the current state of catalyst development likewise facilitates the development of scientific research. However, the review of the literature on catalysts used for TDA synthesis does not provide sufficient information to achieve this aim. The comparison of the catalysts examined so far provides a much more comprehensive picture of the latest developments on their effectiveness. Therefore, the MIRA21 model was used to execute the catalyst's characterization, comparison, and qualification [24].

3.1. Catalyst Library

The results of the literature research are surprising because there are relatively few published scientific results about the dinitrotoluene hydrogenation process. They were mostly prepared before the 2000s. Based on Google Scholar searches for the keywords *dinitrotoluene hydrogenation*, we obtained 2210 matches, however, if we added *toluenediamine*, there were only 212 hits. In total, 92 pieces of these included scientific results obtained after 2010. To demonstrate this, the keyword *kinetic* was added to the initial search, which then yielded 120 articles. Overall, only a few research groups have studied TDA synthesis and have prepared catalysts for this reaction. On one hand, a smaller database reduces the reliability of the MIRA21 results. On the other hand, a smaller dataset makes it easier to delineate the possible research pathways on the topic.

After the first selection, 56 articles remained. During the data analysis, we concluded that it is justified to change the publication year selection criteria (after 2000) and we also worked with previous articles. The left panel of Figure 8 shows the distribution of the scientific publications according to the publication date. The right panel of the figure presents the studied articles based on its Q-index in 2021 after the primary article selection (relevance, publication year, Q-index). The figure shows that the data used to analyze the catalysts mainly came from Q1 articles. A few publications whose publisher has since ceased to exist were also included in the analysis because they had previously provided space for the publication of high-quality scientific works.

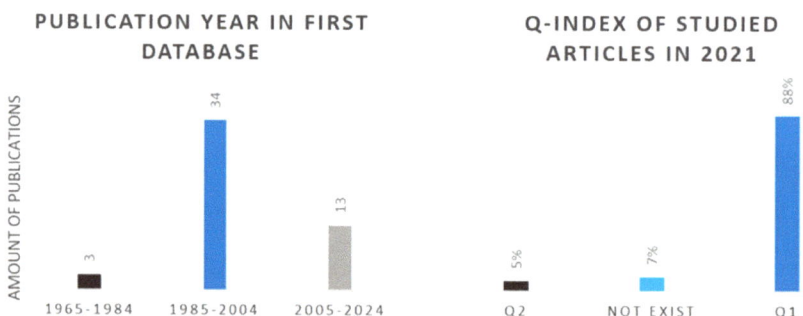

Figure 8. Publication year distribution of 56 articles after first selection (**left**) and Q-index distribution of 15 articles after second selection (**right**).

The 58 qualified catalysts selected from 15 articles were mostly supported catalysts (Figure 9 left) [14,16,40,44–55]. Most of the produced catalysts contained one active component on the support (middle of the figure). The catalysts with two active components generally applied palladium-platinum, palladium-iron combinations. The catalyst systems containing three active components were composed of either iridium-manganese-iron, iridium-iron-cobalt, or nickel-lanthanum-boron. The frequency of the active metal components was in the order of Pd > Pt > Ni. In addition to palladium and platinum, nickel was also seen, which is used as a common catalyst in industrial practice (Figure 9 right). Regarding the catalyst carrier, we mainly identified metal oxides (zirconium, chromium, titanium, aluminum, and silicon), ferrites, maghemites, zeolites, and activated carbon as typical in the chemical industry. Occasionally, PVP-based catalysts were also investigated [14].

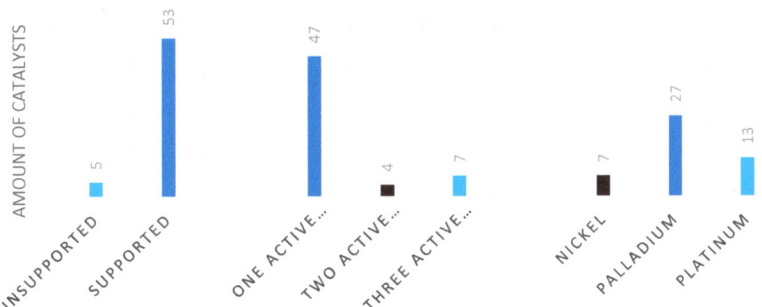

Figure 9. Composition of studied catalysts according to support, active component.

3.2. Catalyst Ranking and Characterization

Based on the available library of the catalyst data, it is difficult to get a consistent picture of DNT hydrogenation catalysts. However, these catalysts can be well-qualified and comparable, according to the MIRA21 model. A total of 58 catalysts from 15 articles reporting research results were successfully analyzed [14,16,40,44–55].

The catalysts were characterized in detail, as 10 or more known parameters could be collected in each case (from 15). The tested reaction conditions are in the range 295–393 K and 1–50 atm, with the exception of two cases (98 and 150 atm). The time required for maximum conversion ranged from a few minutes to a 1-day interval, which therefore shows a large standard deviation. The average reaction time for 100% conversion is 60 min (Figure 10). This shows that the reaction times of the best catalysts were under 40 min.

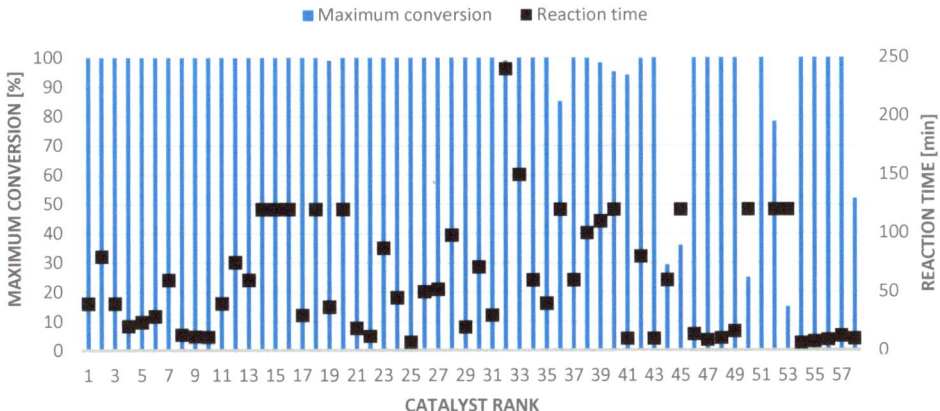

Figure 10. Maximum conversion with required reaction time.

The amount of initial dinitrotoluene was in the range of 0.002 and 0.3 mol. The amount of active metal in the catalyst also showed a large deviation from 5.13×10^{-7} mol to 0.034 mol. Despite the low amount of the catalyst, as mentioned above, 100% conversion was achieved [54]. The increased amount of the material was typical for nickel-type catalysts.

Furthermore, Figure 11 shows the catalytic performance results for the selected, studied, characterized, ranked, and classified catalysts. The conversion of the studied catalysts in classes D1-Q1-Q2 is over 99 n/n%, however, the product selectivity is much more differentiated. Based on these results, it can be said that achieving the pure TDA product produced during hydrogenation is a serious challenge for researchers. The worst-performing catalysts (class Q4) worked below 50 n/n%.

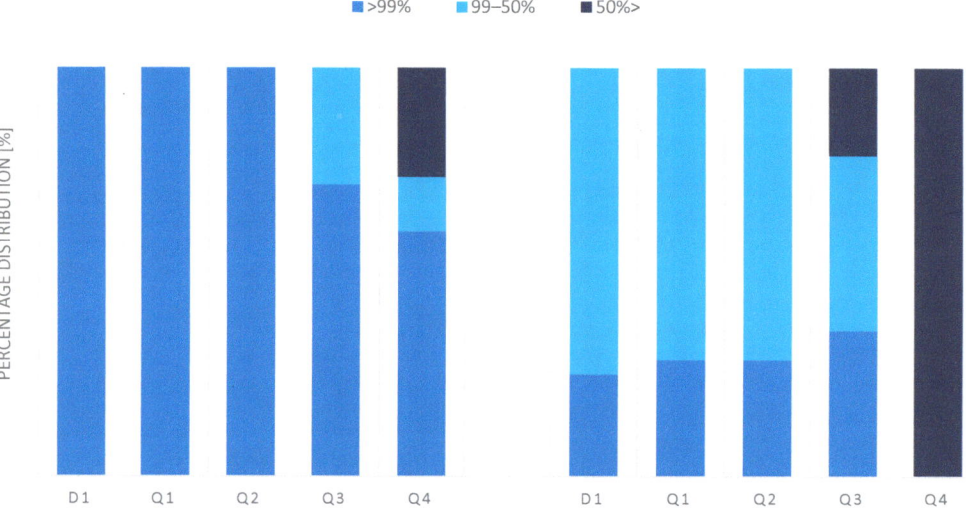

Figure 11. Catalytic performance of catalysts.

The catalyst composition changed according to the ranking of the MIRA21 model. In addition, Figure 12 shows the active components and support types of the catalyst systems based on their classification. The best-performing catalysts (class D1) consist of palladium or platinum and transition metal oxide supports. Although nickel is more commonly used in the industry, these types of catalysts are in the lower half of the ranking. Iridium as an active component in the catalyst also obtained a relatively good MIRA21 number. Most of the unsupported, carbon black, Al_2O_3 and SiO_2-supported catalysts are in the lower half of the ranking.

Practically, the catalyst carrier of the system differed according to the MIRA21 classes. Mainly activated carbon supports can be found in Q2 and Q4 classes. The catalysts with the transition metal supports are at the top of the ranking.

The eight best D1-rated catalysts are listed in Table 1. The columns contain the ID code and the designation of the catalysts, the type of catalyst support and active component, the number of known parameters, and the calculated MIRA21 number. The best MIRA catalysts consist of only one active component and transition metal oxide supports. Based on the results, the platinum-containing catalysts produced better results than their competitors did. The synergistic effect of the combination of the active components is difficult to assess because there is not enough information available. Class D1 includes the catalysts that are studied according to sustainability considerations, such as stability and reactivation capabilities. These catalysts are at the beginning of the innovation pathway and are not yet suitable for industrial application. If the results were compared with the ranking of the catalysts analyzed in the case of the nitrobenzene hydrogenation reaction, it can be found that the best MIRA21-ranked catalyst was similar to the Pt/ZrO_2 catalyst, which is one of the most effective catalyst systems in the first class. Zhang et al. prepared a $Pt/ZrO_2/SBA$-

15 hybrid nanostructure catalyst that showed an excellent catalytic performance at 313 K, 7 atm in 50 min for the hydrogenation of nitrobenzene to aniline [56]. They found that the dispersion of ZrO_2 in SBA-15 improved the performance of the catalyst due to its mesoporous structure. Therefore, it would be worthwhile to try this catalyst for the synthesis of TDA as well.

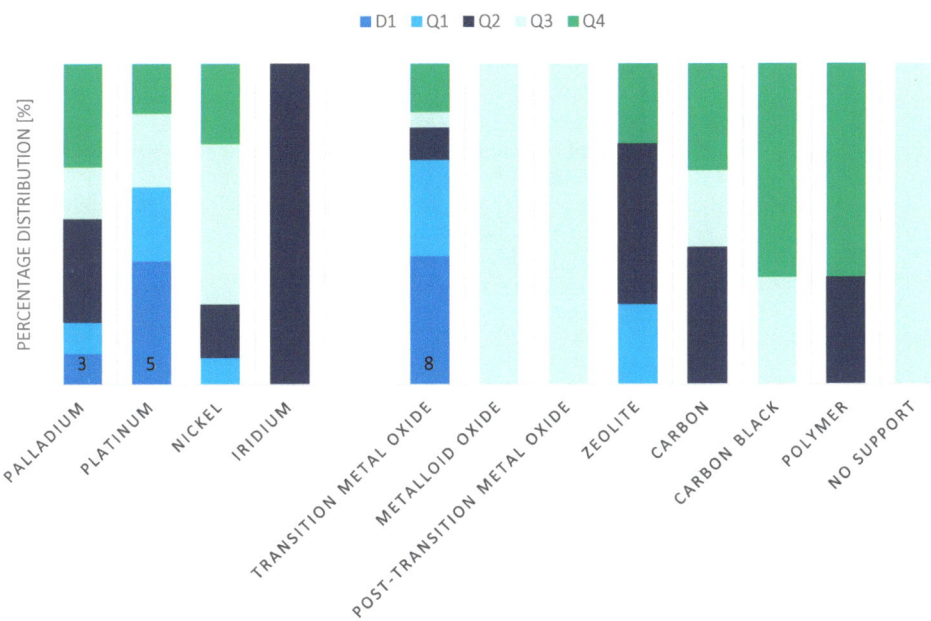

Figure 12. Distribution and active components of catalysts according to MIRA21 ranking and classification (D1-best, Q4-worst qualification, according to MIRA21 coloring).

Table 1. TOP10 catalyst of MIRA21 ranking.

			D1 CATALYSTS			
No.	CATALYST ID	Catalyst Name	Support	Active Component	Number of Known Parameters	MIRA21 Number
1	HDNT/MIS/2021/2/2	Pt/CrO$_2$	Chromium(IV)-dioxide	Platinum	15	11.50
2	HDNT/MIS/2021/2/1	Pd/CrO$_2$	Chromium(IV)-dioxide	Palladium	15	11.49
3	HDNT/MIS/2021/3/1	Pd/NiFe$_2$O$_4$	Nickel ferrite	Palladium	15	11.45
4	HDNT/TIA/2020/1/3	15Pt/ZrO$_2$-300	Zirconium-dioxide	Platinum	13	11.44
5	HDNT/TIA/2020/1/4	15Pt/ZrO$_2$-400	Zirconium-dioxide	Platinum	13	11.43
6	HDNT/TIA/2020/1/2	15Pt/ZrO$_2$-200	Zirconium-dioxide	Platinum	13	11.42
7	HDNT/MIS/2021/1/2	Pd/maghemite	Maghemite	Palladium	15	11.35
8	HDNT/TIA/2020/1/5	45Pt/ZrO$_2$-300	Zirconium-dioxide	Platinum	13	11.06

The work of Hajdu et al. focused on the development of new magnetic catalysts for the hydrogenation of DNT to TDA [44,53,55]. One of the catalysts is Pd/NiFe2O4, which has achieved 99 n/n% TDA yield at 333 K and 20 atm. In this work, they synthetized the nickel ferrite spinel nanoparticles to solve the problem of separating the catalyst from the products by magnetization. Another magnetic catalyst with good catalytic performance is Pd/maghemite, which is made by a combustion method with a sonochemical step. Palladium on a maghemite support resulted in a high catalytic activity for TDA synthesis at about 60 min and under the same reaction conditions as ferrite hydrogenation. The first and the fifth place of the MIRA rankings were the chromium oxide platinum and palladium catalysts. These innovative systems yielded excellent performing catalysts. It was prepared with chromium (IV) oxide nanowires that were decorated with platinum and palladium nanoparticles. These catalysts showed high catalytic activity at 333 K and 20 atm. If a Pt/CrO_2 catalyst was used, 304.8 mol of TDA was produced under these conditions, while only 1 mol of the precious metal catalyst was used. When palladium is used as an active component, only 60.14 mol TDA was produced, but it is also a relatively large amount. From an industrial point of view, it is important that this type of catalyst could be easily separated from the reaction mixture due to its magnetic properties. The stability of the catalyst was studied, and it was found that the catalyst could be used at least four times without regeneration.

Ren and his colleagues made half of the D1 class catalysts, and these catalytic systems consisted of zirconium oxide supports and platinum precious metal [54]. Ren et al. prepared the ZrO_2-supported platinum catalysts with different Pt concentrations and at different reduction temperatures. They found that the 0.156% Pt-containing zirconium oxide catalyst has the highest catalytic performance at 353 K and 20 atm. According to their results, the use of this catalyst reached an initial hydrogen consumption of 4583 mol H2 mol Pt-1 min-1. In this work, they investigated the interaction between the precious metal and the oxide support. It was found that zirconium oxide had the highest adsorption capacity for platinum ions due to its ability to be protonated and deprotonated.

4. MIRA21 Method

In our previous work, we successfully developed the Miskolc Ranking 2021 (MIRA21) system as a multistep process for the identification of new and useful patterns in the catalyst data sets to provide a standard algorithm for catalyst characterization and to compare and rank catalysts with minimal bias [24]. It is a practical and functional mathematical model of exact catalyst qualification with four classes of descriptors: catalyst performance, reaction conditions, catalyst conditions, and catalyst sustainability. The comparison of TDA catalysts could enable the supporting design of catalysts and the monitoring of research and development trends. The model facilitates the determination of the direction of catalyst development by establishing a system for ranking and classifying the catalysts. Furthermore, the standardization of the data in scientific publications could also benefit from accurate and coherent data. Figure 13 illustrates the process of the MIRA21 method from the literature sources to useful knowledge.

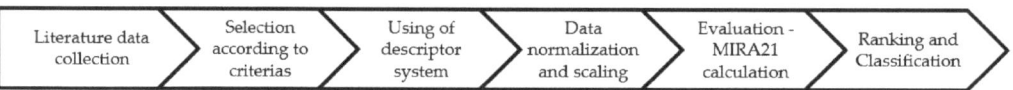

Figure 13. Process steps of MIRA21 model.

Table 2 shows the descriptor system of the model. The parameters can be divided into four classes with different weighting coefficients. The catalyst performance class includes conversion, selectivity, yield, and turnover number attributes. The second group contains the reaction conditions of the laboratory or large-scale experiments with the prepared catalysts. The third group is the catalyst conditions, with these two easily described parameters included in it. The last group addresses the sustainability and industrial application of the catalysts.

Table 2. Descriptor system of MIRA21 for TDA synthesis [24].

	Categories		No.	Notation	Name	Unit	Definition
Quantifiable parameters	Catalyst performance	I.	1.	X_{PRmax}	Maximum conversion	n/n%	Maximum desired product conversion achieved on a given catalyst
			2.	Y_{PR}	Product yield	n/n%	Product yield for maximum conversion
			3.	S_{PR}	Product selectivity	n/n%	Product selectivity for maximum conversion
			4.	TON_{PR}	Turnover number	-	Number of moles of product formed per 1 mol active metal when the maximum conversion reached
	Reaction conditions	II.	5.	$Tmax_{.conv.}$	Temperature	K	Reaction temperature for maximum conversion
			6.	$P_{max.conv.}$	Pressure	atm	Reaction pressure for maximum conversion
			7.	$t_{max.conv.}$	Time	min	Time required to reach maximum conversion
			8.	$n_{cat.}$	Molar amount of initial catalyst	mol	The molar amount of the active metal involved in the reaction; in case of several metals, the sum of molar numbers
			9.	n_{start}	Molar amount of starting reagent	mol	The initial amount of starting reagent involved in the reaction
	Catalyst conditions	III.	10.	CPZ	Catalyst particle size	nm	Average particle size of the catalyst
			11.	CSA	Catalyst surface area	m^2/g	Catalyst (active metal + support) surface area
			Does the publication contain information about these subjects?				
Non-quantifiable parameters	Sustainable parameters	IV.	12.	Rea	Information about reactivation	-	Reactivation means the physical process by which the activity of the catalyst used returns to or near the original activity level.
			13.	Stab	Information about stability of catalyst	-	Stability means preservation of catalytic activity
			14.	Care	Information about catalyst carrier effect	-	Carrier effect means the catalyst support influences the catalytic reaction
			15.		Catalyst carrier effect	-	Nature of the effect (positive, no effect, negative)

Table 2 describes the parameters used to characterize the performance of the catalyst, as well as the main mathematical equations used in the calculation of the MIRA21 number and ranking.

Figure 14 lists the equations used in the MIRA21 model, where n_{DNT}, n_{TDA}, and $n_{catalyst}$ are the corresponding molar amounts of the compounds; A is the value of the attribute; A^t is the transformed attribute value; min_A and max_A are the corresponding calculated minimum and maximum values of the attribute in the data set; MIN is the minimum scoring point; MAX is the maximum scoring point; i = 1 ... 15 is the number of attributes; MAXrank is the highest and MINrank is the lowest score of the MIRA21 ranking.

	Used Equations
Maximum Conversion	$X_{Prmax}\% = \dfrac{consumed\ n_{DNT}}{initial\ n_{DNT}} * 100$
Product Yield	$Y_{PR}\% = \dfrac{synthetized\ n_{TDA}}{initial\ n_{DNT}} * 100$
Product Selectivity	$S_{PR}\% = \dfrac{synthetized\ n_{TDA}}{consumed\ n_{DNT}} * 100$
TurnOver Number	$TON_{PR} = \dfrac{synthetized\ n_{TDA}}{n_{catalyst}}$
Normalization	$A^t = MIN + \dfrac{(MAX - MIN) * A - min_A}{max_A - min_A}$
MIRA21	$MIRA21 = \log \prod_{i=1}^{n} A_i^t \quad i = 1 \ldots 15$
Classification	$SLofD1class = MAXrank - \left(\dfrac{MAXrank - MINrank}{10}\right)$

Figure 14. MIRA21 used equations [24].

5. Summary and Conclusions

In summary, Table 3 includes the MIRA21 results and the classification of selected and studied catalysts for the hydrogenation of DNT. The aim of this work is to make an overview of the hydrogenation of dinitrotoluene to toluenediamine. The chemical technology, development of reaction mechanism, and previous catalyst research were summarized by a quantitative comparison method called MIRA21. In total, 58 catalysts from 15 research articles were selected and studied with the MIRA21 model, which covered the complete scientific literature of the catalytic hydrogenation DNT. According to the ranking and classification, eight catalysts were ranked in the highest class (D1).

Table 3. Catalysts for dinitrotoluene hydrogenation to toluenediamine in liquid phase [14,16,40,44–55].

	CATALYSTS						
No.	CATALYST ID	Catalyst Name	Catalyst Support	Main Active Component	Known Par.	MIRA21 Number	Class
1	HDNT/MIS/2021/2/2	Pt/CrO$_2$	Chromium(IV)-dioxide	Platinum	15	11.50	D1
2	HDNT/MIS/2021/2/1	Pd/CrO$_2$	Chromium(IV)-dioxide	Palladium	15	11.49	D1
3	HDNT/MIS/2021/3/1	Pd/NiFe$_2$O$_4$	Nickel ferrite	Palladium	15	11.45	D1
4	HDNT/TIA/2020/1/3	15Pt/ZrO$_2$-300	Zirconium-dioxide	Platinum	13	11.44	D1
5	HDNT/TIA/2020/1/4	15Pt/ZrO$_2$-400	Zirconium-dioxide	Platinum	13	11.43	D1
6	HDNT/TIA/2020/1/2	15Pt/ZrO$_2$-200	Zirconium-dioxide	Platinum	13	11.42	D1
7	HDNT/MIS/2021/1/2	Pd/maghemite	Maghemite	Palladium	15	11.35	D1
8	HDNT/TIA/2020/1/5	45Pt/ZrO$_2$-300	Zirconium-dioxide	Platinum	13	11.06	D1

Table 3. Cont.

			CATALYSTS				
No.	CATALYST ID	Catalyst Name	Catalyst Support	Main Active Component	Known Par.	MIRA21 Number	Class
9	HDNT/TIA/2020/1/6	60Pt/ZrO$_2$-300	Zirconium-dioxide	Platinum	13	11.01	Q1
10	HDNT/TIA/2020/1/7	85Pt/ZrO$_2$-300	Zirconium-dioxide	Platinum	13	11.00	Q1
11	HDNT/MIS/2021/3/2	Pd/CoFe$_2$O$_4$	Cobalt ferrite	Palladium	15	10.84	Q1
12	HDNT/SHA/2012/1/1	Ni/HY catalyst	HY molecular sieve	Nickel	15	10.77	Q1
13	HDNT/MIS/2021/1/1	Pt/maghemite	Maghemite	Platinum	15	10.67	Q1
14	HDNT/MIS/2021/3/3	Pd/CuFe$_2$O$_4$	Copper ferrite	Palladium	15	10.48	Q1
15	HDNT/MIS/2022/1/3	Pd/NiFe$_2$O$_4$-NH2	Nickel-ferrite	Palladium	13	10.31	Q1
16	HDNT/MIS/2022/1/1	Pd/CoFe$_2$O$_4$-NH2	Cobalt-ferrite	Palladium	13	10.28	Q2
17	HDNT/MIS/2021/1/3	Pd-Pt/maghemite	Maghemite	Palladium	14	10.14	Q2
18	HDNT/PUN/1999/1/5	20% Ni/HY	HY zeolite	Nickel	14	9.50	Q2
19	HDNT/DAL/1997/1/2	PVP-Pd-1/4 Pt	PVP	Palladium	13	9.47	Q2
20	HDNT/PUN/1999/1/6	10% Ni/HY	HY zeolite	Nickel	14	9.44	Q2
21	HDNT/MES/2001/1/1	MGPd05	Chemviron SC XII active carbon	Palladium	13	9.24	Q2
22	HDNT/MES/2001/1/3	MGPd1b	Chemviron SC XII active carbon	Palladium	13	9.23	Q2
23	HDNT/MES/2001/1/8	MGPd5a	Chemviron SC XII active carbon	Palladium	13	9.20	Q2
24	HDNT/HAN/2001/1/2	B	Chemically activated carbon	Iridium	13	9.19	Q2
25	HDNT/MES/2001/1/4	MGPd1c	Chemviron SC XII active carbon	Palladium	13	9.19	Q2
26	HDNT/HAN/2001/1/1	A	Chemically activated carbon	Iridium	13	9.19	Q2
27	HDNT/MES/2001/1/7	MGPd5	Chemviron SC XII active carbon	Palladium	13	9.17	Q2
28	HDNT/MES/2001/1/5	MGPd1d	Chemviron SC XII active carbon	Palladium	13	9.14	Q2
29	HDNT/MES/2001/1/2	MGPd1a	Chemviron SC XII active carbon	Palladium	13	9.13	Q2
30	HDNT/MES/2001/1/6	MGPd3	Chemviron SC XII active carbon	Palladium	13	9.06	Q3
31	HDNT/HAN/2001/1/3	C	Steam activated carbon	Palladium	13	9.03	Q3
32	HDNT/MIS/2022/1/2	Pd/CdFe$_2$O$_4$-NH$_2$	Cadmium-ferrite	Palladium	13	8.95	Q3
33	HDNT/ZUR/1987/1/1	0.5 % Pt/Al$_2$O$_3$	Al$_2$O$_3$	Platinum	13	8.88	Q3
34	HDNT/HAN/2001/1/4	D	Steam activated carbon	Palladium	13	8.79	Q3
35	HDNT/HAN/2001/1/5	E	Oleophilic carbon black	Palladium	13	8.78	Q3
36	HDNT/PUN/1999/1/2	20% Ni/SiO$_2$	SiO$_2$	Nickel	14	8.63	Q3
37	HDNT/SHA/2012/1/4	Ni-La6-B		Nickel	12	8.36	Q3
38	HDNT/SHA/2012/1/3	Ni-La4-B		Nickel	12	8.34	Q3
39	HDNT/SHA/2012/1/2	Ni-La2-B		Nickel	12	8.33	Q3
40	HDNT/SHA/2012/1/1	Ni-La0-B		Nickel	12	8.32	Q3
41	HDNT/SAP/2004/1/3	Pt/C in ethanol	Active carbon	Platinum	13	8.02	Q3
42	HDNT/SHA/2012/1/5	Ni-La8-B		Nickel	12	7.92	Q3
43	HDNT/SAP/2004/1/2	Pt/C in ethanol	Active carbon	Platinum	13	7.90	Q3
44	HDNT/TIA/2020/1/1	15Pt/ZrO$_2$-100	Zirconium-dioxide	Platinum	13	7.85	Q4
45	HDNT/PUN/1999/1/4	20% Ni/HZSM-5	HZSM-5	Nickel	14	7.75	Q4
46	HDNT/TAE/1993/1/1	SA	Activated carbon	Palladium	11	7.58	Q4
47	HDNT/TAE/1993/1/3	DA	Activated carbon	Palladium	11	7.52	Q4
48	HDNT/TAE/1993/1/2	SAON	Activated carbon	Palladium	11	7.50	Q4
49	HDNT/TAE/1993/1/4	DAON	Activated carbon	Palladium	11	7.49	Q4
50	HDNT/PUN/1999/1/1	20% Ni/Al$_2$O$_3$	Al$_2$O$_3$	Nickel	14	7.43	Q4
51	HDNT/BAR/2000/1/1	Pd(AAEMA)2/EMA/EGDMA	Polymer-supported complex	Palladium	13	7.27	Q4
52	HDNT/DAL/1997/1/1	PVP-PdCl2	PVP	Palladium	10	6.99	Q4
53	HDNT/PUN/1999/1/3	20% Ni/TiO$_2$	TiO$_2$	Nickel	14	6.95	Q4
54	HDNT/TAE/1993/1/7	VB	Carbon black	Palladium	10	6.80	Q4
55	HDNT/TAE/1993/1/8	VON	Carbon black	Palladium	10	6.80	Q4
56	HDNT/TAE/1993/1/6	DAOH	Activated carbon	Palladium	10	6.80	Q4
57	HDNT/TAE/1993/1/5	DAOS	Activated carbon	Palladium	10	6.80	Q4
58	HDNT/SAP/2004/1/1	Pt/C in scCO$_2$	Active carbon	Platinum	13	6.68	Q4

The number of catalysts developed for TDA synthesis is low, since the scientific research focused mostly on the reaction mechanism and reaction kinetic. Despite this fact, many different catalyst systems have been developed.

More than 80% of the 58 types of catalysts produced and tested had excellent conversions, but only 45% of them demonstrated a selectivity above 90% n/n%. More than 80% of the produced catalysts consisted of only one active component. Since the combination of catalysts has not been scarcely investigated, one recommended direction of research is the multi-component catalyst. Catalyst development represents a new trend that has led to the establishment of many high-performance catalysts. Based on the analyzed catalysts, compared to the traditional carbon-based supports, catalysts with oxide and/or magnetic supports showed better results in laboratory conditions. Carbon-supported nickel catalysts are primarily used in the industry, but nickel catalysts did not yield the best results. The advantage of well-performing magnetic catalysts due to their ability to be repaired is indisputable, but the economic implications of their industrial application must also be considered.

Author Contributions: Conceptualization, A.J.-N. and B.V.; methodology, A.J.-N., B.V., V.H. and L.V.; writing—original draft preparation, A.J.-N., V.H., L.F. and L.V.; writing—review and editing, A.J.-N., L.F. and B.V.; supervision, B.V.; funding acquisition, B.V. All authors have read and agreed to the published version of the manuscript.

Funding: This research was supported by the Ministry of Innovation and Technology-financed 2020-1.1.2-PICI-KFI-2020-00121 project. This research was prepared with the professional support of the Doctoral Student Scholarship Program of the Cooperative Doctoral Program of the Ministry of Innovation and Technology financed from the National Research, Development and Innovation Fund.

Acknowledgments: We thank Tamás Purzsa from Wanhua-BorsodChem for his helpful contributions. We also acknowledge the opportunity provided by Wanhua-BorsodChem to conduct this study.

Conflicts of Interest: The authors declare no conflict of interest.

References

1. Ashida, K. Polyurethane and Related Foams, Polyurethane Relat. Foam. 2006. Available online: https://www.taylorfrancis.com/books/mono/10.1201/9780203505991/polyurethane-related-foams-kaneyoshi-ashida (accessed on 15 December 2022).
2. Polyurethane Market Size | Global Research Report [2021–2028], (n.d.). Available online: https://www.fortunebusinessinsights.com/industry-reports/polyurethane-pu-market-101801 (accessed on 29 December 2021).
3. Dinitrotoluene Hydrogenation Toluenediamine—Google Patents, (n.d.). Available online: https://patents.google.com/?q=dinitrotoluene+hydrogenation+toluenediamine&oq=dinitrotoluene+hydrogenation+toluenediamine (accessed on 15 December 2022).
4. CH372683A—Process for the Catalytic Hydrogenation of Dinitro Derivatives of Benzene and Toluene to Give the Corresponding Diamines—Google Patents, (n.d.). Available online: https://patents.google.com/patent/CH372683A/en?q=dinitrotoluene+hydrogenation+toluenediamine&oq=dinitrotoluene+hydrogenation+toluenediamine&sort=old&page=1 (accessed on 15 December 2022).
5. DE1092478B—Process for the Catalytic Reduction of Dinitrobenzenes and Dinitrotoluenes to the Corresponding Diamine—Google Patents, (n.d.). Available online: https://patents.google.com/patent/DE1092478B/en?q=dinitrotoluene+hydrogenation+toluenediamine&oq=dinitrotoluene+hydrogenation+toluenediamine&sort=old&page=1 (accessed on 15 December 2022).
6. US2894036A—Catalytic Reduction of Aromatic Polynitro Compounds—Google Patents, (n.d.). Available online: https://patents.google.com/patent/US2894036A/en?q=dinitrotoluene+hydrogenation+toluenediamine&oq=dinitrotoluene+hydrogenation+toluenediamine&sort=old (accessed on 15 December 2022).
7. DE951930C—Process for the Preparation of Tolylenediamine—Google Patents, (n.d.). Available online: https://patents.google.com/patent/DE951930C/en?q=dinitrotoluene+hydrogenation+toluenediamine&oq=dinitrotoluene+hydrogenation+toluenediamine&sort=old (accessed on 15 December 2022).
8. DE1044099B—Process for the Preparation of Tolylenediamine—Google Patents, (n.d.). Available online: https://patents.google.com/patent/DE1044099B/en?q=dinitrotoluene+hydrogenation+toluenediamine&oq=dinitrotoluene+hydrogenation+toluenediamine&sort=old (accessed on 15 December 2022).
9. CN114289034A—Noble Metal Catalyst, Preparation Method and Application Thereof in Preparation of Toluenediamine by Catalyzing Dinitrotoluene Hydrogenation—Google Patents, (n.d.). Available online: https://patents.google.com/patent/CN114289034A/en?q=dinitrotoluene+hydrogenation+toluenediamine&oq=dinitrotoluene+hydrogenation+toluenediamine&sort=new (accessed on 15 December 2022).
10. CN114682288A—Supported Catalyst for Hydrogenation of Aromatic Nitro Compound and Preparation Method Thereof—Google Patents, (n.d.). Available online: https://patents.google.com/patent/CN114682288A/en?q=dinitrotoluene+hydrogenation+toluenediamine&oq=dinitrotoluene+hydrogenation+toluenediamine&sort=new (accessed on 15 December 2022).
11. WO2022234021A1—A Catalytic Material Suitable for Hydrogenation Reactions Comprising ni, One or More Additional Metals m, and a Specific Oxidic Support Material—Google Patents, (n.d.). Available online: https://patents.google.com/patent/WO2022234021A1/en?q=dinitrotoluene+hydrogenation+toluenediamine&oq=dinitrotoluene+hydrogenation+toluenediamine&sort=new (accessed on 15 December 2022).
12. Jackson, S.D.; AlAsseel, A.K.A.; Allgeier, A.M.; Hargreaves, J.S.J.; Kelly, G.J.; Kirkwood, K.; Lok, C.M.; Schauermann, S.; Schmidt, R.S.; Sengupta, S.K. *Hydrogenation: Catalysts and Processes*; De Guyter: Berlin, Germany; Boston, MA, USA, 2018. [CrossRef]

13. Dinitrotoluene Hydrogenation Toluenediamine—Google Tudós, (n.d.). Available online: https://scholar.google.com/scholar?hl=hu&as_sdt=0%2C5&q=dinitrotoluene+hydrogenation+toluenediamine&btnG= (accessed on 16 December 2022).
14. Yu, Z.; Liao, S.; Xu, Y.; Yang, B.; Yu, D. Hydrogenation of nitroaromatics by polymer-anchored bimetallic palladium-ruthenium and palladium-platinum catalysts under mild conditions. *J. Mol. Catal. A Chem.* **1997**, *120*, 247–255. [CrossRef]
15. Mathew, S.; Rajasekharam, M.; Chaudhari, R. Hydrogenation of p-isobutyl acetophenone using a Ru/Al2O3 catalyst: Reaction kinetics and modelling of a semi-batch slurry reactor. *Catal. Today* **1999**, *49*, 49–56. [CrossRef]
16. Kut, O.M.; Yücelen, F.; Gut, G. Selective liquid-phase hydrogenation of 2,6-dinitrotoluene with platinum catalysts. *J. Chem. Technol. Biotechnol.* **1987**, *39*, 107–114. [CrossRef]
17. Song, J.; Huang, Z.-F.; Pan, L.; Li, K.; Zhang, X.; Wang, L.; Zou, J.-J. Review on selective hydrogenation of nitroarene by catalytic, photocatalytic and electrocatalytic reactions. *Appl. Catal. B Environ.* **2018**, *227*, 386–408. [CrossRef]
18. Anand, S.; Pinheiro, D.; Devi, K.R.S. Recent Advances in Hydrogenation Reactions Using Bimetallic Nanocatalysts: A Review. *Asian J. Org. Chem.* **2021**, *10*, 3068–3100. [CrossRef]
19. Qi, S.-C.; Wei, X.-Y.; Zong, Z.-M.; Wang, Y.-K. Application of supported metallic catalysts in catalytic hydrogenation of arenes. *RSC Adv.* **2013**, *3*, 14219–14232. [CrossRef]
20. Kamalzare, M. *Recovery and Reusability of Catalysts in Various Organic Reactions*; Micro and Nano Technologies: Rajasthan, India, 2022; pp. 149–165. [CrossRef]
21. Sheldon, R.A. Engineering a more sustainable world through catalysis and green chemistry. *J. R. Soc. Interface* **2016**, *13*, 20160087. [CrossRef]
22. Sciscenko, I.; Arques, A.; Escudero-Oñate, C.; Roccamante, M.; Ruiz-Delgado, A.; Miralles-Cuevas, S.; Malato, S.; Oller, I. A Rational Analysis on Key Parameters Ruling Zerovalent Iron-Based Treatment Trains: Towards the Separation of Reductive from Oxidative Phases. *Nanomaterials* **2021**, *11*, 2948. [CrossRef]
23. The Value of Basic Scientific Research (Dec 2004)—ICSU, (n.d.). Available online: https://web.archive.org/web/20170506184955/http://www.icsu.org/publications/icsu-position-statements/value-scientific-research (accessed on 16 December 2022).
24. Jakab-Nácsa, A.; Sikora, E.; Prekob, Á.; Vanyorek, L.; Szőri, M.; Boros, R.Z.; Nehéz, K.; Szabó, M.; Farkas, L.; Viskolcz, B. Comparison of Catalysts with MIRA21 Model in Heterogeneous Catalytic Hydrogenation of Aromatic Nitro Compounds. *Catalysts* **2022**, *12*, 467. [CrossRef]
25. BorsodChem Zrt. *Sustainability Report 2019–2020*; BorsodChem Zrt: Kazincbarcika, Hungary, 2020.
26. Charra, F.; Gota-Goldmann, S.; Warlimont, H. Nanostructured Materials. In *Springer Handbook of Materials Data*; Springer: Cham, Switzerland, 2018. [CrossRef]
27. SMahood, A.; Schaffner, P.V.L. 2,4-DIAMINOTOLUENE. *Org. Synth.* **1931**, *11*, 32. [CrossRef]
28. Saboktakin, M.R.; Tabatabaie, R.M.; Maharramov, A.; Ramazanov, M. Hydrogenation of 2,4-Dinitrotoluene to 2,4-Diaminotoluene over Platinum Nanoparticles in a High-Pressure Slurry Reactor. *Synth. Commun.* **2011**, *41*, 1455–1463. [CrossRef]
29. Westerterp, K.; Molga, E.; van Gelder, K. Catalytic hydrogenation reactors for the fine chemicals industries. Their design and operation. *Chem. Eng. Process.-Process. Intensif.* **1997**, *36*, 17–27. [CrossRef]
30. Westerterp, K.; Janssen, H.; van der Kwast, H. The catalytic hydrogenation of 2,4-dinitrotoluene in a continuous stirred three-phase slurry reactor with an evaporating solvent. *Chem. Eng. Sci.* **1992**, *47*, 4179–4189. [CrossRef]
31. Molga', E.; Westerterpf, K. Kinetics of the hydrogenation of 2,4-dinitrotoluene over a palladium on alumina catalyst. *Chem. Eng. Sci.* **1992**, *47*, 1733–1749. [CrossRef]
32. Rajashekharam, M.V.; Nikalje, D.D.; Jaganathan, R.; Chaudhari, R.V. Hydrogenation of 2,4-Dinitrotoluene Using a Pd/Al2O3 Catalyst in a Slurry Reactor: A Molecular Level Approach to Kinetic Modeling and Nonisothermal Effects. *Ind. Eng. Chem. Res.* **1997**, *36*, 592–604. [CrossRef]
33. Dovell, F.S.; Ferguson, W.E. Greenfield Kinetic Study of Hydrogenation of dinitrotoluene and m-Dinitrobenzene. *Ind. Eng. Chem. Prod. Res. Develop.* **1970**, *9*, 224–229. [CrossRef]
34. Kreutzer, M.T.; Kapteijn, F.; Moulijn, J.A. Fast gas-liquid-solid reactions in monoliths: A case study of nitro-aromatic hydrogenation. *Catal. Today* **2005**, *105*, 421–428. [CrossRef]
35. Janssen, H.; Kruithof, A. Kinetics of the catalytic hydrogenation of 2,4-dinitrotoluene. 1. Experiments, Reaction Scheme and Catalyst Activity. *Ind. Eng. Chem. Res.* **1990**, *29*, 754–766. [CrossRef]
36. Janssen, H.; Kruithof, A. Kinetics of the catalytic hydrogenation of 2,4-dinitrotoluene. 2. Modeling of the Reaction Rates and Catalyst Activity. *Ind. Eng. Chem. Res.* **1990**, *29*, 1822–1829. [CrossRef]
37. Neri, G.; Rizzo, G.; Milone, C.; Galvagno, S.; Musolino, M.G.; Capanelli, G. Microstructural characterization of doped-Pd/C catalysts for the selective hydrogenation of 2,4-dinitrotoluene to arylhydroxylamines. *Appl. Catal. A Gen.* **2003**, *249*, 303–311. [CrossRef]
38. Neri, G.; Musolino, M.G.; Milone, C.; Visco, A.M.; Di Mario, A. Mechanism of 2,4-dinitrotoluene hydrogenation over Pd/C. *J. Mol. Catal. A Chem.* **1995**, *95*, 235–241. [CrossRef]
39. Neri, G.; Musolino, M.G.; Rotondo, E.; Galvagno, S. Catalytic hydrogenation of 2,4-dinitrotoluene over a Pd/C catalyst: Identification of 2-(hydroxyamino)-4-nitrotoluene (2HA4NT) as reaction intermediate. *J. Mol. Catal. A Chem.* **1996**, *111*, 257–260. [CrossRef]
40. Neri, G.; Musolino, M.G.; Milone, C.; Pietropaolo, D.; Galvagno, S. Particle size effect in the catalytic hydrogenation of 2,4-dinitrotoluene over Pd/C catalysts. *Appl. Catal. A Gen.* **2001**, *208*, 307–316. [CrossRef]

41. Neri, G.; Musolino, M.G.; Bonaccorsi, L.; Donato, A.; Mercadante, L.; Galvagno, S. Catalytic Hydrogenation of 4-(Hydroxyamino)-2-nitrotoluene and 2,4-Nitroaminotoluene Isomers: Kinetics and Reactivity. *Ind. Eng. Chem. Res.* **1997**, *36*, 3619–3624. [CrossRef]
42. Barone, G.; Duca, D. Hydrogenation of 2,4-Dinitro-toluene on Pd/C Catalysts: Computational Study on the Influence of the Molecular Adsorption Modes and of Steric Hindrance and Metal Dispersion on the Reaction Mechanism. *J. Catal.* **2002**, *211*, 296–307. [CrossRef]
43. Barone, G.; Duca, D. New time-dependent Monte Carlo algorithm designed to model three-phase batch reactor processes: Applications on 2,4-dinitro-toluene hydrogenation on Pd/C catalysts. *Chem. Eng. J.* **2003**, *91*, 133–142. [CrossRef]
44. Hajdu, V.; Jakab-Nácsa, A.; Muránszky, G.; Kocserha, I.; Fiser, B.; Ferenczi, T.; Nagy, M.; Viskolcz, B.; Vanyorek, L. Precious-Metal-Decorated Chromium(IV) Oxide Nanowires as Efficient Catalysts for 2,4-Toluenediamine Synthesis. *Int. J. Mol. Sci.* **2021**, *22*, 5945. [CrossRef]
45. Rajashekharam, V.; Malyala, V.; Chaudhari, V.R. Hydrogenation of 2,4-Dinitrotoluene Using a Supported Ni Catalyst: Reaction Kinetics and Semibatch Slurry Reactor Modeling. *Ind. Eng. Chem. Res.* **1999**, *38*, 906–915. [CrossRef]
46. Yan, S.; Fan, H.; Liang, C.; Li, Z.; Yu, Z. Preparation and Characterization of Ni-La-B Amorphous Alloy Catalyst for Low-Pressure Dinitrotoluene Hydrogenation. *Chin. J. Catal. (Chin. Version)* **2012**, *33*, 1374–1382. [CrossRef]
47. Zhang, X.M.; Niu, F.X. Liquid-Phase Hydrogenation to 2, 4-Tolylenediamine over Supported HY Catalysts. *Adv. Mater. Res.* **2012**, *512–515*, 2381–2385. [CrossRef]
48. Suh, D.J.; Tae-Jin, P.; Son-Ki, I. Effect of surface oxygen groups of carbon supports on the characteristics of Pd/C catalysts. *Carbon* **1993**, *31*, 427–435. [CrossRef]
49. Dell'Anna, M.M.; Gagliardi, M.; Mastrorilli, P.; Suranna, G.P.; Nobile, C. Hydrogenation reactions catalysed by a supported palladium complex. *J. Mol. Catal. A Chem.* **2000**, *158*, 515–520. [CrossRef]
50. Hajdu, V.; Sikora, E.; Kristály, F.; Muránszky, G.; Fiser, B.; Viskolcz, B.; Nagy, M.; Vanyorek, L. Palladium Decorated, Amine Functionalized Ni-, Cd- and Co-Ferrite Nanospheres as Novel and Effective Catalysts for 2,4-Dinitrotoluene Hydrogenation. *Int. J. Mol. Sci.* **2022**, *23*, 13197. [CrossRef]
51. Zhao, F.; Fujita, S.-I.; Sun, J.; Ikushima, Y.; Arai, M. Hydrogenation of nitro compounds with supported platinum catalyst in supercritical carbon dioxide. *Catal. Today* **2004**, *98*, 523–528. [CrossRef]
52. Auer, E.; Gross, M.; Panster, P.; Takemoto, K. Supported iridium catalysts—A novel catalytic system for the synthesis of toluenediamine. *Catal. Today* **2001**, *65*, 31–37. [CrossRef]
53. Hajdu, V.; Varga, M.; Muránszky, G.; Karacs, G.; Kristály, F.; Fiser, B.; Viskolcz, B.; Vanyorek, L. Development of magnetic, ferrite supported palladium catalysts for 2,4-dinitrotoluene hydrogenation. *Mater. Today Chem.* **2021**, *20*, 100470. [CrossRef]
54. Ren, X.; Li, J.; Wang, S.; Zhang, D.; Wang, Y. Preparation and catalytic performance of ZrO_2-supported Pt single-atom and cluster catalyst for hydrogenation of 2,4-dinitrotoluene to 2,4-toluenediamine. *J. Chem. Technol. Biotechnol.* **2020**, *95*, 1675–1682. [CrossRef]
55. Hajdu, V.; Muránszky, G.; Hashimoto, M.; Kristály, M.; Szőri, M.; Fiser, B.; Kónya, Z.; Viskolcz, B.; Vanyorek, L. Combustion method combined with sonochemical step for synthesis of maghemite-supported catalysts for the hydrogenation of 2,4-dinitrotoluene. *Catal. Commun.* **2021**, *159*, 106342. [CrossRef]
56. Yanji, Z.; Zhou, J. Synergistic-catalysis-by-a-hybrid-nanostructure-Pt-catalyst-for-high-efficiency-selective-hydrogenation-of-nitroarenes. *J. Catal.* **2021**, *395*, 445–456.

Disclaimer/Publisher's Note: The statements, opinions and data contained in all publications are solely those of the individual author(s) and contributor(s) and not of MDPI and/or the editor(s). MDPI and/or the editor(s) disclaim responsibility for any injury to people or property resulting from any ideas, methods, instructions or products referred to in the content.

Article

Preformed Pd(II) Catalysts Based on Monoanionic [N,O] Ligands for Suzuki-Miyaura Cross-Coupling at Low Temperature

Matthew J. Andrews, Sebastian Brunen, Ruaraidh D. McIntosh * and Stephen M. Mansell *

Institute of Chemical Sciences, School of Engineering and Physical Sciences, Heriot-Watt University, Edinburgh EH14 4AS, UK
* Correspondence: r.mcintosh@hw.ac.uk (R.D.M.); s.mansell@hw.ac.uk (S.M.M.)

Abstract: This paper describes the synthesis and catalytic testing of a palladium complex with a 5-membered chelating [N,O] ligand, derived from the condensation of 2,6-diisopropylphenyl aniline and maple lactone. This catalyst was active towards the Suzuki-Miyaura cross-coupling reaction, and its activity was optimised through the selection of base, solvent, catalytic loading and temperature. The optimised conditions are mild, occurring at room temperature and over a short timescale (1 h) using solvents considered to be 'green'. A substrate scope was then carried out in which the catalyst showed good activity towards aryl bromides with electron-withdrawing groups. The catalyst was active across a broad scope of electron-donating and high-withdrawing aryl bromides with the highest activity shown for weak electron-withdrawing groups. The catalyst also showed good activity across a range of boronic acids and pinacol esters with even boronic acids featuring strong electron-withdrawing groups showing some activity. The catalyst was also a capable catalyst for the cross-coupling of aryl chlorides and phenylboronic acid. This more challenging reaction requires slightly elevated temperatures over a longer timescale but is still considered mild compared to similar examples in the literature.

Keywords: palladium; cross-coupling; Suzuki-Miyaura; aryl Halide; sustainable

Citation: Andrews, M.J.; Brunen, S.; McIntosh, R.D.; Mansell, S.M. Preformed Pd(II) Catalysts Based on Monoanionic [N,O] Ligands for Suzuki-Miyaura Cross-Coupling at Low Temperature. *Catalysts* **2023**, *13*, 303. https://doi.org/10.3390/catal13020303

Academic Editor: Carl Redshaw

Received: 20 December 2022
Revised: 9 January 2023
Accepted: 18 January 2023
Published: 29 January 2023

Copyright: © 2023 by the authors. Licensee MDPI, Basel, Switzerland. This article is an open access article distributed under the terms and conditions of the Creative Commons Attribution (CC BY) license (https://creativecommons.org/licenses/by/4.0/).

1. Introduction

Suzuki-Miyaura cross-coupling is an important and popular reaction with many applications across a wide field, including the synthesis of ligands for catalysis, natural products and pharmaceuticals [1,2]. This wide applicability is due to the high tolerance of the reaction to a wide variety of functional groups and the mild reaction conditions that are required [1,3–6]. The reaction typically occurs through the reaction of an aryl boronic acid with a haloarene in the presence of a base with many different catalysts available [7]. These catalysts are typically Pd and range from heterogeneous nanoparticles to homogeneous catalysts [4–14]. Early catalysts for Suzuki-Miyaura coupling reactions were based on phosphorus ligands, of which [Pd(PPh$_3$)$_4$] is a typical example, with the choice of Pd(II) pre-catalysts based on their air stability and their ready reduction to form the active Pd(0) species [1,3,14]. The choice of ligand was also observed to have a major effect. This led to much work being carried out into monoligated species through the development of bulky, electron-rich phosphines or sterically demanding N–Heterocyclic carbene (NHC) ligands [15,16]. An important advance in Suzuki-Miyaura cross-coupling reactions is the activation of aryl chlorides due to their decreased cost compared to aryl bromides, which was first achieved by Shen et al. using Pd(OAc)$_2$ with 1,3 bis(diphenylphosphino)propane (dppp) [17]. Through tuning the phosphorus ligands, it has been possible to cross-couple a wide variety of coupling partners with aryl chlorides [18]. Whilst the yields for aryl chlorides are typically poor, high reaction temperatures, such as refluxing *ortho*-xylene for 5–20 h, can be used to improve the yields [19].

Another main class of ligands in this field are NHCs, first utilised for the Suzuki–Miyaura coupling of 4–chloroacetophenone with arylboronic acid by Herrmann and co-workers [20]. Following this work, a wide range of NHCs have been utilized as ligands for catalytic cross-coupling reactions owing to the strong σ-donating properties and steric shielding [16]. A recent development in cross-coupling catalysis is the utilization of well-defined pre-catalysts that do not require the addition of ancillary ligands to metal salts. These catalysts can allow for transformations under milder conditions or at lower catalytic loadings [21]. These advantages have been demonstrated by Beller and co-workers where the use of a pre-formed catalyst improved performance over the same catalytic system generated in situ [22].

There is a large drive within the field of chemistry to work towards 'greener' reactions. These are summarized as the twelve principles of green chemistry introduced by Anastas and Warner in 1998 [23]. One of these principles is the choice of safer solvents and auxiliaries [24]. With many cross-coupling reactions requiring DMF, DMA or toluene heated under reflux, there is space here to improve these conditions and move to more sustainable alternatives since these solvents are classified by several companies as undesirable or in need of substitution where available following environmental, health and safety property analysis [25,26]. The use of better media for Suzuki-Miyaura reactions was reviewed recently across a range of solvents and catalyst types [27,28]. The switch away from solvents such as DMF is also beneficial when solvents such as methanol are used, as a simple extraction is required rather than a lengthy aqueous workup.

Schiff base-derived ligands have been shown to provide alternatives to classic phosphine-based catalysts with similar easy tuning of their steric and electronic properties [8,29–32]. One catalyst design of note was developed by Liu et al. using salicylaldimine ligands with $PdCl_2$ to form catalytic species in situ, which were capable of coupling a wide range of aryl bromides between room temperature and 60 °C under 6 h (Figure 1) [33]. The use of pre-formed neutral Pd [N,O] catalysts in green solvents has also been observed in recent work by Muthumari et al. which was capable of coupling alkyl bromides with a range of boronic acids in yields of approximately 90% after 3 h at 100 °C in water. The catalyst was also capable of coupling 4–chloroacetophenone to phenyl boronic acid, but this required 8 h at 100 °C [34].

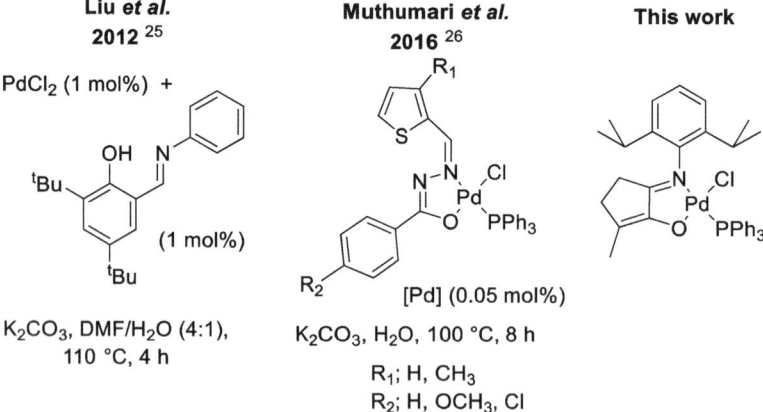

Figure 1. Selected [N,O] catalysts capable of Suzuki-Miyaura cross-coupling reactions [25,26].

We have previously developed [N,O]-chelating ligands based on maple lactone, 2-hydroxy–3–methylcyclopent–2–enone, also called cyclotene [35,36], which when condensed with a substituted aniline, gave ligands featuring a 5-membered chelate ring and a correspondingly smaller bite-angle than salicylaldimines [37]. Complexation with [Ni(Ar)(PPh$_3$)] fragments gave catalysts for ethylene polymerization generating different polyethylene properties depending upon the initiator [38]. These ligands are based on maple lactone, a cheap, bio-sourced material that is useful in the development of catalysts with inexpensive ligand systems. With limited work into the development of [N,O]-ligated catalysts in cross-coupling, we set out to investigate the use of a Pd catalyst with a 5-membered chelating ring in Suzuki-Miyaura cross-coupling reactions using benign conditions. The work presented herein includes the synthesis, complexation and characterization of an active catalyst, optimization studies for the Suzuki–Miyaura cross-coupling reaction and a substrate scope for both cross-coupling partners.

2. Results and Discussion

2.1. Catalyst Synthesis

Synthesis of the proligand was carried out as previously detailed through the condensation of maple lactone with 2,6–diisopropylaniline using catalytic amounts of *p*-toluenesulfonic acid (Scheme 1) with single substitution seen to occur selectively at the ketone position [38]. The reaction with NaH gave the corresponding sodium imino–enolate (**2**) that reacted smoothly with [PdCl$_2$(COD)] and PPh$_3$ to give the desired Pd complex **3** in 94% yield.

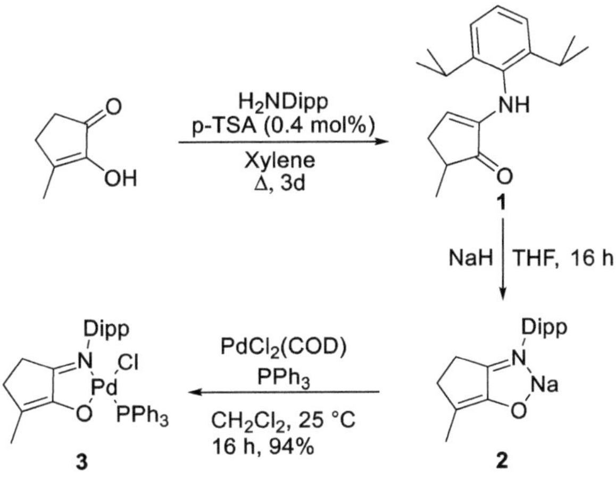

Scheme 1. Synthesis of the Pd complex **3**. Dipp = 2,6–diisopropylphenyl.

3 was characterized using multinuclear NMR spectroscopy, single-crystal X-ray diffraction (SCXRD), elemental analysis and high-resolution mass spectrometry. The NMR spectroscopic studies showed free rotation of the aniline group around the metal centre with a pair of doublets at 1.20 ppm and 1.44 ppm, each with an integration equivalent to 6H, and a septet at 3.46 ppm with an integration of 2H, indicative of two distinct environments for the Dipp protons. Alongside this, signals corresponding to the PPh$_3$ group were also seen, including a singlet at 26.6 ppm in the ^{31}P{^{1}H} NMR spectrum.

The molecular structure of **3** was determined using SCXRD and showed a Pd(II) complex with square planar geometry (Figure 2). The N-Pd-O bite angle was 82.74(7)° with the imine donor *trans* to PPh$_3$. This is similar to the other reported 5-membered chelating [N,O] Pd complexes displaying bite angles between 78.46(6)° and 85.27(4)°. Similar bond lengths to those in the literature were also observed with bond lengths of Pd-O; 2.018(2) Å and Pd–N; 2.099(2) Å for compound **3** [34,39,40]. The imino–enolate tautomer was identified from the short C1–C5 bond length (1.359(4) Å) and the longer C2–C3 bond length (1.497(3) Å). We note that **3** is air and moisture stable over a 3-month period, which is of significant importance for use in Suzuki-Miyaura reactions, as this allows for easier handling and the use of hydrous solvents.

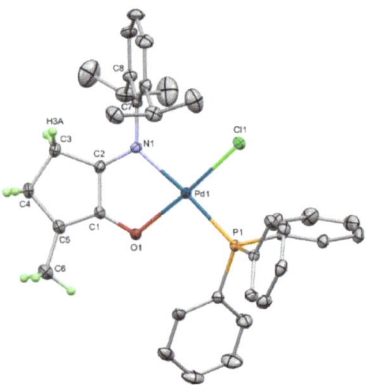

Figure 2. Molecular structure for compound **3** with 50% ellipsoids (all H atoms, except for selected H atoms on the imino-enolate ligand, are omitted for clarity). Selected bond lengths (Å) and bond angles (°): Pd1-Cl1 2.2905(6), Pd1-P1 2.2530(6), Pd1-O1 2.018(2), Pd1-N1 2.099(2), O1-C1 1.328(3), C1-C5 1.359(4), C1-C2 1.442(3), N1-C2 1.294(3), C2-C3 1.497(3), C3-C4 1.535(4), C4-C5 1.502(4), O1-Pd1-N1 82.74(7).

2.2. Cross-Coupling Reactions

The initial reactions between 4–bromoacetophenone and phenylboronic acid were carried out using methanol as the solvent under both atmospheric conditions and in an anaerobic N_2 environment at either room temperature or 40 °C (Table 1). These substrates were chosen as they represent common reactions within the literature allowing for the benchmarking of results; in addition, the acetophenone group on the halogen coupling partner acts as an unambiguous ^{1}H NMR spectroscopic handle for reaction monitoring through comparative integration of the respective signals in the starting material at 2.44 ppm and the product at 2.50 ppm. Assignment of these signals were confirmed through both starting material characterization and doping of additional starting material into the final reaction mixture.

Table 1. Initial optimization studies carried out between 4–bromoacetophenone and phenylboronic acid.

Entry	Air/N_2	Time (hr)	Temperature (°C)	NMR Conversion [a]
1	N_2	2	r.t.	79%
2	Air	2	r.t.	90%
3	N_2	1	r.t.	73%
4	Air	1	r.t.	85%
5	N_2	1	40	83%
6	Air	1	40	99%

[a] Conversion was determined with ^{1}H NMR spectroscopy via integration against starting material of the aryl halide and the product, as outlined above.

From these initial studies, it was determined that carrying the reaction out in air was not detrimental to the yield with a conversion of 90% (Table 1, entry 2) compared to 79%

(Table 1, entry 1) when the reaction was carried out under N_2 over a two-hour period at room temperature. The difference between a one-hour and two-hour reaction time was small with an 85% completion over 1 h in air (Table 1, entry 4) compared to 90% completion after 2 h in air at room temperature (Table 1, entry 2). Room temperature was selected for further reactions, incorporating 'green' chemistry principles and benign conditions, and to allow for easier observations of the different optimization studies.

Following this initial screen of reaction time and temperature, a base screen was carried out using standardised conditions of 1 h at room temperature with a 1 mol% catalytic loading in methanol. The choice of base has an important effect on the rate of the reaction with the base playing various roles within the catalytic cycle [11,41–45]. The hydroxide ion is known to promote the transmetallation and the reductive elimination steps; however, a concentration of hydroxide that is too high has a negative impact, enhancing the formation of unreactive boronates. The different bases screened can be seen in Table 2 and were chosen to give a range of anions and cations.

Table 2. Base optimization screen with the range of bases trialled and the NMR spectroscopic conversion obtained.

Entry	Base	NMR Conversion
7	Et$_3$N	No reaction observed
8	NaOH	80%
9	K$_3$PO$_4$	65%
10	K(OAc)	45%
11	K$_2$CO$_3$	81%
12	Na$_2$CO$_3$	82%
13	Cs$_2$CO$_3$	71%
14	No base	2%

When the neutral base Et$_3$N (Table 2, entry 7) was used, no reaction was observed. The classic ionic bases enabled the catalytic reaction to proceed; acetates and phosphates displayed moderate conversions (Table 2, entries 9, 10), whilst hydroxides and carbonates showed good conversion (Table 2, entries 8 and 11). Following these experiments, carbonate anions were chosen, and through comparison of the cations (Table 2, entries 11–13), it was seen that sodium carbonate and potassium carbonate were similar in performance; therefore, sodium carbonate was chosen for future reactions. For the control studies, the reaction was also trialled with no base added, which showed very little reaction (2% conversion, Table 2, entry 14).

A solvent screen was then carried out on this reaction at room temperature with 1 mol% catalyst and Na$_2$CO$_3$ as the base.

Acetone is a poor solvent for this reaction with a low conversion of 23% (Table 3, entry 15) along with dimethyl carbonate (Table 3, entry 16), trialled due to its growing relevance as a green solvent, with a poor conversion of 4%. Other typical laboratory solvents, including toluene, DCM and DMF, also showed very poor conversion of below 8% (Table 3, entries 17–20). Whilst surprising due to the common use of these solvents within the literature for Suzuki-Miyaura cross-coupling reactions [4,14,15], this could be due to the low solubility of Na_2CO_3 in these solvents at room temperature with the literature examples requiring refluxing conditions. Further evidence of this low reactivity being caused by poor base solubility can be seen in entry 21 when a 50:50 DMF/H_2O solvent mixture was utilised. In this entry, the NMR conversion increased from 4% to 14%, which is likely due to the increased solubility of the reagents. Following the success seen with methanol in the initial reactions, a screen of common alcohols was then carried out (Table 3, entries 22–24). This showed an increase in catalytic activity with lower molecular weight alcohols, with methanol enabling the highest catalytic activity. Methanol was then trialled with the addition of water (Table 3, entries 25–26). Entry 25 shows that the addition of water was beneficial for an improved conversion, possibly due to the increased solubility of the reagents, with the conversion increasing from 80% to 88%. There is a balance with the number of equivalents of water though, as by decreasing the proportion of water to 25% (entry 26), a further increase in conversion is seen to 97%.

Table 3. Solvent screen results with solvent used and NMR conversion.

Entry	Solvent	NMR Conversion
15	Acetone	23%
16	Dimethylcarbonate	4%
17	Toluene	2%
18	THF	7%
19	DCM	2%
20	DMF	4%
21	DMF/H_2O (50:50)	14%
22	IPA	21%
23	EtOH	52%
24	MeOH	80%
25	$MeOH/H_2O$ (50:50)	88%
26	$MeOH/H_2O$ (75:25)	97%

From the solvent scope, a mixed solvent system of $MeOH/H_2O$ (75:25) was then chosen for further reactions. This result contributes towards the initial aim of improving the reaction with the use of 'green' solvents since methanol has a low environment, health and safety (EHS) indicator of 2.67 and is, therefore, classified as a 'green' solvent alongside water [12,16].

Following these optimisations, the catalytic loading of the system was investigated to try to lower the catalytic loading from 1 mol%, using sodium carbonate as the base, with a 75:25 $MeOH:H_2O$ solvent mixture at room temperature for an hour, as outlined in Table 4.

Table 4. Results from the catalytic loading screen with NMR conversion and calculated TON from the conversion.

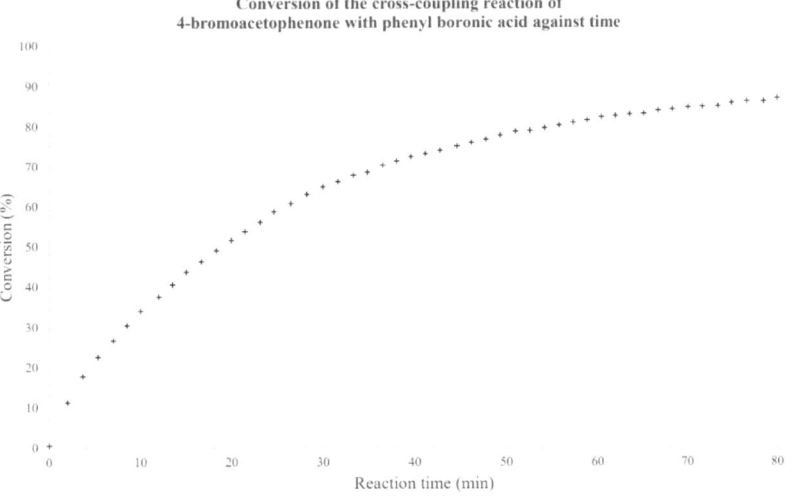

Entry	Catalyst Loading	NMR Conversion	TON
27	0.001 mol%	0.3%	300
28	0.01 mol%	2%	200
29	0.1 mol%	43%	430
30	1 mol%	97%	97

From these results, it can be seen that a very low catalyst loading of 0.001 mol% results in very little conversion (Table 4, entry 27). This also can act as a control run with very little catalyst effectively shutting off the reaction. Increasing the catalytic loading can be seen to increase the conversion. This TON is lower than other catalysts reported in the literature [8] and may be due to the mild conditions utilised in this screen. However, the TON is higher than other reported catalysts in the literature in mild conditions with the work by Wang et al. exhibiting a TON of 316. This experiment was carried out over 8 h, however, with a more comparable TON calculated after 1 h of 160 [17]. Comparisons with phosphine complexes at room temperatures are still low with a comparative TON calculated at 9900 at room temperature, but this was again calculated over 4 h [18].

The rate of conversion over time was investigated through the use of an NMR scale reaction and monitored using the addition of mesitylene to act as an internal standard with MeOD as the solvent (Figure 3). This graph shows that no induction period was observed with the fastest rate of reaction occurring at the start of the experiment (Table 4).

Figure 3. Reaction profile for the NMR conversion of the cross-coupling reaction of 4–bromoacetophenone with phenyl boronic acid against time.

The temperature of the reaction was then screened to investigate how this affected the NMR spectroscopic conversion. Due to the high conversion obtained under 1 mol% loading in mild conditions and the relative short time scale of the reaction, these experiments were carried out using 0.01 mol% of catalyst to allow for the effects to be seen (Table 5).

Table 5. Temperature and NMR conversion obtained from the temperature screen.

Entry	Temperature (°C)	NMR conv.
39	20	2%
40	40	10%
41	60	99%
42	Reflux	99%

From this screen, a dramatic increase in reactivity was observed above 40 °C. This increase could be due to catalyst degradation resulting in the formation of Pd(0) nanoparticles changing, or in fact producing, the active catalytic species. In order to test for the formation of Pd(0) nanoparticles, a mercury drop test was carried out [7]. Whilst this is not a completely infallible test, we did observe a decrease in the NMR conversion from 97% to an average of 40%, leaving the role of nanoparticles as a possibility [32].

2.3. Substrate Scope

With the catalytic conditions optimized, the study then progressed to establishing the substrate scope. This was first carried out across a range of aryl halides. The conditions used were a 1 mol% [Pd] loading at room temperature for an hour utilizing MeOH/H_2O (75:25) as the solvent (Table 5). These conditions were chosen to maintain the 'green' chemistry of the system and limit the possible catalyst degradation through heating of the system.

An initial screen of groups with varying electron-donating/withdrawing properties showed that the cross-coupling reaction proceeded well when strong or moderate electron-withdrawing groups were used (Table 6, entries 46 and 45, respectively). When weak electron-donating groups were used at the *ortho–* and *para–* positions (Table 6, entries 47 and 48), the conversion decreased to 82% and 75%, respectively. Comparing the *ortho–* and *para–*bromotoluene to *meta–*bromotoluene (Table 6, entry 49), the conversion dropped significantly to 15%. This effect continued further when a strong electron-donating group was used in entry 50 when only trace amounts of product were obtained. Entry 45 was subjected to column chromatography to check for homo-coupling of either partner. This resulted in an isolated yield of 70% with all other components isolated relating to trace starting materials. These trends have been seen in the literature; previous salicylaldimine ligated catalysts showed decreasing activity with increasing electron-donating substituents [4]. Electron-donating groups were also seen to slow the catalysis down in the work by Ref. [23].

Table 6. Substrate scope investigating the effect of changing the aryl bromide coupling to phenylboronic acid.

Entry	Aryl Bromide	NMR conv.
45	4-bromoacetophenone	96%
46	4-nitrobromobenzene	97%
47	4-bromotoluene	82%
48	2-bromotoluene	75%
49	3-bromotoluene	15%
50	4-bromoaniline	Trace
51 [1]	4-bromoaniline	29% [a]
52 [2]	4-bromoaniline	87% [a]

[1]: Reaction carried out at 40 °C for 24 h. [2]: Reaction carried out at 60 °C for 24 h. [a]: Isolated yield obtained.

The reaction was carried out at increased temperatures over a longer reaction time with the isolated yield recorded for added clarity because no convenient characteristic signals were observed through ^{1}H NMR spectroscopy. This showed an increase in reactivity with an isolated yield of 29% obtained at 40 °C after 24 h. This was further improved with heating at 60 °C for 24 h instead, resulting in an 87% isolated yield (Table 6, entries 51 and 52).

A similar study was then carried out for the boronic acid scope with a range of electronic properties present on the aryl ring. Phenylboronic acid pinacol ester was also included in this trial for tolerance of boronic acid derivatives.

The catalyst showed a broad reaction scope tolerating many substituents except for a *para*–nitro substituent (a strong electron-withdrawing group), which displayed a low conversion of 17% (Table 7, entry 58). Phenylboronic acid pinacol ester (Table 7, entry 59) displayed quantitative conversion, indicating that the catalyst is tolerant towards pinacol esters. This trend has also been seen in the literature with electron deficient boronic acids showing reduced catalytic activity [8,23].

Table 7. Substrate scope investigating the effect of changing the boronic acid coupling to 4–bromoacetophenone.

Entry	R'	NMR Conversion
53	4-methoxybenzene boronic acid	99%
54	4-tolylboronic acid	97%
55	2-tolylboronic acid	94%
56	3-tolylboronic acid	99%
57	4-methoxycarbonylbenzene boronic acid	Quant.
58	4-nitrobenzene boronic acid	17%
59	Phenylboronic acid pinacol ester	Quant.

There is an incentive to be able to use aryl chlorides as the coupling partner in these reactions due to their decreased cost compared to aryl bromides. With that in mind, compound **3** was trialled for catalytic activity towards aryl chlorides using the conditions previously optimised (Table 8).

Table 8. Cross-coupling reaction between 4–chloroacetophenone and phenyl boronic acid in varying conditions.

Entry	Temperature	Time (hr)	NMR Conversion
60	Room temperature	1 hr	trace
61	40 °C	24 hr	62%
62	60 °C	1 hr	42%
63	60 °C	6 hr	74%
64	60 °C	24 hr	95%

The initial conditions of room temperature for one hour proved insufficient for the reaction to proceed with only trace quantities of product observed. Because of this, the temperature was increased to 40 °C for 24 h (Table 8, entry 61). This increase in temperature and prolonged reaction time was sufficient for the reaction to proceed with an NMR spectroscopic conversion of 62% observed. A further increase in temperature to 60 °C allowed for a higher rate of conversion with 42% observed over 1 h, 74% over 6 h, and 95% after 24 h (Table 8, entries 62–64).

Through comparison of these results to other [N,O] catalysts in the field, compound **3** shows good catalytic activity. Work by Muthumari et al. required high temperatures of 100 °C for 8 h for the coupling of aryl chlorides when using 4-methoxychlorobenzene with phenyl boronic acid with a comparative isolated yield of 58% [8]. Cui et al. utilised an [N,O]-ligated palladium catalyst for the reaction of 4–chloroacetophenone with phenyl boronic acid, offering a more direct comparison between the catalytic systems [24]. These catalysts were capable of completing this reaction to high conversion (95%) with a lower catalytic loading of 0.5 mol% but required DMF at 110 °C for 4 h. Similar high conversions were also observed by Liu et al. with [N,O] catalysts generated in situ [4]. These reaction conditions were also forcing with a DMF:H_2O (80:20) solvent heated to 110 °C for 4 h. In comparison to this, catalyst **3** was capable of cross-coupling 4–chloroacetophenone to phenyl boronic acid to moderate yields in 6 h and high yields over 24 h at 60 °C.

3. Experimental

Reactions were either performed under an oxygen-free (H_2O, O_2 < 0.5 ppm) nitrogen atmosphere using standard Schlenk line techniques and an MBRAUN UNIlab Plus glovebox or in the open laboratory, as indicated. Anhydrous toluene, anhydrous DCM and anhydrous THF were obtained from an MBRAUN SPS–800,(MBRAUN, Munich, Germany) and petroleum ether 40–60 was distilled from sodium wire; benzene and benzene-d^6 were dried over molten potassium and distilled. All anhydrous solvents were degassed before use and stored over activated molecular sieves.

The following compounds were prepared according to the literature methods: 2–((2,6-diisopropylphenyl)amino)–5–methylcyclopent–2–en–1–one (**1**), **2**, [38] and $PdCl_2$(COD) [46]. The following were purchased from commercial suppliers and used without further purification: maple lactone, p–toluenesulfonic acid, sodium hydride (95%), palladium chloride, 1,5–cyclooctadiene and triphenylphosphine. 2,6–Diisopropylaniline was distilled under reduced pressure before use. Air-sensitive samples for NMR spectroscopy were prepared in NMR tubes equipped with a J. Young tap. The NMR spectra were recorded on a Bruker AV300 (Bruker, MA, USA) (300 MHz), AVI400 (400 MHz), AVIII400 (400 MHz) or AVHDIII (400 MHz) spectrometer at 25 °C unless specified. Chemical shifts δ are noted in parts per million (ppm). ^{1}H and ^{13}C spectra were calibrated to the residual proton resonances of the deuterated solvent. ^{31}P NMR spectra were referenced to external samples of 85% H_3PO_4 at 0 ppm. The elemental analyses were performed by Brian Hutton at Heriot-Watt University using an Exeter CE440 elemental analyser. (Exeter Analytical, Coventry, UK)The mass spectrometry analysis was performed at the UK National Mass Spectrometry Facility at Swansea University using an Atmospheric Solids Analysis Probe interfaced to a Waters Xevo G2–S instrument (Waters, MA, USA).

[Pd(Cl)(N,O^{Dipp})(PPh$_3$)] (3)

[PdCl$_2$(cod)] (0.371 g, 1.30 mmol), **2** (0.400 g, 1.365 mmol, 1.05 eq) and PPh$_3$ (0.341 g, 1.30 mmol) were weighed into a Schlenk tube in a glovebox before CH_2Cl_2 (0.8 mL) was added. The solution was then stirred at room temperature overnight, and the solvent was removed via rotatory evaporation yielding a dark-brown solid. The solid was then dissolved in toluene and filtered before evaporation of the solvent under reduced pressure. Recrystallisation was then carried out using THF/petroleum ether 40–60 to yield the desired product as dark red crystals (0.826 g, 1.22 mmol, 94%).

^{1}**H NMR (400 Hz, 298 K, C$_6$D$_6$):** δ 7.73 (6H, m, H$_{20}$), 7.46 (m, 3H, H$_{22}$), 7.38 (m, 6H, H$_{21}$), 7.16 (3H, m, H$_{9-11}$), 6.24 (1H, td, J = 7.3, 1.2 Hz, H$_{Ni-o-tolyl}$), 3.46 (2H, sept, J = 7.3 Hz, H$_{13,16}$), 2.57 (bs, 2H, H$_{4/3}$), 2.02 (bs, 2H, H$_{3/4}$), 1.67 (s, 3H, H$_6$), 1.44 (6H, d, J = 6.8 Hz, H$_{iPr}$), 1.20 (6H, d, J = 6.8 Hz, H$_{iPr}$). ^{13}C{^{1}H} **NMR (100.6 Hz, 298 K, C$_6$D$_6$):** δ 194.06 (C), 166.65 (C), 141.25 (C), 140.84 (C), 137.84 (C), 135.10 (CH), 135.00 (CH), 132.32 (CH), 130.80 (CH), 129.16 (C), 129.03 (CH), 128.62 (C), 128.21 (CH), 128.03 (CH), 127.92 (CH), 126.56 (CH), 125.29 (CH), 123.15 (CH), 34.44 (CH$_2$), 28.31 (CH), 24.82 (CH$_2$), 23.99 (CH$_3$), 23.86 (CH$_2$), 21.41 (CH$_3$), 13.15 ppm (CH$_3$). ^{31}P{^{1}H} **NMR (162 Hz, 298 K, C$_6$D$_6$):** δ 26.6 ppm. **HRMS (ASAP):** Calculated for [M+H]: 674.1581, Found: 674.1575 m/z; calculated for [M-Cl]: 638.1818, Found: 638.1816 m/z. **Elemental analysis** calcd (%) for C$_{15}$H$_{19}$NO: C; 64.10, H; 5.83, N; 2.08. Found: C; 63.82, H; 5.86, N; 2.21.

3.1. Crystallographic Details

The X-ray diffraction experiments were performed on single crystals of **3** with the data collected on a Bruker X8 APEX II (Bruker, MA, USA) four-circle diffractometer. The crystal was kept at 100 K during the data collection. Indexing, data collection and absorption corrections were performed, and the structures were solved using direct methods (SHELXT [47] and refined with full-matrix least-squares (SHELXL) interfaced with the programme OLEX2 [48,49].

Crystal data for C$_{36}$H$_{39}$ClNOPPd (M = 674.50 g/mol): orthorhombic, space group P2$_1$2$_1$2$_1$ (no. 19), a = 8.9925(4) Å, b = 16.4896(8) Å, c = 22.0909(10) Å, V = 3275.7(3) Å^3, Z = 4, T = 100.15 K, µ(synchrotron) = 0.765 mm^{-1}, Dcalc = 1.368 g/cm^3, 103,238 reflections measured (4.552° ≤ 2Θ ≤ 62.764°), 9994 unique (R$_{int}$ = 0.0781, R$_{sigma}$ = 0.0398), which were used in all calculations. The final R_1 was 0.0298 (I > 2σ(I)), and wR_2 was 0.0726 (all data). CCDC deposition number: 2102831.

3.2. Suzuki-Miyaura Cross-Coupling Reactions

A typical cross-coupling reaction was carried out as follows:

4–bromoacetophenone (0.100 g, 0.5 mmol, 1 eq.), phenyl boronic acid (0.092 g, 0.754 mmol, 1.5 eq.), Na$_2$CO$_3$ (0.1065 g, 1.005 mmol, 2 eq.) and [PdCl(N,O^{Dipp})(PPh$_3$)] (3.4 mg, 0.005 mmol, 1 mol%) were added to a 100 mL round-bottomed flask equipped with a stirrer bar. A solvent mixture of methanol:water 3:1 (10 mL) was added, and the mixture was stirred for 1 h at room temperature. Water (10 mL) was then added to the flask followed by diethyl ether (10 mL), and the organic phase was collected by separation. This was then dried over MgSO$_4$, filtered and the solvent removed by rotatory evaporation. ^{1}H NMR spectroscopic analysis was then carried out in CDCl$_3$, and the conversion was calculated through integration of the resonances at 2.60 ppm (starting material) and 2.65 ppm (product). The product was then isolated using column chromatography (199:1 petroleum ether 40–60:ethyl acetate) as an off-white solid (0.069 g, 0.35 mmol, 70%). **^{1}H NMR (400 Hz, 298 K, C$_6$D$_6$):** δ 7.92 (2H, d, *J* = 8.8 Hz), 7.57 (2H, d, *J* = 8.5 Hz), 7.51 (2H, m), 7.36 (2H, m), 7.29 (1H, m), 2.52 (3H, s). The data matches the literature values [10].

3.3. Cross-Coupling Reaction of 4–Bromoaniline at Elevated Temperatures

4–bromoaniline (0.0864 g, 0.5 mmol, 1 eq.), phenyl boronic acid (0.092 g, 0.754 mmol, 1.5 eq.), Na$_2$CO$_3$ (0.1065 g, 1.005 mmol, 2 eq.) and [PdCl(N,O^{Dipp})(PPh$_3$)] (3.4 mg, 0.005 mmol, 1 mol%) were added to a 100 mL round-bottomed flask equipped with a stirrer bar. A solvent mixture of methanol:water 3:1 (10 mL) was added, and the mixture was stirred at 60 °C for 24 h. The mixture was then cooled to room temperature before water (10 mL) was then added to the flask followed by diethyl ether (10 mL), and the organic phase was collected by separation. This was then dried over MgSO$_4$, filtered and the solvent removed by rotatory evaporation. The product was then isolated using column chromatography (1:1 petroleum ether 40–60:ethyl acetate) with a small amount of DCM added to improve solubility to yield a brown solid (0.0736 g, 0.43 mmol, 87%). **^{1}H NMR (400 Hz, 298 K, C$_6$D$_6$):** δ 7.47 (2H, d, *J* = 8.3 Hz), 7.34 (4H, m), 7.20 (1H, m), 6.69 (2H, d, *J* = 8.4 Hz), 3.66 (2H, br. s, NH$_2$). The data matches the literature values [10].

4. Conclusions

In conclusion, a Pd(II) complex with a 5-membered chelating [N,O] ligand derived from maple lactone was synthesized and characterized. This complex is a rare example of a pre-formed [N,O] Pd catalyst for the Suzuki-Miyaura cross-coupling reaction, and screening revealed good catalytic activity even at low temperatures. The catalytic activity was optimized through the selection of base, solvent, catalytic loading and temperature to yield conditions that involve a mild, room-temperature reaction at a short timescale (1 h) using solvents that are typically considered to be 'green'. Using these optimised conditions, a substrate scope identified that the catalyst showed good activity towards aryl bromides with electron-withdrawing groups. This activity decreased with weak electron-donating groups, and only trace quantities of products were observed with strong electron-donating groups. However, these less reactive substrates were capable of being coupled through increased temperature and timescale with an 87% isolated yield of the cross-coupled product derived from 4–bromoaniline obtained after 24 h at 60 °C. The catalyst also showed good activity across a range of boronic acids and pinacol esters with only boronic acids featuring strong electron-withdrawing groups showing limited activity. The catalyst was also a capable catalyst for the cross-coupling of aryl chlorides and phenylboronic acid. This reaction, however, required slightly elevated temperatures of 60 °C over a longer timescale but is still considered mild compared to other examples of these compounds in the literature.

Author Contributions: M.J.A. and S.B. synthesised and characterised the proligands and catalyst. M.J.A. carried out catalytic studies. M.J.A., S.M.M. and R.D.M. undertook the analysis of the results. M.J.A., S.M.M. and R.D.M. conceived the project and contributed to writing the paper. All authors have read and agreed to the published version of the manuscript.

Funding: This research was funded by EPSRC, grant number EP/L016419/1.

Data Availability Statement: Additional research data supporting this publication are available from Heriot-Watt University's research data repository at DOI: 10.17861/68f1c4a3-630a-4e7d-ac72-15ed00dbc68e.

Acknowledgments: The authors thank the UK National Mass Spectrometry Facility at Swansea University for assistance with sample analysis. David Ellis (HWU) is thanked for assistance with NMR spectroscopy, and Brian Hutton (HWU) for CHN analysis. We would like to thank the Engineering and Physical Sciences Research Council and the EPSRC Centre for Doctoral Training in Critical Resource Catalysis (CRITICAT) for financial support [M.A; Grant code: EP/L016419/1].

Conflicts of Interest: The authors declare no conflict of interest.

References

1. Miyaura, N.; Suzuki, A. Palladium-Catalyzed Cross-Coupling Reactions of Organoboron COMPOUNDS. *Chem. Rev.* **1995**, *95*, 2457–2483. [CrossRef]
2. Suzuki, A. Cross-Coupling Reactions of Organoboranes: An Easy Way to Construct C-C Bonds (Nobel Lecture). *Angew. Chem. Int. Edit.* **2011**, *50*, 6722–6737. [CrossRef]
3. Suzuki, A. Recent advances in the cross-coupling reactions of organoboron derivatives with organic electrophiles, 1995–1998. *J. Organomet. Chem.* **1999**, *576*, 147–168. [CrossRef]
4. Seechurn, C.; Kitching, M.O.; Colacot, T.J.; Snieckus, V. Palladium-Catalyzed Cross-Coupling: A Historical Contextual Perspective to the 2010 Nobel Prize. *Angew. Chem. Int. Edit.* **2012**, *51*, 5062–5085. [CrossRef]
5. Farhang, M.; Akbarzadeh, A.R.; Rabbani, M.; Ghadiri, A.M. A retrospective-prospective review of Suzuki-Miyaura reaction: From cross-coupling reaction to pharmaceutical industry applications. *Polyhedron* **2022**, *227*, 116214. [CrossRef]
6. D'Alterio, M.C.; Casals-Cruañas, E.; Tzouras, T.V.; Talarico, G.; Nolan, S.P.; Poater, A. Mechanistic Aspects of the Palladium-Catalyzed Suzuki-Miyaura Cross-Coupling Reaction. *Chem. Eur. J.* **2021**, *27*, 13481–13493. [CrossRef]
7. Phan, N.T.S.; Van Der Sluys, M.; Jones, C.W. On the nature of the active species in palladium catalyzed Mizoroki-Heck and Suzuki-Miyaura couplings–Homogeneous or heterogeneous catalysis, a critical review. *Adv. Synth. Catal.* **2006**, *348*, 609–679. [CrossRef]
8. Das, P.; Linert, W. Schiff base-derived homogeneous and heterogeneous palladium catalysts for the Suzuki-Miyaura reaction. *Coord. Chem. Rev.* **2016**, *311*, 1–23. [CrossRef]
9. Maluenda, I.; Navarro, O. Recent Developments in the Suzuki-Miyaura Reaction: 2010–2014. *Molecules* **2015**, *20*, 7528–7557. [CrossRef]
10. Shaaban, M.R.; Farghaly, T.A.; Khormi, A.Y.; Farag, A.M. Recent Advances in Synthesis and Uses of Heterocycles-based Palladium(II) Complexes as Robust, Stable, and Low-cost Catalysts for Suzuki- Miyaura Cross-couplings. *Curr. Org. Chem.* **2019**, *23*, 1601–1662. [CrossRef]
11. Sain, S.; Jain, S.; Srivastava, M.; Vishwakarma, R.; Dwivedi, J. Application of Palladium-Catalyzed Cross-Coupling Reactions in Organic Synthesis. *Curr. Org. Synth.* **2019**, *16*, 1105–1142. [CrossRef]
12. Kotha, S.; Lahiri, K.; Kashinath, D. Recent applications of the Suzuki-Miyaura cross-coupling reaction in organic synthesis. *Tetrahedron* **2002**, *58*, 9633–9695. [CrossRef]
13. Scheuermann, G.M.; Rumi, L.; Steurer, P.; Bannwarth, W.; Mulhaupt, R. Palladium Nanoparticles on Graphite Oxide and Its Functionalized Graphene Derivatives as Highly Active Catalysts for the Suzuki-Miyaura Coupling Reaction. *J. Am. Chem. Soc.* **2009**, *131*, 8262–8270. [CrossRef]
14. Stanforth, S.P. Catalytic cross-coupling reactions in biaryl synthesis. *Tetrahedron* **1998**, *54*, 263–303. [CrossRef]
15. Christmann, U.; Vilar, R. Monoligated palladium species as catalysts in cross-coupling reactions. *Angew. Chem. Int. Edit.* **2005**, *44*, 366–374. [CrossRef] [PubMed]
16. Marion, N.; Nolan, S.P. Well-Defined N-Heterocyclic Carbenes-Palladium(II) Precatalysts for Cross-Coupling Reactions. *Acc. Chem. Res.* **2008**, *41*, 1440–1449. [CrossRef] [PubMed]
17. Shen, W. Palladium catalyzed coupling of aryl chlorides with arylboronic acids. *Tetrahedron Lett.* **1997**, *38*, 5575–5578. [CrossRef]
18. Littke, A.F.; Fu, G.C. Palladium-catalyzed coupling reactions of aryl chlorides. *Angew. Chem. Int. Edit.* **2002**, *41*, 4176–4211. [CrossRef]
19. Bei, X.H.; Crevier, T.; Guram, A.S.; Jandeleit, B.; Powers, T.S.; Turner, H.W.; Uno, T.; Weinberg, W.H. A convenient palladium/ligand catalyst for Suzuki cross-coupling reactions of arylboronic acids and aryl chlorides. *Tetrahedron Lett.* **1999**, *40*, 3855–3858. [CrossRef]
20. Herrmann, A.W.; Reisinger, C.-P.; Spiegler, M. Chelating N-heterocyclic carbene ligands in palladium-catalyzed heck-type reactions. *J. Organomet. Chem.* **1998**, *557*, 93–96. [CrossRef]
21. Hazari, N.; Melvin, P.R.; Beromi, M.M. Well-defined nickel and palladium precatalysts for cross-coupling. *Nat. Rev. Chem.* **2017**, *1*, 16. [CrossRef]
22. Andreu, M.G.; Zapf, A.; Beller, M. Molecularly defined palladium (0) monophosphine complexes as catalysts for efficient cross-coupling of aryl chlorides and phenylboronic acid. *Chem. Commun.* **2000**, *24*, 2475–2476. [CrossRef]
23. Anastas, P.; Eghbali, N. Green Chemistry: Principles and Practice. *Chem. Soc. Rev.* **2010**, *39*, 301–312. [CrossRef]

24. Jimenez-Gonzalez, C.; Ponder, C.S.; Broxterman, Q.B.; Manley, J.B. Using the Right Green Yardstick: Why Process Mass intensity Is Used in the Pharmaceutical Industry to Drive More Sustainable Processes. *Org. Process Res. Dev.* **2011**, *15*, 912–917. [CrossRef]
25. Byrne, F.P.; Jin, S.; Paggiola, G.; Petchey, T.H.M.; Clark, J.H.; Farmer, T.J.; Hunt, A.J.; Robert McElroy, C.; Sherwood, J. Tools and techniques for solvent selection: Green solvent selection guides. *Sustain. Chem. Process.* **2016**, *4*, 7. [CrossRef]
26. Capello, C.; Fischer, U.; Hungerbuhler, K. What is a green solvent? A comprehensive framework for the environmental assessment of solvents. *Green Chem.* **2007**, *9*, 927–934. [CrossRef]
27. Hooshmand, S.E.; Heidari, B.; Sedghi, R.; Varma, R.S. Recent advances in the Suzuki-Miyaura cross-coupling reaction using efficient catalysts in eco-friendly media. *Green Chem.* **2019**, *21*, 381–405. [CrossRef]
28. Santoro, S.; Ferlin, F.; Luciani, L.; Ackermann, L.; Vaccaro, L. Biomass-derived solvents as effective media for cross-coupling reactions and C-H functionalization processes. *Green Chem.* **2017**, *19*, 1601–1612. [CrossRef]
29. Nehzat, F.; Grivani, G. Two efficient ligand-assisted systems of two different ionic Schiff base ligands for palladium chloride catalyzed in Suzuki-Miyaura reaction. *J. Chem. Sci.* **2020**, *132*, 12. [CrossRef]
30. Kumar, S. Recent Advances in the Schiff Bases and N-Heterocyclic Carbenes as Ligands in the Cross-Coupling Reactions: A Comprehensive Review. *J. Heterocycl. Chem.* **2019**, *56*, 1168–1230. [CrossRef]
31. Manna, C.K.; Naskar, R.; Bera, B.; Das, A.; Mondal, T.K. A new palladium (II) phosphino complex with ONS donor Schiff base ligand: Synthesis, characterization and catalytic activity towards Suzuki-Miyaura cross-coupling reaction. *J. Mol. Struct.* **2021**, *1237*, 130322. [CrossRef]
32. Lin, W.; Zhang, L.; Ma, Y.; Liang, T.; Sun, W.-H. Sterically enhanced 2-iminopyridylpalladium chlorides as recyclable ppm-palladium catalyst for Suzuki-Miyaura coupling in aqueous solution. *Appl. Organomet. Chem.* **2022**, *36*, e6474. [CrossRef]
33. Liu, F.S.; Huang, Y.T.; Lu, C.; Shen, D.S.; Cheng, T. Efficient salicylaldimine ligands for a palladium-catalyzed Suzuki-Miyaura cross-coupling reaction. *Appl. Organomet. Chem.* **2012**, *26*, 425–429. [CrossRef]
34. Muthumari, S.; Ramesh, R. Highly efficient palladium(II) hydrazone based catalysts for the Suzuki coupling reaction in aqueous medium. *RSC Adv.* **2016**, *6*, 52101–52112. [CrossRef]
35. Wozniak, B.; Spannenberg, A.; Li, Y.H.; Hinze, S.; de Vries, J.G. Cyclopentanone Derivatives from 5-Hydroxymethylfurfural via 1-Hydroxyhexane-2,5-dione as Intermediate. *Chem. Sus. Chem* **2018**, *11*, 356–359. [CrossRef] [PubMed]
36. Ball, D.W. The chemical composition of maple syrup. *J. Chem. Educ.* **2007**, *84*, 1647–U6. [CrossRef]
37. Mansell, S.M. Catalytic applications of small bite-angle diphosphorus ligands with single-atom linkers. *Dalton Trans.* **2017**, *46*, 15157–15174. [CrossRef]
38. Andrews, M.J.; Ewing, P.; Henry, M.C.; Reeves, M.; Kamer, P.C.J.; Muller, B.H.; McIntosh, R.D.; Mansell, S.M. Neutral Ni(II) Catalysts Based on Maple-Lactone Derived N,O Ligands for the Polymerization of Ethylene. *Organometallics* **2020**, *39*, 1751–1761. [CrossRef]
39. Fild, M.; Thone, C.; Totos, S. Synthetic and structural studies of Pd-II and Pt-II complexes with quincorine and quincoridine derivatives. *Eur. J. Inorg. Chem.* **2004**, *2004*, 749–761. [CrossRef]
40. Vignesh, A.; Shalini, C.; Dharmaraj, N.; Kaminsky, W.; Karvembu, R. Delineating the Role of Substituents on the Coordination Behavior of Aroylhydrazone Ligands in Pd-II Complexes and their Influence on Suzuki-Miyaura Coupling in Aqueous Media. *Eur. J. Inorg. Chem.* **2019**, *2019*, 3869–3882. [CrossRef]
41. Maegawa, T.; Kitamura, Y.; Sako, S.; Udzu, T.; Sakurai, A.; Tanaka, A.; Kobayashi, Y.; Endo, K.; Bora, U.; Kurita, T.; et al. Heterogeneous Pd/C-Catalyzed Ligand-Free, Room-Temperature Suzuki-Miyaura Coupling Reactions in Aqueous Media. *Chem. Eur. J.* **2007**, *13*, 5937–5943. [CrossRef]
42. Amatore, C.; Jutand, A.; le Duc, G. Kinetic Data for the Transmetalation/Reductive Elimination in Palladium-Catalyzed Suzuki-Miyaura Reactions: Unexpected Triple Role of Hydroxide Ions Used as Base. *Chem. Eur. J.* **2011**, *17*, 2492–2503. [CrossRef]
43. Carrow, B.P.; Hartwig, J.F. Distinguishing Between Pathways for Transmetalation in Suzuki−Miyaura Reactions. *J. Am. Chem. Soc.* **2011**, *133*, 2116–2119. [CrossRef]
44. Schmidt, A.F.; Kurokhtina, A.A.; Larina, E.V. Role of a base in Suzuki-Miyaura reaction. *Russ. J. Gen. Chem.* **2011**, *81*, 1573–1574. [CrossRef]
45. Schmidt, A.F.; Kurokhtina, A.A. Distinguishing between the homogeneous and heterogeneous mechanisms of catalysis in the Mizoroki-Heck and Suzuki-Miyaura reactions: Problems and prospects. *Kinet. Catal.* **2012**, *53*, 714–730. [CrossRef]
46. Bailey, C.T.; Lisensky, G.C. Synthesis of Organometallic Palladium Complexes–An Undergraduate Experiment. *J. Chem. Educ.* **1985**, *62*, 896–897. [CrossRef]
47. Sheldrick, G.M. Crystal structure refinement with Shelxl. *Acta Crystallogr. Sect. C Struct. Chem.* **2015**, *71*, 3–8. [CrossRef]
48. Dolomanov, O.V.; Bourhis, L.J.; Gildea, R.J.; Howard, J.A.K.; Puschmann, H. OLEX2: A complete structure solution, refinement and analysis program. *J. Appl. Crystallogr.* **2009**, *42*, 339–341. [CrossRef]
49. Bourhis, L.J.; Dolomanov, O.V.; Gildea, R.J.; Howard, J.A.K.; Puschmann, H. The anatomy of a comprehensive constrained, restrained refinement program for the modern computing environment-Olex2 dissected. *Acta Crystallogr. Sect. A* **2015**, *71*, 59–75. [CrossRef]

Disclaimer/Publisher's Note: The statements, opinions and data contained in all publications are solely those of the individual author(s) and contributor(s) and not of MDPI and/or the editor(s). MDPI and/or the editor(s) disclaim responsibility for any injury to people or property resulting from any ideas, methods, instructions or products referred to in the content.

Review

Photocatalyzed Oxygenation Reactions with Organic Dyes: State of the Art and Future Perspectives

Mattia Forchetta, Francesca Valentini, Valeria Conte *, Pierluca Galloni and Federica Sabuzi *

Department of Chemical Science and Technologies, University of Rome Tor Vergata, 00133 Rome, Italy
* Correspondence: valeria.conte@uniroma2.it (V.C.); federica.sabuzi@uniroma2.it (F.S.)

Abstract: Oxygen atom incorporation into organic molecules is one of the most powerful strategies to increase their pharmacological activity and to obtain valuable intermediates in organic synthesis. Traditional oxidizing agents perform very well, but their environmental impact and their low selectivity constitute significant limitations. On the contrary, visible-light-promoted oxygenations represent a sustainable method for oxidizing organic compounds, since only molecular oxygen and a photocatalyst are required. Therefore, photocatalytic oxygenation reactions exhibit very high atom-economy and eco-compatibility. This mini-review collects and analyzes the most recent literature on organo-photocatalysis applications to promote the selective oxygenation of organic substrates. In particular, acridinium salts, Eosin Y, Rose Bengal, cyano-arenes, flavinium salts, and quinone-based dyes are widely used as photocatalysts in several organic transformations as the oxygenations of alkanes, alkenes, alkynes, aromatic compounds, amines, phosphines, silanes, and thioethers. In this context, organo-photocatalysts proved to be highly efficient in catalytic terms, showing similar or even superior performances with respect to their metal-based counterparts, while maintaining a low environmental impact. In addition, given the mild reaction conditions, visible-light-promoted photo-oxygenation processes often display remarkable selectivity, which is a striking feature for the late-stage functionalization of complex organic molecules.

Keywords: organic dyes; photocatalysis; photooxygenation; metal-free photocatalysts; acridinium; Eosin Y; Rose Bengal; 4CzIPN; quinones; flavinium

Citation: Forchetta, M.; Valentini, F.; Conte, V.; Galloni, P.; Sabuzi, F. Photocatalyzed Oxygenation Reactions with Organic Dyes: State of the Art and Future Perspectives. *Catalysts* 2023, *13*, 220. https://doi.org/10.3390/catal13020220

Academic Editor: Carl Redshaw

Received: 17 December 2022
Revised: 13 January 2023
Accepted: 16 January 2023
Published: 18 January 2023

Copyright: © 2023 by the authors. Licensee MDPI, Basel, Switzerland. This article is an open access article distributed under the terms and conditions of the Creative Commons Attribution (CC BY) license (https:// creativecommons.org/licenses/by/ 4.0/).

1. Introduction

Considering the climate emergency and, recently, the surge in oil prices, the research of alternative sources of renewable energy is becoming an increasingly urgent issue to be addressed [1]. At the beginning of the 20th century, Giacomo Ciamician, one of the pioneers of photochemistry, with shrewd foresight, noticed that modern societies needed an energy transition from fossil fuels to solar energy [2]. The Italian chemist claimed this concept as a possibility for his time and a necessity for the future. Later, in 1998, Anastas and Warner coined the 12 principles of green chemistry, based on the concept of minimizing the environmental footprint of chemical processes by reducing or eliminating the use or formation of hazardous substances [3]. Photocatalysis, given the capability to convert visible light to chemical energy, thoroughly embraces green chemistry culture [4]. In particular, photoredox catalysis ranks as an effective sustainable choice in organic synthesis, also allowing the formation of challenging carbon–heteroatom bonds in mild conditions [5]. Metal-based photocatalysts, mainly Ru(II) and Ir(III) bipyridyls complexes, have been extensively studied and described in several reviews [5–8]. However, due to the environmental issues related to rare metals applications, in recent years, the consideration of organic dyes as metal-free photocatalysts has grown rapidly, and, in some cases, even better photocatalytic performances with respect to their metal counterparts have been observed (Scheme 1) [9–13].

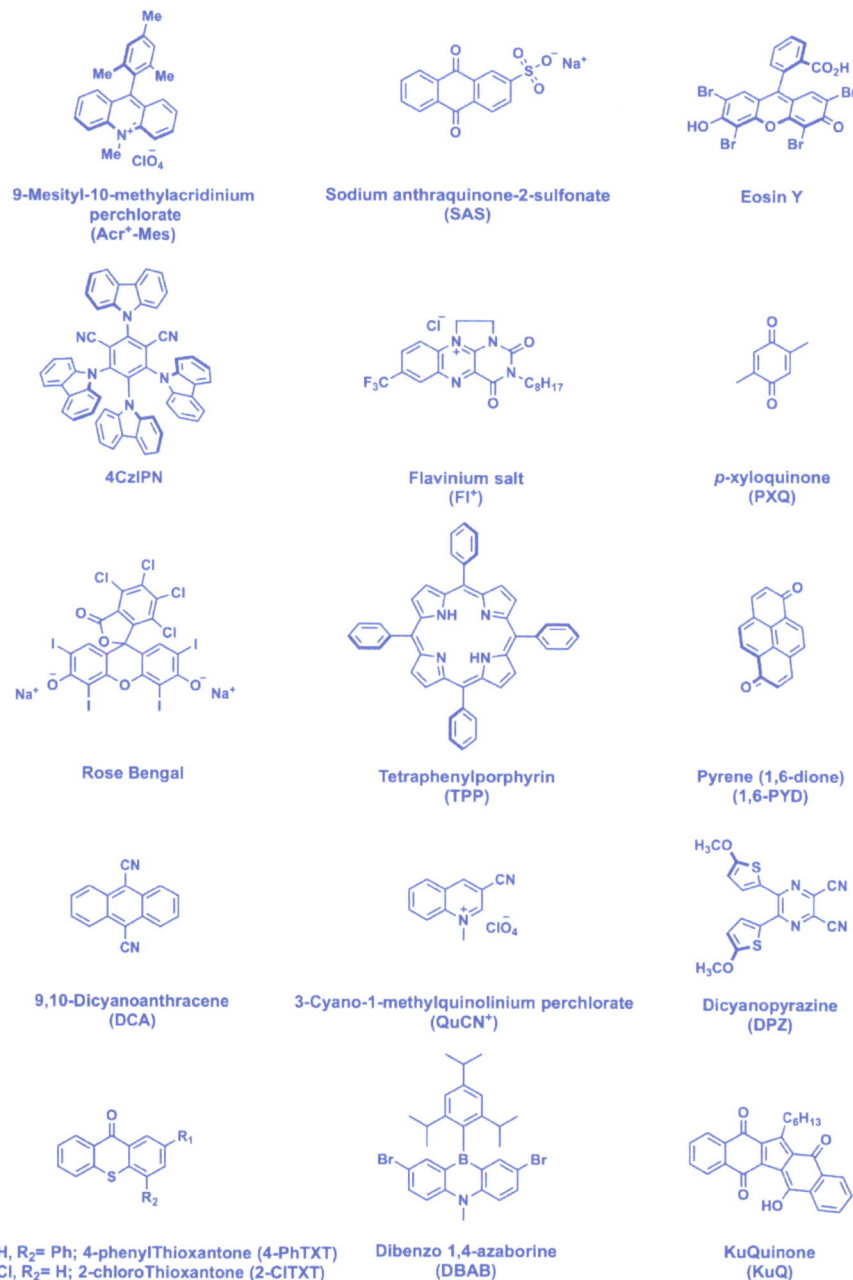

Scheme 1. Organic dyes applied as metal-free photocatalysts in oxygenation reactions.

In particular, 9-mesityl-10-methylacridinium perchlorate (**Acr⁺-Mes**) and cyanoarenes [14,15], **Eosin-Y** [16], **Rose Bengal** [17], flavins [18], anthraquinones [19], and thioxanthones [20] have shown very high efficiency.

In the framework of metal-free photocatalysts, graphitic carbon nitride (g-C_3N_4) is emerging as an innovative alternative in heterogeneous catalysis [21,22]. However, the engineering of such photoresponsive metal-free materials needs multiple time-consuming modifications. In fact, pure g-C_3N_4 exhibits poor photocatalytic efficiency because of its low specific surface area and its high charge carrier recombination [23]. Therefore, morphology control, metal/metal-free doping, or dye sensitization are usually required to enhance g-C_3N_4 performances in visible-light-driven catalytic processes [24]. On the contrary, the advantage in using organic dyes as photocatalysts lies in the variegated heterogeneity of the photochemical and photophysical catalytic pathways that can be investigated. Moreover, the possibility to tailor their structure through specific functionalizations is an advantageous opportunity to further implement catalytic efficiency [25]. In this context, photocatalytic oxygenations represent a vivid example of the potential of merging sustainable chemistry and photochemistry, revealing a green, effective alternative compared to the traditional oxidative methods [26]. In biological systems, mono- or di-oxygenase metalloenzymes are essential to promote molecular oxygen activation leading to oxygen atom incorporation into organic substrates [27,28]. Following a bio-mimetic approach, oxygenations play an important role in chemistry for the synthesis of valuable synthons in mild and green conditions, using O_2 as a bio-available and bio-compatible oxidizing agent [29]. In fact, the development of sustainable oxidative processes arises primarily from the choice of the oxidizing agent: reactivity, cost, and eco-friendliness are the main parameters to consider. Obviously, molecular oxygen represents the greenest choice in oxygenation reactions given its excellent atom economy and its low environmental impact [30]. However, O_2, owing to the spin-forbidden reaction [31], needs to be activated to a reactive oxygen species (ROS), such as singlet oxygen, superoxide anion, hydroxyl radical, or hydrogen peroxide, in order to be used as an oxidant [32]. Such activated species can be generated through biological, physical, or chemical traditional methods, with low efficiency and energy consuming procedures [33–35]. On the other hand, visible light photocatalysis has been affirmed as one of the most promising and best performing tools for molecular oxygen activation in order to carry out sustainable oxygenation processes.

The aim of this mini-review is to provide a critical analysis of recent applications of organic dyes as photocatalysts to promote light-driven oxygenation reactions in an aerobic environment.

2. Organic-Dye-Promoted Photooxygenation Processes

2.1. C(sp³) and Alkyl Arene (Benzyl) Oxygenation

Selective C(sp^3-H) oxygenation in the benzylic position is one of the most interesting tools to synthesize carbonyl intermediates, which are relevant molecules for biomedical or agrochemical applications [36–38]. Several organic photocatalysts have been recently examined to perform such a reaction. To this purpose, the 2,3-dichloro-5,6-dicyano-1,4-benzoquinone (DDQ)/*tert*-butyl nitrite couple proved to be a powerful organo-photocatalytic system for aerobic visible-light-enabled C-H oxygenation [39]. Fukuzumi et al. in 2011 reported for the first time the use of the **Acr⁺-Mes** photocatalyst to perform the selective oxygenation of *p*-xylene in acetonitrile, using molecular oxygen as an oxidant under visible light irradiation (Scheme 2) [40].

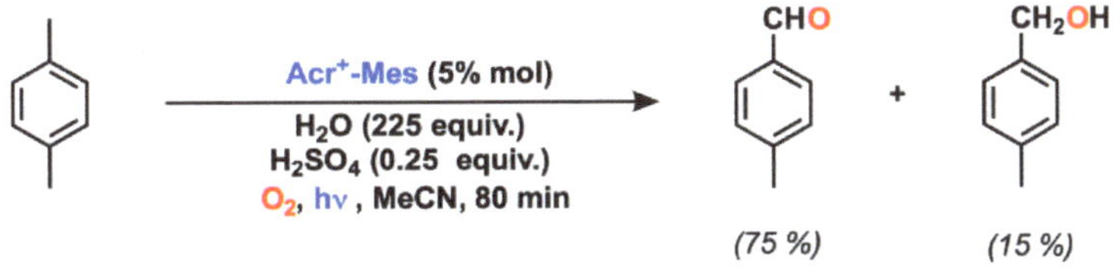

Scheme 2. *p*-xylene photooxygenation promoted by **Acr⁺-Mes**.

p-methyl benzaldehyde and *p*-methylbenzyl alcohol were obtained in mild conditions in only 80 min with satisfactory yields. The photocatalytic efficiency of such systems was then enhanced with the addition of aqueous sulfuric acid that allowed the reaching of good benzaldehyde yields (>70%). Remarkably, no over-oxidation to benzoic acid occurred, demonstrating the high selectivity of the reaction. A further improvement was achieved using the 9-mesityl-2,7,10-trimethylacridinium derivative (**Me$_2$Acr$^+$–Mes**) as a photocatalyst that accomplished the quantitative *p*-xylene conversion to *p*-methylbenzaldehyde in 80 min. Mechanistic studies proved a radical pathway, in which the alkyl arene was oxidized by the excited photocatalyst, and then it reacted with molecular oxygen to generate an activated alkyl peroxyradical, which collapsed into the oxidation products (Scheme 3).

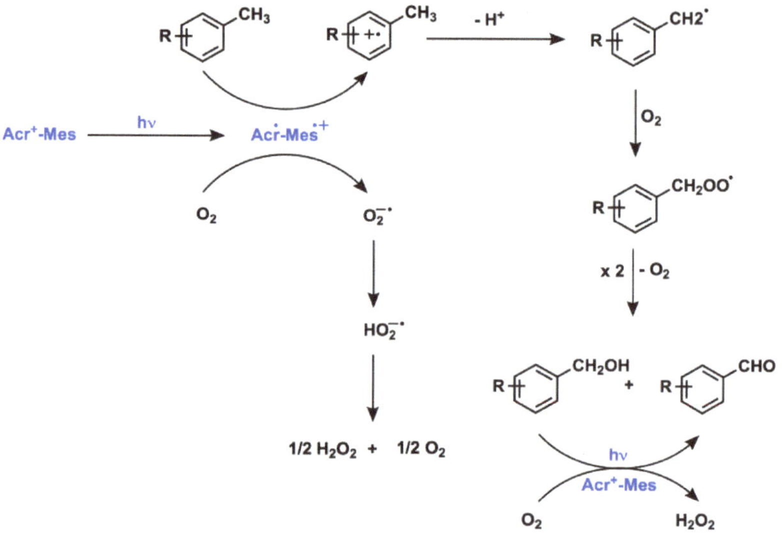

Scheme 3. Mechanistic insights of *p*-xylene photooxygenation promoted by **Acr$^+$-Mes**.

The authors demonstrated that photooxygenation simultaneously occurred with molecular oxygen reduction to hydrogen peroxide to restore the catalytic cycle. Drawbacks of such procedures are related to the substrate scope. In fact, reaction efficiently occurred for only three similar substrates, and no photooxidation of complex compounds was accomplished. Similarly, the water-soluble sodium anthraquinone sulfonate (**SAS**) proved to be a promising photocatalyst for the oxygenation of alkyl arenes in biphasic systems [41,42]. Hollmann et al. studied the neat photooxygenation of benzylic and allylic C-H bonds in alkane:water (3:7 v/v) (Scheme 4a) [43].

SAS displayed good catalytic efficiency, reaching in 24 h a TON of 37 in the toluene-selective oxygenation to benzaldehyde. However, long reaction times and photocatalyst deactivation were the resulting major drawbacks of such a system. Later, **Eosin Y** metal-free photocatalyst was investigated for aerobic benzylic C-H oxygenation in water (Scheme 4b) [44]. Such a sustainable protocol exhibited good photocatalytic efficiency in 24 h (32–89% yields) and a broad substrate scope (26 derivatives). However, 365 nm of UV irradiation and tetrabutylammonium borohydride (**TBABH**) as a phase-transfer catalyst were required to promote such a biphasic oxygenation reaction. Sing et al. recently developed an oxidative system, made up of **4CzIPN** organic photocatalyst (2%), tetrabutylammonium azide (**TBAN$_3$** 40%), and air/oxygen for the selective oxygenation of alkylarenes to the corresponding carbonyl compounds (Scheme 4c) [45]. The use of TBAN$_3$ was necessary to trigger the hydrogen atom transfer (HAT) mechanism, as no oxygenation occurred in the absence of the azido radical precursor. This technique efficiently accomplished the oxygenation of 23 structurally different substrates, exhibiting high functional group tolerance.

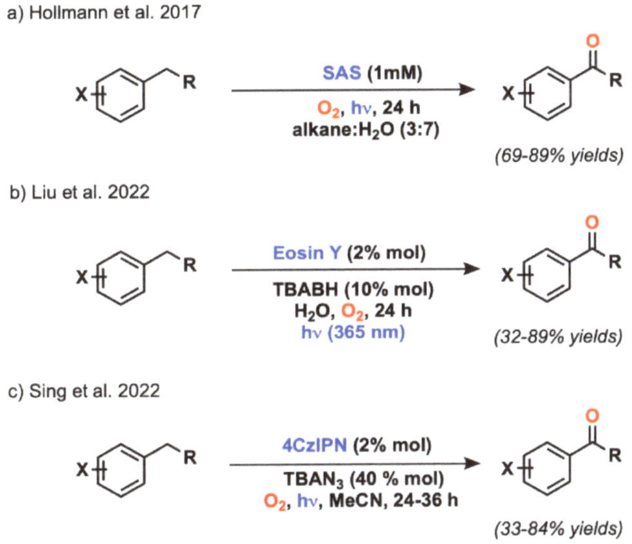

Scheme 4. Selective benzyl C(sp³-H) oxygenation enabled by (**a**) **SAS** [43], (**b**) **Eosin Y** [44], and (**c**) **4CzIPN** [45] organic photocatalysts.

In recent years, Cibulka's group has been involved in the study of ethylene-bridged flavinium salts (**Fl⁺**), which demonstrated enhanced photostability and improved catalytic properties compared with other flavinium derivatives in the oxidation of electron-deficient benzylic substrates to carboxylic acids (Scheme 5) [46,47].

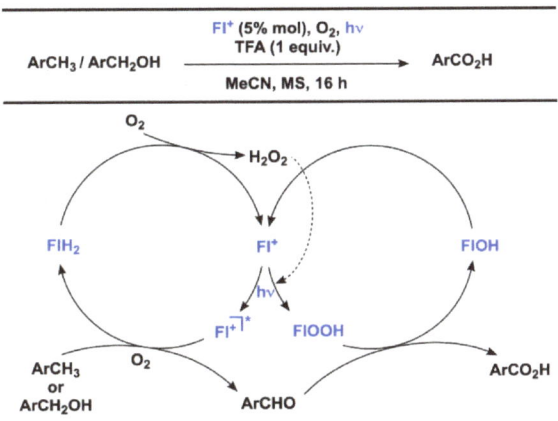

Scheme 5. Photo/organocatalytic oxygenation of alkyl arenes promoted by the **Fl⁺** photocatalyst.

Fl⁺ acts with a tandem photo-organo catalytic mechanism. Indeed, upon excitation, **Fl⁺** oxidizes the substrate to the corresponding aldehyde in the presence of O_2. Then, the reduced photocatalyst is restored by O_2, generating H_2O_2 as a byproduct. However, **Fl⁺** can react with H_2O_2 to generate the corresponding flavin hydroperoxides with a monooxygenase-like behavior. The broad substate scope was examined, affording satisfactory to good yields in 16 h.

The direct oxygenation of saturated hydrocarbons to synthesize fine materials in the chemical industry is a challenging process and it generally requires harsh conditions [48,49].

In 2011, Fukuzumi et al. reported the first example of the aerobic metal-free-promoted oxygenation of cyclohexane with visible light irradiation, using **Acr+-Mes** in acetonitrile amongst a sub-stoichiometric amount of HCl (Scheme 6a) [50].

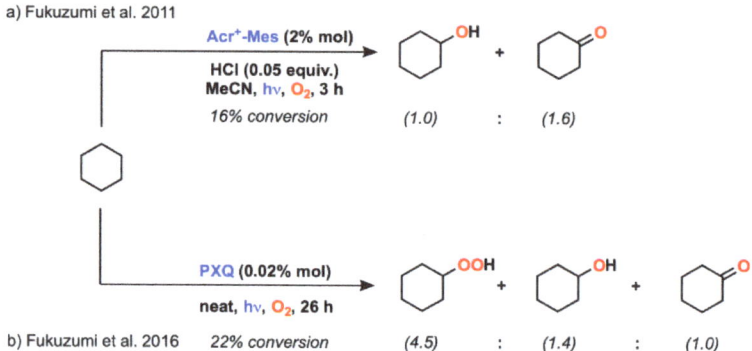

Scheme 6. Cyclohexane oxygenation promoted by (**a**) **Acr+-Mes** [50] and (**b**) **PXQ** [51].

Cyclohexanol and cyclohexanone were obtained with good selectivity in 3 h with a 7% quantum yield. More recently, the same authors performed the reaction solventless and with no inorganic acid, using 0.02% of *p*-xyloquinone (**PXQ**) photocatalyst (Scheme 6b) [51]. In these conditions, 22% of the cyclohexane was converted in 26 h, and cyclohexyl hydroperoxide, cyclohexanol, and cyclohexanone were obtained with a 4.5:1.4:1 ratio. A high quantum yield and good photocatalytic performances were obtained for the linear and branched alkanes. Although long reaction times were needed and unsatisfactory conversions and selectivity were achieved, this method is surely promising in terms of sustainability, since it requires only the use of cyclohexane, dioxygen, visible light, and photocatalyst to afford valuable oxygenated products in mild conditions.

2.2. Alkene Oxygenation

The dihydroxylation of alkenes is one of the classical methods to achieve 1,2-diols, biologically active molecules with pharmaceutical interest [52,53]. **Acr+-Mes** proved to be a powerful photocatalyst to promote the selective aerobic oxygenation of alkenes to 1,2-diols (Scheme 7) [54].

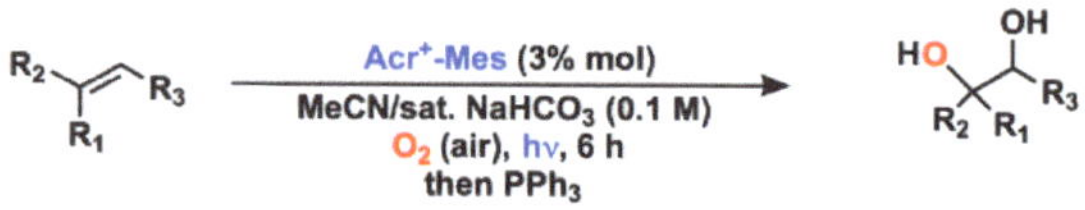

Scheme 7. Visible-light-induced alkene dihydroxylation, catalyzed by **Acr+-Mes**.

Reactions carried out under visible light irradiation in acetonitrile in the presence of water and a weakly basic medium led to good yields (46–90%) in 6 h with a broad substrate scope. Scavenging experiments indicated an electron-transfer radical mechanism, and studies performed with $H_2^{18}O$ suggested that one hydroxy group comes from water and the other from molecular oxygen. Similarly, Chen's group proposed **Rose Bengal** as a photocatalyst for alkene oxidation in the presence of pyridine and N-hydroxyphthalimide (Scheme 8) [55].

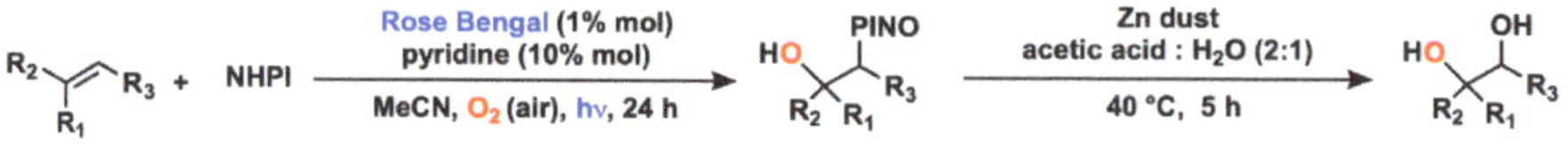

Scheme 8. Two-step visible-light-induced alkene dihydroxylation, catalyzed by **Rose Bengal**.

Despite the good efficiency (53–83% yield), such a method requires an additional hydrolysis step to obtain 1,2-diol; moreover, it is not suitable for the dioxygenation of inactivated alkenes or in the presence of an ester group. In 2018, Oliveira's group studied the continuous aerobic endoperoxidation of conjugated dienes, promoted by tetraphenylporphyrin (**TPP**) (Scheme 9) [56].

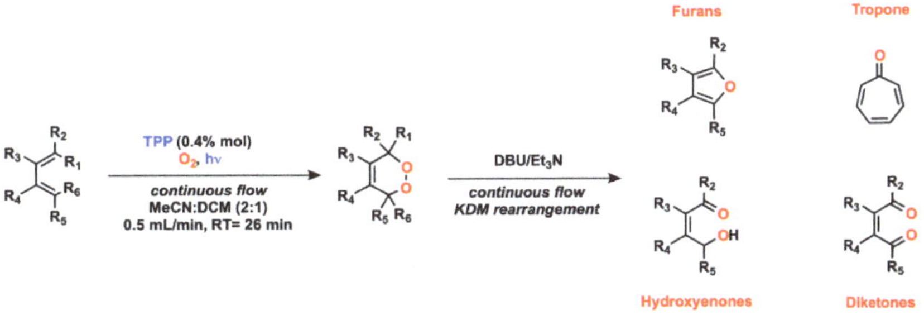

Scheme 9. Photocatalytic aerobic endoperoxidation of conjugated dienes, promoted by **TPP**, followed by Kornblum–DeLaMare (KDM) rearrangement to afford essential synthons.

Reactions were carried out in flow by irradiating the reaction mixture with a white LED under an oxygen atmosphere. In this process, **TPP** in its triplet excited state interacted with O_2 through an energy transfer process, generating singlet oxygen as an active oxidizing species. The latter reacted with the conjugated diene to form the corresponding endoperoxides. Photooxygenation showed good yields (ca. 60%) with 0.4% mol of **TPP** loading and a throughput of 7.7 g/day. Endoperoxidation was followed by the Kornblum–DeLaMare (KDM) rearrangement in the presence of 1,8-diazabiciclo[5.4.0]undec-7-ene (DBU) or triethylamine to synthesize C-H-oxidized relevant synthons, such as furans, diketones, enones, and tropones. The efficiency of the 1,6-pyrenedione (**1,6-PYD**) metal-free photocatalyst for the selective oxygenation of 1,1-dicyanoalkenes to epoxides was recently explored (Scheme 10) [57].

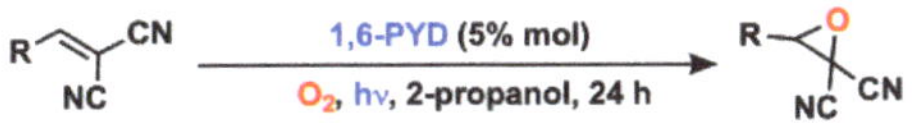

Scheme 10. Electron-deficient alkene epoxidation promoted by the **1,6-PYD** photocatalyst.

Reactions carried out in 2-propanol with 5% of **1,6-PYD** and O_2, under blue LED irradiation proceeded with good to excellent yields (51–98%) and high functional group tolerance. Scavenging experiments proved the presence of superoxide anion as an oxidizing active species. However, only electron-deficient alkenes were examined, and no examples of unsubstituted or activated alkenes were provided. Alkene oxidation to allylic alcohols or α,β-unsaturated carbonyl compounds is a remarkable synthetic organic transformation [58]. Recently, **Rose Bengal** was exploited as a metal-free photocatalyst to perform allylic C-H bond photooxygenation (Scheme 11).

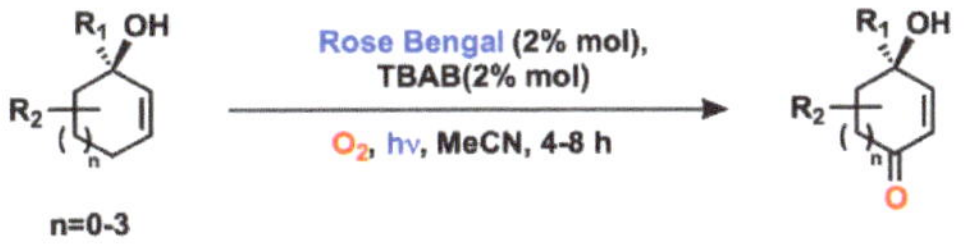

Scheme 11. Rose-Bengal-promoted allylic C-H bond photooxygenation.

This method used oxygen as a green oxidizing agent, tetrabutylammonium bromide (TBAB) as a HAT co-catalyst, and visible light to yield functionalized enones. Good to excellent yields (45–91%) were achieved in 4–8 h on differently substituted substrates, including pharmaceutically relevant compounds.

2.3. Alkyne Oxygenation to 1,2-Diketones

The synthons 1,2-diketones are essential precursors in the synthesis of azacyclic molecules [59,60] and bioactive compounds [61]. Alkyne oxygenation is considered one of the best performing strategies to afford 1,2-diketones. In the context of metal-free visible-light-promoted oxygenations, in 2016, Sun's group reported the aerobic photooxygenation of alkynes to 1,2-diketones, catalyzed by **Eosin Y** (Scheme 12) [62].

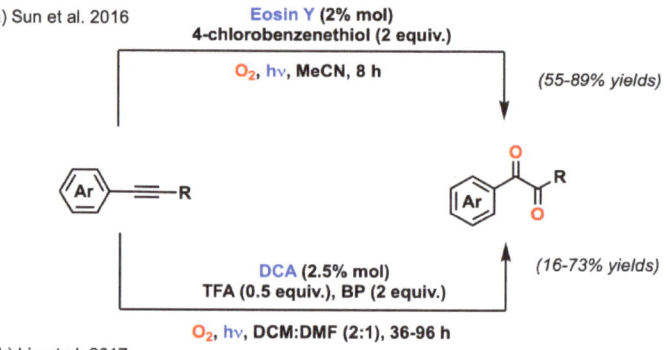

Scheme 12. Aerobic photooxygenation to 1,2-diketones promoted by (**a**) **Eosin Y** [62] and (**b**) **DCA** [63].

Reactions were carried out in acetonitrile for 8 h in the presence of two equivalents of 4-chlorobenzenethiol. Good to excellent yields were achieved on a series of different substrates; interestingly, the photocatalytic efficiency of such a system was higher than the traditional metal-based Ru(bpy)$_3$Cl·6H$_2$O photocatalyst. Mechanistic studies pointed out the presence of a radical-type pathway, in which superoxide anion was involved. The only limitation of such an oxidative protocol is related to the use of 4-chlorobenzenethiol as an additive, which has an essential role in triggering the radical chain. More recently, dicyanoanthracene (**DCA**) was investigated as a photocatalyst for the alkyne photooxygenation to 1,2-diketones [63]. The reactions were performed in DCM:DMF (2:1 v/v) in presence of 0.5 equivalents of trifluoroacetic acid (TFA) as an additive, and two equivalents of biphenyl redox mediator as a co-sensitizer (BP). TFA proved to be essential in increasing product yield, while BP was useful in reducing reaction time. However, very long reaction times were required to afford low to moderate yields.

2.4. Aromatic Oxygenation

Fukuzumi et al. in 2004 reported anthracene selective oxidation to anthraquinone, promoted by visible light under O$_2$, using metal-free **Acr$^+$-Mes** photocatalyst (Scheme 13) [64].

Mechanistic studies proved the presence of an electron transfer mechanism, in which the photocatalyst in the excited state oxidized anthracene, generating anthracene radical cations. Consequently, the reduced photocatalyst interacted with molecular oxygen, generating superoxide anion and restoring the catalytic cycle. A direct reaction between superoxide anion and anthracene radical cation caused the generation of epidioxyanthracene intermediate, which was further oxidized to quinone by the excited photocatalyst, producing hydrogen peroxide. The reactions were performed for three anthracene derivatives in acetonitrile with 10% of the photocatalyst, and 75–99% yields of epidioxyanthracene were achieved in 10 min. Thus, with this interesting approach, by using light, oxygen, and an organic catalyst, it was possible to obtain valuable anthraquinones.

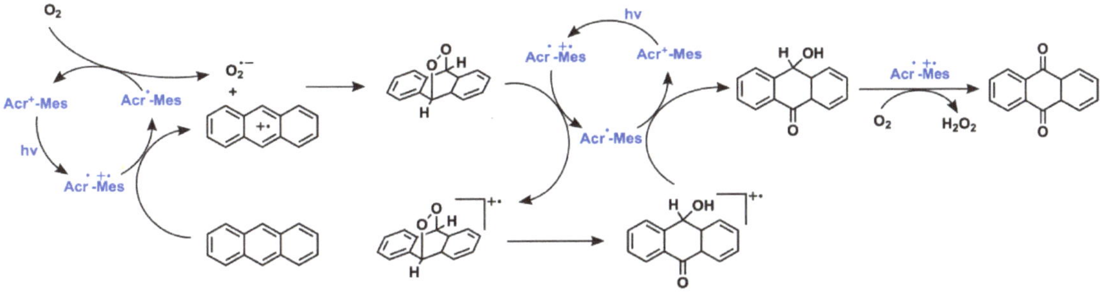

Scheme 13. Mechanism of aerobic photooxygenation of anthracenes to quinones promoted by **Acr⁺-Mes**.

Later, the same group investigated the selective oxidation of benzene to phenol, promoted by 3-cyano-1-methyl quinolinium (**QuCN⁺**) (Scheme 14) [65].

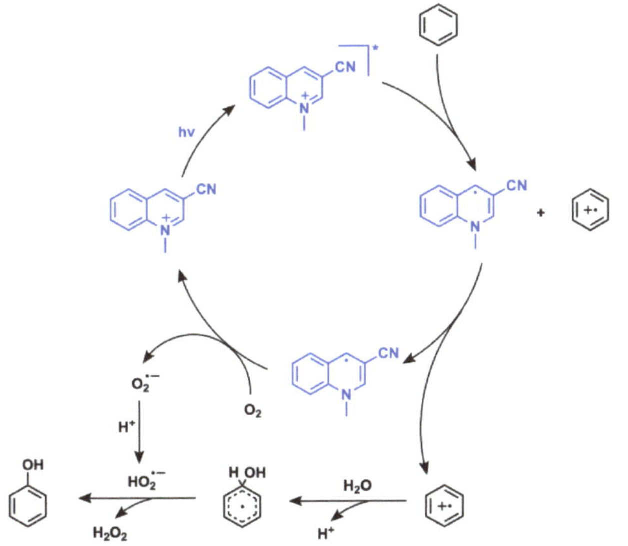

Scheme 14. Benzene oxidation to phenol promoted by **QuCN⁺** in an oxygen atmosphere.

Photooxidation was carried out on a gram-scale with benzene (2.3 g, 29 mmol), 3% **QuCN⁺**, and water (0.2 mmol) in O_2-saturated acetonitrile under visible light irradiation. In these conditions, 1.1 g of phenol (a 41% yield) was obtained after 48 h, an appreciable result considering the mild reaction conditions applied and the absence of overoxidation products. Importantly, mechanistic investigations pointed out that the hydroxyl group comes from water, while molecular oxygen re-oxidizes the reduced photocatalyst to restore the catalytic cycle. **TPP** was also used as a photocatalyst for the aerobic endoperoxidations of α-naphthols to yield naphthoquinones through singlet oxygen generation (Scheme 15) [66].

Importantly, this procedure can be used also to produce valuable compounds on a gram-scale. Reactions were performed on 11 derivatives in batch (a 7–20% yield) and continuous-flow conditions (an up to 82% yield). Continuous-flow reactions showed improved selectivity since minor by-products were obtained, thus simplifying product isolation.

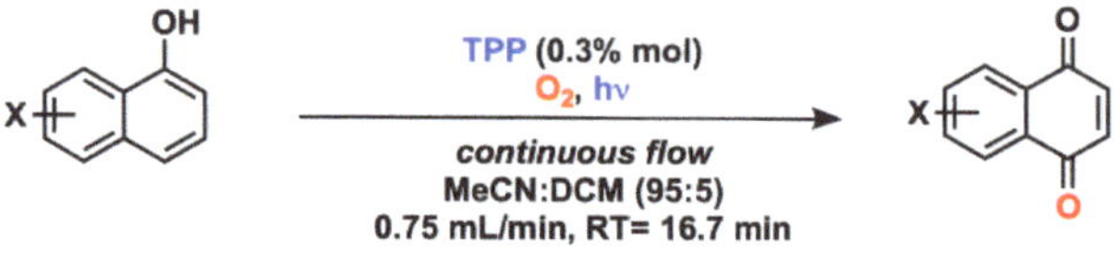

Scheme 15. Continuous-flow photocatalytic naphthol endoperoxidation promoted by **TPP** in oxygen atmosphere.

2.5. Oxidative C-C and C=C Cleavage

The selective cleavage and late-stage functionalization of C–C bonds have a substantial impact in the organic synthesis for the production of complex molecules [67], as well as in medicinal chemistry [68]. However, the high dissociation bond energy and the stability of C-C bonds make this process challenging, and, in general, the use of a metal-based catalyst is needed [69]. **4CzIPN** was recently studied as an organic photocatalyst for the selective C-C bond oxidative cleavage (Scheme 16) [70].

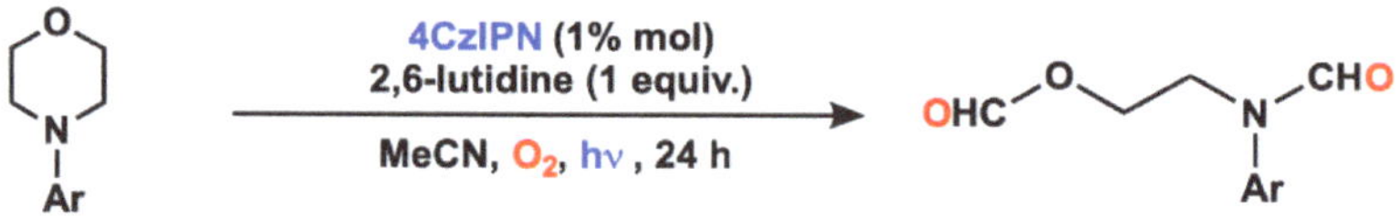

Scheme 16. *N*-aryl morpholine aerobic C(sp^3)-C(sp^3) oxidative cleavage promoted by **4CzIPN**.

In this study, 29 *N*-aryl morpholine derivatives were tested, showing low to good yields (13–83%) but significant functional group tolerance. Nevertheless, this protocol was not successful for *N*-alkyl morpholine derivatives because C$_{alkyl}$–N bond cleavage was obtained instead of the ring opening product. Olefin C=C bond oxidative cleavage is a hot topic in photochemistry. In fact, recently, several visible-light-promoted systems for alkene C=C cleavage have been developed, using riboflavin tetraacetate, **Eosin Y**, disulfide charge transfer, sodium benzene sulfinate, or nitroarene [71–75]. In 2021, Zhang et al. reported a mild and efficient procedure for the aerobic oxidative cleavage of olefines to ketones using **Rose Bengal** metal-free photocatalyst under visible light irradiation in the presence of 0.5 equivalents of acetic acid (Scheme 17) [76].

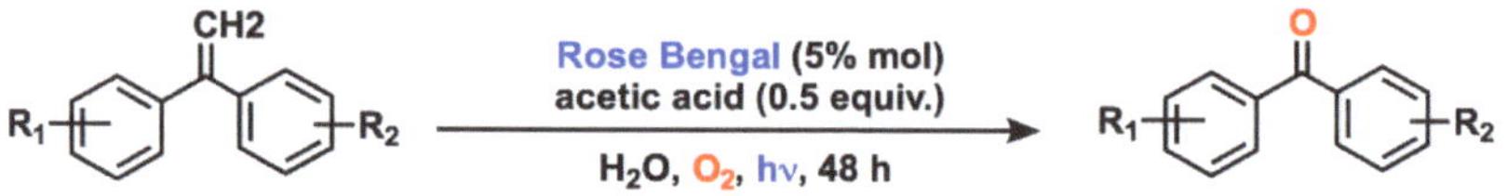

Scheme 17. Alkene aerobic oxidative cleavage to ketones promoted by **Rose Bengal** photocatalyst.

Reactions were carried out in water, and added-value carbonyl products were obtained with good to excellent yields (40–95%) in 48 h. In addition, reactions could be efficiently scaled up on a gram-scale; 1,1-diphenylethylene was converted to benzophenone with a 64% yield after 48 h, thus demonstrating the potential industrial applicability of such a protocol.

2.6. Amine Oxygenation

Amine α-oxygenation products are valuable building blocks in organic synthesis and interesting molecules for pharmaceutical, agrochemical, and photovoltaic applications [77–81]. In recent years, several metal-free photocatalysts have been evaluated for the aerobic photooxygenation of tertiary amines to amides. In 2019, Singh's group described the use of **Eosin Y** to promote the aerobic oxidation of tertiary amines conjugated with pyridine functionality through a single-electron-transfer mechanism (Scheme 18a) [82].

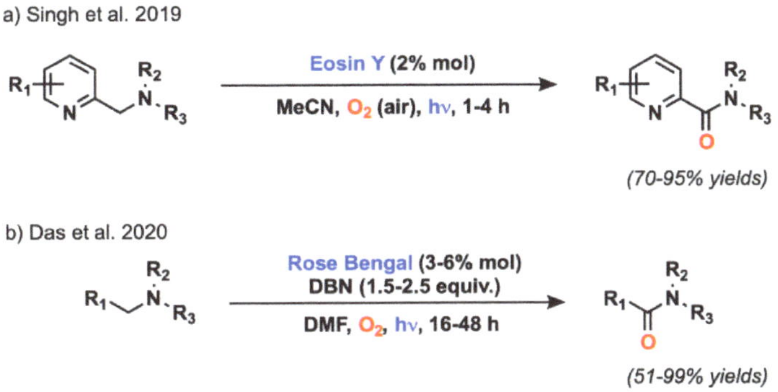

a) Singh et al. 2019

Scheme 18. Aerobic photooxygenation of tertiary amines to amides promoted by (**a**) **Eosin Y** [82] and (**b**) **Rose Bengal** [83,84].

This procedure showed good catalytic efficiency (16 substrates, 70–95% yields, in 1–4 h). The use of atmospheric oxygen as an oxidant, **Eosin Y** as a metal-free photocatalyst, and visible light irradiation outlined the sustainability of such an approach. Later, in 2020, Das and co-workers reported the use of **Rose Bengal** photocatalyst in the α-oxygenation of differently substituted 1-benzylpiperidines [83,84], showing a broad substrate scope (Scheme 18b). This methodology was suitable for the synthesis of natural products and for the late-stage selective oxygenation of drugs. Additionally, the oxygenation of N-substituted piperidines, 1-benzylpyrrolidine, N,N-dimethylbenzylamine, N-substituted tetrahydroquinoline, and N-substituted tetrahydroisoquinoline was investigated, obtaining the corresponding amides in good to excellent yields (51–99%). This protocol, despite its versatility, required the use of 1,5-diazabicyclo(4.3.0)non-5-ene (DBN) as a base and longer reaction times (16–48 h) if compared with the **Eosin-Y**-catalyzed protocol [82]. Recently, Anandhan's group reported, for the first time, the direct α-oxygenation of amines to imides using the **Acr⁺-Mes** organic photocatalyst (Scheme 19) [85].

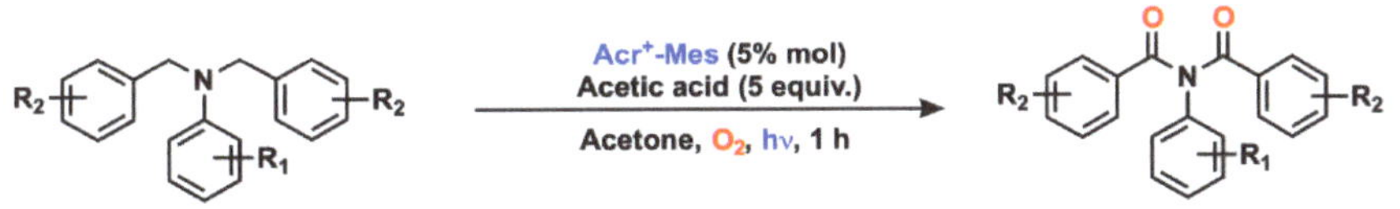

Scheme 19. Acr⁺-Mes-photocatalyzed aerobic photooxygenation of N,N-dibenzylanilines to imides.

The photooxygenation of more than 20 N,N-dibenzylanilines was carried out in acetone in the presence of five equivalents of acetic acid and 5% of the photocatalyst under blue LED irradiation in just 1 h with satisfactory to good yields (30–92%). Mechanistic studies pointed out the presence of an electron transfer mechanism, which is typical for acridinium-promoted oxygenation processes. Despite its potential practical applicability, such a method is not effective for the oxygenation of heteroaryl and aliphatic amines and electron-deficient N,N-dibenzylamines.

9,10-dicyanoanthracene-2,6-disulfonamide (**DCAS**) was recently investigated as a metal-free photocatalyst to perform the late-stage photooxygenation of trialkylamines to N-formamides in continuous flow under blue LED irradiation (Scheme 20) [86].

Indeed, the direct C–H oxidation of the N–CH$_3$ group of trialkylamines to the N-formyl group is an interesting process, since these products are relevant in biomedical applications and valuable intermediates in organic synthesis [87–89]. Generally, metal-based catalysts are required in such transformations [90,91]. However, in this procedure, the direct N–CH$_3$ oxygenation to N-formyl was accomplished using a metal-free photocatalyst. The reaction could be performed on several substrates in continuous-flow conditions in acetonitrile with 5% of the photocatalyst, affording high selectivity and good yields (15–68%). This protocol is unsuitable for the oxygenation of benzylic amines and trialkylamines containing benzylic alcohols or free carboxylic acids. Mechanistic studies proved an energy transfer mechanism, in which singlet oxygen was the active oxidizing species. Lu et al. in 2016 described the visible-light-driven oxygenation of primary amines to carboxylic acids and lactones, promoted by **Rose Bengal** (Scheme 21) [92].

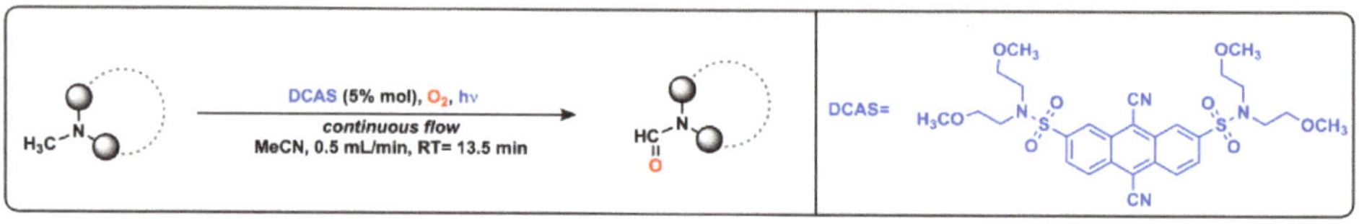

Scheme 20. Continuous-flow-**DCAS**-promoted photocatalytic oxygenation of trialkylamines to *N*-formamides.

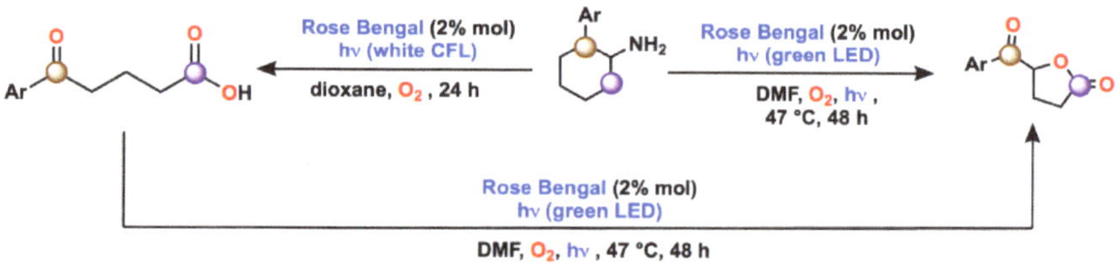

Scheme 21. Primary amine photooxygenation to carboxylic acids or lactones, catalyzed by **Rose Bengal**.

This process occurred via an electron transfer and energy transfer cooperative mechanism, through an oxidative ring opening followed by C=C oxidation. The reactions were performed with 2% of the photocatalyst, obtaining satisfactory yields in 24–48 h and an interesting tunability. In fact, the aerobic oxidation of primary amines led to the corresponding carboxylic acids working with an 18 W compact fluorescent light bulb (CFL) in dioxane, while lactones were obtained using an 8 W green LED in DMF at 47 °C.

2.7. Indole Oxygenation

The oxidative de-aromatization of *N*-heteroaromatic compounds is a smart tool for the synthesis of added-value bio-active products [93]. A dicyanopyrazine photocatalyst (**DPZ**) was recently investigated for the aerobic oxygenation of indoles to afford interesting bio-active compounds (Scheme 22) [94].

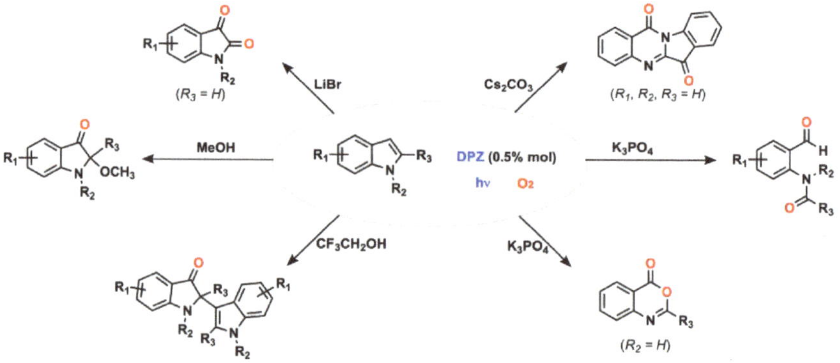

Scheme 22. pH and inorganic salt-tunable **DPZ**-promoted aerobic indole photooxygenation.

The authors demonstrated that the electrochemical properties of **DPZ** could be tuned by switching the reaction pH and using diverse inorganic salts as additives. Reactions proceeded through an electron transfer mechanism, with the generation of superoxide anion or through an energy transfer oxidative pathway via singlet oxygen. This protocol showed high photocatalytic performances, converting indoles and double-substituted indoles to added-value *N*-heterocyclic products, including isatins, tryptanthrin, 2-methoxy-3-oxoindoles, and benzo-oxazinones. Similar performances were obtained by Das and co-workers using **Rose Bengal** (Scheme 23) [95].

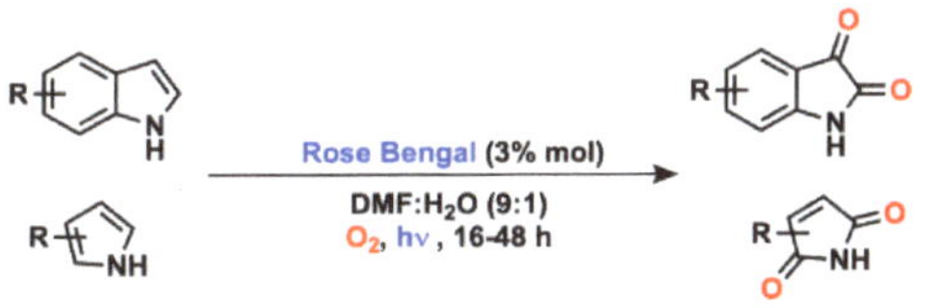

Scheme 23. Indole and pyrrole light-driven oxygenation catalyzed by **Rose Bengal**.

Reactions were carried out in an oxygen atmosphere in DMF:H$_2$O (9:1 v/v) in 16–48 h with 3% of the photocatalyst. Such a procedure was suitable for pyrrole oxidative de-aromatization to afford cyclic imides in a one-pot reaction. A broad substrate scope was reported (38 derivatives), obtaining pharmaceutically relevant compounds, such as 7-azaisatin, an anticancer drug, which was synthetized on a gram-scale in 45 h. More recently, Singh et al. investigated the light-driven aerobic oxidative coupling of indole and activated methylene compounds (e.g., malononitrile, ethyl acetoacetate, dimedone, and barbituric acid) to afford valuable building blocks for the synthesis of spiro-oxindoles, biologically active compounds (Scheme 24) [96].

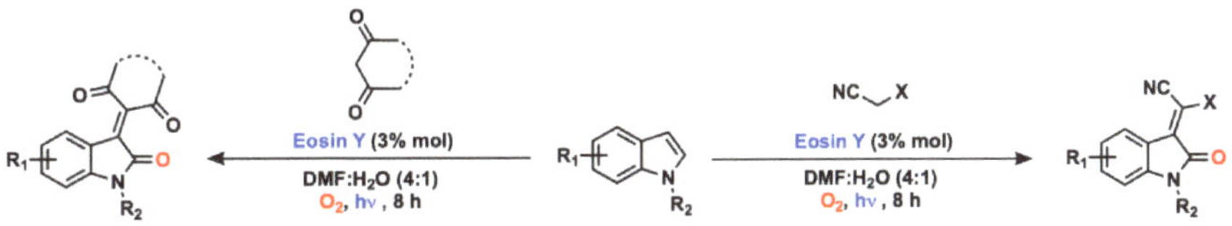

Scheme 24. Eosin-Y-promoted photooxidative coupling of indoles with activated methylene compounds.

Reactions performed with 3% **Eosin Y** in DMF:H$_2$O (4:1 v/v) displayed remarkable selectivity and good yields in mild conditions.

2.8. Triaryl Phosphine Oxygenation

Phosphine oxides are relevant intermediates in organic synthesis and valuable building blocks for the synthesis of drugs [97–99]. In 2017, Guo and co-workers described the visible-light-enabled photooxidation of triaryl phosphines to phosphine oxides, promoted by **Eosin Y** (Scheme 25) [100].

Reactions were performed in DCM:MeOH (5:1 v/v) using 1% of the photocatalyst. Scavenging experiments proved the presence of an energy transfer mechanism, where singlet oxygen was the active oxidizing species. More recently, the same group investigated such reactions by using 4-phenylthioxanthone (**4-PhTXT**) as a photocatalyst in methanol, obtaining a quantitative conversion to the oxygenated product in a shorter reaction time (e.g., triphenylphosphine oxide was achieved in 40 min compared with 3.5 h required for **Eosin Y**) [101]. Mechanistic investigations pointed out the presence of a merged energy- and electron-transfer mechanism. Remarkably, the **4-PhTXT** photocatalyst was suitable in the oxidation of different mono-, di- and tri-alkyl phosphines, and the reactions were accomplished on a gram-scale in 9 h. Agou et al. recently studied the triaryl phosphine aerobic photooxygenation promoted by a dibenzo-fused 1,4-azaborine (**DBAB**) photocatalyst, which occurred through singlet oxygen formation [102]. Compared with **Eosin Y** and **4-PhTXT**, the reactions catalyzed by **DBAB** required longer reaction times (8–10 h); in addition, such a catalyst was unsuitable for the oxidation of alkyl phosphine derivatives, but it was effective in the oxidation of 2,2′-bis(diphenylphosphino)-1,1′-binaphthyl (BINAP) substrates.

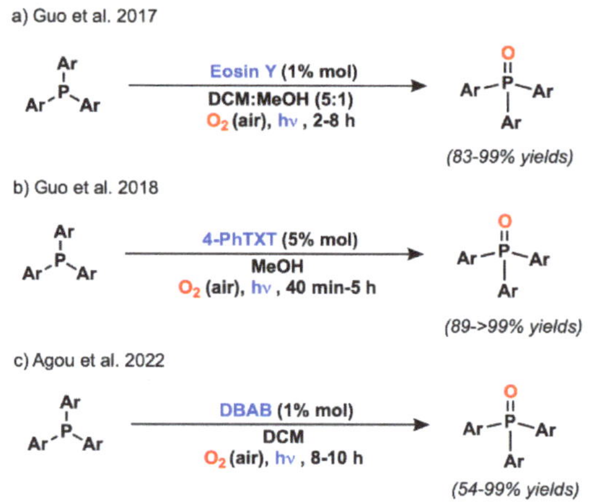

Scheme 25. Triaryl phosphine aerobic oxygenation to phosphine oxides, promoted by (**a**) **Eosin Y** [100], (**b**) **Rose Bengal** [101], and (**c**) **DBAB** [102].

2.9. Silane Oxygenation

Silanols are essential synthons in organic synthesis [103–105] and potential bioactive fragments in pharmaceutical applications [106,107]. The direct oxidation of silanes is the most effective and sophisticated strategy to produce silanols with high selectivity [108,109]. However, traditional methods require the use of stoichiometric oxidizing agents [110]. Recently, metal-based photocatalysts have been explored to perform aerobic silane oxygenation in sustainable conditions [111–113]. However, only two examples of the organo-photocatalyst-enabling silane oxidation to silanols have been reported (Scheme 26) [114,115].

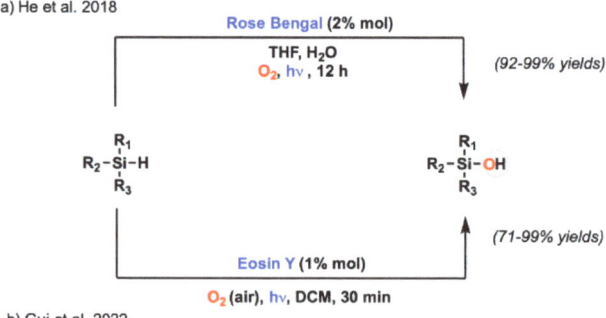

Scheme 26. Visible-light-driven aerobic silane oxygenation to silanols, promoted by (**a**) **Rose Bengal** [114] and (**b**) **Eosin Y** [115].

In that respect, He et al. developed for the first time a metal-free protocol using a 2% **Rose Bengal** photocatalyst (Scheme 26a) [114]. Silane oxygenation was performed in THF using water as an additive via an energy-transfer- and electron-transfer-combined mechanism. A wide substrate scope was analyzed, showing appreciable functional group tolerance with excellent yields (92–99%) in 12 h. Additionally, triphenylsilane was quantitatively converted to the corresponding silanol on a gram-scale in 12 h. Later, Gui and co-workers developed a mild and fast procedure to oxygenate silanes to silanols using 1% **Eosin Y** (Scheme 26b) [115]. Photooxygenation was performed under an air atmosphere in a chlorinated solvent, obtaining excellent yields (71–99%) in just 30 min. Moreover, triphenylsilane oxygenation was performed on a gram-scale, obtaining the corresponding silanol with a 92% yield, just prolonging the reaction time to 1 h. The promising photocatalytic efficiency together with the mild reaction conditions make this protocol also suitable for industrial application.

2.10. Thioether Oxygenation to Sulfoxides

Sulfoxides are bioactive molecules of pharmaceutical interest [116–118] and valuable intermediates in organic synthesis [119,120]. Thioether oxidation using the traditional oxidants in a stoichiometric amount is the most straightforward method to synthetize sulfoxides [121–123]. Recently, several metal-free photocatalysts have been explored to promote thioether oxygenation [124–134]; among them, good performances were achieved using thioxanthone derivatives as organic photocatalysts [132,133]. In particular, in 2018, Guo and co-workers described the selective oxidation of thioethers to sulfoxides promoted by **4-PhTXT** in methanol, obtaining good yields in 5–40 h (Scheme 27a) [132].

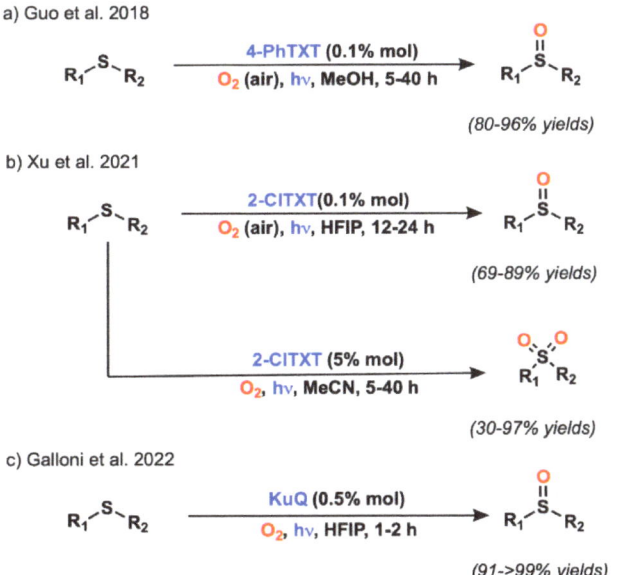

Scheme 27. Selective thioether oxidation to sulfoxides, promoted by (a) **4-PhTXT** [132], (b) **2-ClTXT** [133], and (c) **KuQ** [134] organo-photocatalysts.

More recently, Xu et al. reported the 2-chlorothioxanthone (**2-ClTXT**)-catalyzed thioether photooxygenation (Scheme 27b) [133]. Here, reactions performed in 1,1,1-3,3,3-hexafluoro-2-propanol (HFIP) selectively led to the corresponding sulfoxides, while sulfones were obtained using acetonitrile as the solvent. Notwithstanding the wide substrate scope and the good yields obtained, such methods still present significant limitations, such as long reaction times [132,133]. Our group recently proposed the use of KuQuinone (**KuQ**) as a homogeneous photocatalyst for the oxidation of thioethers selectively to sulfoxides in HFIP (Scheme 27c) [134]. Molecular oxygen and visible light were the required reagents to afford aliphatic, cyclic, diaryl, and heteroaromatic sulfoxides with high yields (91–>99%) in short reaction times (1–2 h) using a 0.5% **KuQ** photocatalyst. High functional group tolerance was observed, also in the presence of redox-labile substituents, and no over-oxidation products were detected. Remarkably, the **KuQ** photocatalyst could be recycled and reused for at least 20 runs, reaching a TON > 4000, showing unique robustness, which is a rare feature for organic photocatalysts.

3. Conclusions

Photooxygenation reactions represent a clear example of the effective applicability of sustainable methods in organic synthesis. In this context, organo-photocatalysis is turning out to be an effective alternative to metal catalysis, showing in some cases even better efficiency. Moreover, with respect to their metal counterparts, organic photocatalysts do not require controlled pressure or temperature conditions, and they are not sensitive to moisture. Indeed, **Acr$^+$-Mes**, **Eosin Y**, **4CzIPN**, and **Rose Bengal** are frequently adopted for the oxygenation of alkanes, alkenes, alkynes, aromatic compounds, amines, phosphines, silanes, and thioethers, leading to valuable organic compounds in mild reaction conditions.

However, research is still ongoing in this field, and future perspectives point towards:

- *The heterogenization of organic photocatalysts on inert solid supports or their encapsulation into nano-porous systems.* Heterogenization ensures efficient catalyst recovery, thus improving process efficiency and sustainability. In particular, catalyst recovery and reuse are aimed to enhance TONs, likely allowing for organic photocatalyst applications at the industrial level.
- *The development of additive-free photocatalytic systems.* Photocatalytic reactions often require the use of electron/hole sacrificial additives in order to trigger or boost redox pathways. Considering the high versatility of organic photocatalysts, future challenges are intended to avoid the use of additives through appropriate organic photocatalyst structural modifications.
- *The design of tandem oxidative-reductive processes.* Following the oxidation of an organic substrate, the reduced photocatalyst participates in a subsequent redox process to restore the native photocatalyst.

Such features undoubtedly highlight the organo-photocatalysis potential future for sustainable chemistry synthetic applications, even feasibly with direct solar light irradiation.

Author Contributions: Conceptualization, P.G., M.F., F.S. and V.C.; investigation, M.F. and F.V.; data curation, M.F.; writing—original draft preparation, M.F.; writing—review and editing, F.S., V.C. and P.G.; supervision, V.C. and P.G.; project administration, P.G.; funding acquisition, P.G. and F.S. All authors have read and agreed to the published version of the manuscript.

Funding: This research received no external funding.

Data Availability Statement: The data presented in this study are available upon request from the corresponding author.

Conflicts of Interest: The authors declare no conflict of interest.

References

1. Shaikh, I.R. Organocatalysis: Key Trends in Green Synthetic Chemistry, Challenges, Scope towards Heterogenization, and Importance from Research and Industrial Point of View. *J. Catal.* **2014**, *2014*, 402860. [CrossRef]
2. Ciamician, G. The Photochemistry of the Future. *Science* **1912**, *36*, 385–394. [CrossRef] [PubMed]
3. Anastas, P.T.; Warner, J.C. *Green Chemistry: Theory and Practice*; Oxford University Press: New York, NY, USA, 1998; p. 30.
4. Albini, A.; Fagnoni, M. Green Chemistry and Photochemistry Were Born at the Same Time. *Green Chem.* **2004**, *6*, 1. [CrossRef]
5. Shaw, M.H.; Twilton, J.; MacMillan, D.W.C. Photoredox Catalysis in Organic Chemistry. *J. Org. Chem.* **2016**, *81*, 6898–6926. [CrossRef] [PubMed]
6. Xuan, J.; Xiao, W.J. Visible-Light Photoredox Catalysis. *Angew. Chem. Int. Ed.* **2012**, *51*, 6828–6838. [CrossRef]
7. Prier, C.K.; Rankic, D.A.; MacMillan, D.W.C. Visible Light Photoredox Catalysis with Transition Metal Complexes: Applications in Organic Synthesis. *Chem. Rev.* **2013**, *113*, 5322–5363. [CrossRef]
8. Liang, Y.-F.; Jiao, N. Oxygenation via C–H/C–C Bond Activation with Molecular Oxygen. *Acc. Chem. Res.* **2017**, *50*, 1640–1653. [CrossRef]

9. Sideri, I.K.; Voutyritsa, E.; Kokotos, C.G. PhotoOrganocatalysis, Small Organic Molecules and Light in the Service of Organic Synthesis: The Awakening of a Sleeping Giant. *Org. Biomol. Chem.* **2018**, *16*, 4596–4614. [CrossRef]
10. Zhang, Y.; Schilling, W.; Das, S. Metal-Free Photocatalysts for C−H Bond Oxygenation Reactions with Oxygen as the Oxidant. *ChemSusChem* **2019**, *12*, 2898–2910. [CrossRef]
11. Amos, S.G.E.; Garreau, M.; Buzzetti, L.; Waser, J. Photocatalysis with Organic Dyes: Facile Access to Reactive Intermediates for Synthesis. *Beilstein J. Org. Chem.* **2020**, *16*, 1163–1187. [CrossRef]
12. Nicewicz, D.A.; Nguyen, T.M. Recent Applications of Organic Dyes as Photoredox Catalysts in Organic Synthesis. *ACS Catal.* **2014**, *4*, 355–360. [CrossRef]
13. Ravelli, D.; Fagnoni, M. Dyes as Visible Light Photoredox Organocatalysts. *ChemCatChem* **2012**, *4*, 169–171. [CrossRef]
14. Fukuzumi, S.; Ohkubo, K. Organic Synthetic Transformations Using Organic Dyes as Photoredox Catalysts. *Org. Biomol. Chem.* **2014**, *12*, 6059–6071. [CrossRef] [PubMed]
15. Tlili, A.; Lakhdar, S. Acridinium Salts and Cyanoarenes as Powerful Photocatalysts: Opportunities in Organic Synthesis. *Angew. Chem. Int. Ed.* **2021**, *60*, 19526–19549. [CrossRef] [PubMed]
16. Bosveli, A.; Montagnon, T.; Kalaitzakis, D.; Vassilikogiannakis, G. Eosin: A Versatile Organic Dye Whose Synthetic Uses Keep Expanding. *Org. Biomol. Chem.* **2021**, *19*, 3303–3317. [CrossRef]
17. Sharma, S.; Sharma, A. Recent advances in photocatalytic manipulations of Rose Bengal in organic synthesis. *Org. Biomol. Chem.* **2019**, *17*, 4394–4405. [CrossRef]
18. Srivastava, V.; Singh, P.K.; Srivastava, A.; Singh, P.P. Synthetic Applications of Flavin Photocatalysis: A Review. *RSC Adv.* **2021**, *11*, 14251–14259. [CrossRef]
19. Cervantes-González, J.; Vosburg, D.A.; Mora-Rodriguez, S.E.; Vázquez, M.A.; Zepeda, L.G.; Villegas Gómez, C.; Lagunas-Rivera, S. Anthraquinones: Versatile Organic Photocatalysts. *ChemCatChem* **2020**, *12*, 3811–3827. [CrossRef]
20. Nikitas, N.F.; Gkizis, P.L.; Kokotos, C.G. Thioxanthone: A Powerful Photocatalyst for Organic Reactions. *Org. Biomol. Chem.* **2021**, *19*, 5237–5253. [CrossRef]
21. Iqbal, S.; Amjad, A.; Javed, M.; Alfakeer, M.M.; Rabea, S.; Elkaeed, E.B.; Pashameah, R.A.; Alzahrani, E.; Farouk, A.E. Boosted spatial charge carrier separation of binary $ZnFe_2O_4/S$-g-C_3N_4 heterojunction for visible-light-driven photocatalytic activity and antimicrobial performance. *Front. Chem.* **2022**, *10*, 975355. [CrossRef]
22. Wang, X.; Blechert, S.; Antonietti, M. Polymeric Graphitic Carbon Nitride for Heterogeneous Photocatalysis. *ACS Catal.* **2012**, *2*, 1596–1606. [CrossRef]
23. Ismael, M. A review on graphitic carbon nitride (g-C_3N_4) based nanocomposites: Synthesis, categories, and their application in photocatalysis. *J. Alloys Comp.* **2020**, *846*, 156446. [CrossRef]
24. Zhou, L.; Zhang, H.; Sun, H.; Liu, S.; Tade, M.O.; Wang, S.; Jin, W. Recent advances in non-metal modification of graphitic carbon nitride for photocatalysis: A historic review. *Catal. Sci. Technol.* **2016**, *6*, 7002–7023. [CrossRef]
25. Marin, M.L.; Santos-Juanes, L.; Arques, A.; Amat, A.M.; Miranda, M.A. Organic Photocatalysts for the Oxidation of Pollutants and Model Compounds. *Chem. Rev.* **2012**, *112*, 1710–1750. [CrossRef]
26. Li, Q.; Li, F. Recent Advances in Molecular Oxygen Activation via Photocatalysis and Its Application in Oxidation Reactions. *Chem. Eng. J.* **2021**, *421*, 129915. [CrossRef]
27. Bäckvall, J.E. *Modern Oxidation Methods*, 2nd ed.; Wiley-VCH: Weinheim, Germany, 2010; pp. 1–465. [CrossRef]
28. Poulos, T.L. Heme Enzyme Structure and Function. *Chem. Rev.* **2014**, *114*, 3919–3962. [CrossRef]
29. Tang, C.; Qiu, X.; Cheng, Z.; Jiao, N. Molecular Oxygen-Mediated Oxygenation Reactions Involving Radicals. *Chem. Soc. Rev.* **2021**, *50*, 8067–8101. [CrossRef]
30. Zhang, X.; Rakesh, K.P.; Ravindar, L.; Qin, H.L. Visible-Light Initiated Aerobic Oxidations: A Critical Review. *Green Chem.* **2018**, *20*, 4790–4833. [CrossRef]
31. Metz, M.; Solomon, E.I. Dioxygen Binding to Deoxyhemocyanin: Electronic Structure and Mechanism of the Spin-Forbidden Two-Electron Reduction of O_2. *J. Am. Chem. Soc.* **2001**, *123*, 4938–4950. [CrossRef]
32. Vo, N.T.; Mekmouche, Y.; Tron, T.; Guillot, R.; Banse, F.; Halime, Z.; Sircoglou, M.; Leibl, W.; Aukauloo, A. A Reversible Electron Relay to Exclude Sacrificial Electron Donors in the Photocatalytic Oxygen Atom Transfer Reaction with O_2 in Water. *Angew. Chem. Int. Ed.* **2019**, *58*, 16023–16027. [CrossRef]
33. Romero, E.; Gómez Castellanos, J.R.; Gadda, G.; Fraaije, M.W.; Mattevi, A. Same Substrate, Many Reactions: Oxygen Activation in Flavoenzymes. *Chem. Rev.* **2018**, *118*, 1742–1769. [CrossRef]
34. Qian, K.; Du, L.; Zhu, X.; Liang, S.; Chen, S.; Kobayashi, H.; Yan, X.; Xu, M.; Dai, Y.; Li, R. Directional Oxygen Activation by Oxygen-Vacancy-Rich WO_2 Nanorods for Superb Hydrogen Evolution via Formaldehyde Reforming. *J. Mater. Chem. A* **2019**, *7*, 14592–14601. [CrossRef]
35. Zang, Y.; Gong, L.; Mei, L.; Gu, Z.; Wang, Q. Bi_2WO_6 Semiconductor Nanoplates for Tumor Radiosensitization through High-Z Effects and Radiocatalysis. *ACS Appl. Mater. Interfaces* **2019**, *11*, 18942–18952. [CrossRef] [PubMed]
36. Brugmans, J.P.; Thienpont, D.C.; van Wijngaarden, I.; Vanparijs, O.F.; Schuermans, V.L.; Lauwers, H.L. Mebendazole in Enterobiasis Radiochemical and Pilot Clinical Study in 1,278 Subjects. *JAMA* **1971**, *217*, 313–316. [CrossRef] [PubMed]
37. Sunshine, A. Analgesic Value of Fenbufen in Postoperative Patients. A Comparative Oral Analgesic Study of Fenbufen, Aspirin, and Placebo. *J. Clin. Pharmacol.* **1975**, *15*, 591–597. [CrossRef] [PubMed]

38. European Food Safety Authority. Conclusion on the Peer Review of the Pesticide Risk Assessment of the Active Substance Pyriofenone. *EFS2* **2013**, *11*, 4302. [CrossRef]
39. Rusch, F.; Schober, J.C.; Brasholz, M. Visible-Light Photocatalytic Aerobic Benzylic C(sp^3)−H Oxygenations with the ^{3}DDQ*/tert-Butyl Nitrite Co-Catalytic System. *ChemCatChem* **2016**, *8*, 2881–2884. [CrossRef]
40. Ohkubo, K.; Mizushima, K.; Iwata, R.; Souma, K.; Suzuki, N.; Fukuzumi, S. Simultaneous Production of p-Tolualdehyde and Hydrogen Peroxide in Photocatalytic Oxygenation of p-Xylene and Reduction of Oxygen with 9-Mesityl-10-Methylacridinium Ion Derivatives. *Chem. Commun.* **2010**, *46*, 601–603. [CrossRef] [PubMed]
41. Jiang, D.; Zhang, Q.; Yang, L.; Deng, Y.; Yang, B.; Liu, Y.; Zhang, C.; Fu, Z. Regulating Effects of Anthraquinone Substituents and Additives in Photo-Catalytic Oxygenation of p-Xylene by Molecular Oxygen under Visible Light Irradiation. *Renew. Energy* **2021**, *174*, 928–938. [CrossRef]
42. Morrison, G.; Bannon, R.; Wharry, S.; Moody, T.S.; Mase, N.; Hattori, M.; Manyar, H.; Smyth, M. Continuous Flow Photooxidation of Alkyl Benzenes Using Fine Bubbles for Mass Transfer Enhancement. *Tetrahedron Lett.* **2022**, *90*, 153613. [CrossRef]
43. Zhang, W.; Gacs, J.; Arends, I.W.C.E.; Hollmann, F. Selective Photooxidation Reactions Using Water-Soluble Anthraquinone Photocatalysts. *ChemCatChem* **2017**, *9*, 3821–3826. [CrossRef] [PubMed]
44. Bo, C.; Bu, Q.; Liu, J.; Dai, B.; Liu, N. Photocatalytic Benzylic Oxidation Promoted by Eosin Y in Water. *ACS Sust. Chem. Eng.* **2022**, *10*, 1822–1828. [CrossRef]
45. Shee, M.; Singh, N.D.P. Photogenerated Azido Radical Mediated Oxidation: Access to Carbonyl Functionality from Alcohols, Alkylarenes, and Olefins via Organophotoredox. *Adv. Synth. Catal.* **2022**, *364*, 2032–2039. [CrossRef]
46. Zelenka, J.; Svobodová, E.; Tarábek, J.; Hoskovcová, I.; Boguschová, V.; Bailly, S.; Sikorski, M.; Roithová, J.; Cibulka, R. Combining Flavin Photocatalysis and Organocatalysis: Metal-Free Aerobic Oxidation of Unactivated Benzylic Substrates. *Org. Lett.* **2019**, *21*, 114–119. [CrossRef] [PubMed]
47. Pokluda, A.; Anwar, Z.; Boguschová, V.; Anusiewicz, I.; Skurski, P.; Sikorski, M.; Cibulka, R. Robust Photocatalytic Method Using Ethylene-Bridged Flavinium Salts for the Aerobic Oxidation of Unactivated Benzylic Substrates. *Adv. Synth. Catal.* **2021**, *363*, 4371–4379. [CrossRef]
48. Hermans, I.; Spier, E.S.; Neuenschwander, U.; Turrà, N.; Baiker, A. Selective Oxidation Catalysis: Opportunities and Challenges. *Top. Catal.* **2009**, *52*, 1162–1174. [CrossRef]
49. Musser, M.T. Cyclohexanol and Cyclohexanone. In *Ullmann's Encyclopedia of Industrial Chemistry*; Wiley-VCH: Weinheim, Germany, 2010; Volume 11, pp. 49–60. [CrossRef]
50. Ohkubo, K.; Fujimoto, A.; Fukuzumi, S. Metal-Free Oxygenation of Cyclohexane with Oxygen Catalyzed by 9-Mesityl-10-Methylacridinium and Hydrogen Chloride under Visible Light Irradiation. *Chem. Commun.* **2011**, *47*, 8515. [CrossRef]
51. Ohkubo, K.; Hirose, K.; Fukuzumi, S. Photooxygenation of Alkanes by Dioxygen with p-Benzoquinone Derivatives with High Quantum Yields. *Photochem. Photobiol. Sci.* **2016**, *15*, 731–734. [CrossRef]
52. Tate, E.W.; Dixon, D.J.; Ley, S.V. A Highly Enantioselective Total Synthesis of (+)-Goniodiol. *Org. Biomol. Chem.* **2006**, *4*, 1698–1706. [CrossRef]
53. Liu, Y.; Wang, W.; Zeng, A.P. Biosynthesizing Structurally Diverse Diols via a General Route Combining Oxidative and Reductive Formations of OH-Groups. *Nat. Commun.* **2022**, *13*, 1595. [CrossRef]
54. Yang, B.; Lu, Z. Visible Light-Promoted Dihydroxylation of Styrenes with Water and Dioxygen. *Chem. Commun.* **2017**, *53*, 12634–12637. [CrossRef]
55. Zhang, M.Z.; Tian, J.; Yuan, M.; Peng, W.Q.; Wang, Y.Z.; Wang, P.; Liu, L.; Gou, Q.; Huang, H.; Chen, T. Visible Light-Induced Aerobic Dioxygenation of α,β-Unsaturated Amides/Alkenes toward Selective Synthesis of β-Oxy Alcohols Using Rose Bengal as a Photosensitizer. *Org. Chem. Front.* **2021**, *8*, 2215–2223. [CrossRef]
56. De Souza, J.M.; Brocksom, T.J.; McQuade, D.T.; De Oliveira, K.T. Continuous Endoperoxidation of Conjugated Dienes and Subsequent Rearrangements Leading to C–H Oxidized Synthons. *J. Org. Chem.* **2018**, *83*, 7574–7585. [CrossRef] [PubMed]
57. Zhang, Y.; Yang, X.; Tang, H.; Liang, D.; Wu, J.; Huang, D. Pyrenediones as Versatile Photocatalysts for Oxygenation Reactions with in Situ Generation of Hydrogen Peroxide under Visible Light. *Green Chem.* **2020**, *22*, 22–27. [CrossRef]
58. Liu, C.; Liu, H.; Zheng, X.; Chen, S.; Lai, Q.; Zheng, C.; Huang, M.; Cai, K.; Cai, Z.; Cai, S. Visible-Light-Enabled Allylic C–H Oxidation: Metal-Free Photocatalytic Generation of Enones. *ACS Catal.* **2022**, *12*, 1375–1381. [CrossRef]
59. Zhao, Z.; Wisnoski, D.D.; Wolkenberg, S.E.; Leister, W.H.; Wang, Y.; Lindsley, C.W. General Microwave-Assisted Protocols for the Expedient Synthesis of Quinoxalines and Heterocyclic Pyrazines. *Tetrahedron Lett.* **2004**, *45*, 4873–4876. [CrossRef]
60. Samanta, S.; Roy, D.; Khamarui, S.; Maiti, D.K. Ni(II)–Salt Catalyzed Activation of Primary Amine-sp^{3}C$_\alpha$–H and Cyclization with 1,2-Diketone to Tetrasubstituted Imidazoles. *Chem. Commun.* **2014**, *50*, 2477–2480. [CrossRef]
61. Irannejad, H.; Nadri, H.; Naderi, N.; Rezaeian, S.N.; Zafari, N.; Foroumadi, A.; Amini, M.; Khoobi, M. Anticonvulsant Activity of 1,2,4-Triazine Derivatives with Pyridyl Side Chain: Synthesis, Biological, and Computational Study. *Med. Chem. Res.* **2015**, *24*, 2505–2513. [CrossRef]
62. Liu, X.; Cong, T.; Liu, P.; Sun, P. Synthesis of 1,2-Diketones via a Metal-Free, Visible-Light-Induced Aerobic Photooxidation of Alkynes. *J. Org. Chem.* **2016**, *81*, 7256–7261. [CrossRef]
63. Qin, H.T.; Xu, X.; Liu, F. Aerobic Oxidation of Alkynes to 1,2-Diketones by Organic Photoredox Catalysis. *ChemCatChem* **2017**, *9*, 1409–1412. [CrossRef]

64. Kotani, H.; Ohkubo, K.; Fukuzumi, S. Photocatalytic Oxygenation of Anthracenes and Olefins with Dioxygen via Selective Radical Coupling Using 9-Mesityl-10-Methylacridinium Ion as an Effective Electron-Transfer Photocatalyst. *J. Am. Chem. Soc.* **2004**, *126*, 15999–16006. [CrossRef] [PubMed]
65. Ohkubo, K.; Kobayashi, T.; Fukuzumi, S. Direct Oxygenation of Benzene to Phenol Using Quinolinium Ions as Homogeneous Photocatalysts. *Angew. Chem. Int. Ed.* **2011**, *50*, 8652–8655. [CrossRef] [PubMed]
66. De Oliveira, K.T.; Miller, L.Z.; McQuade, D.T. Exploiting Photooxygenations Mediated by Porphyrinoid Photocatalysts under Continuous Flow Conditions. *RSC Adv.* **2016**, *6*, 12717–12725. [CrossRef]
67. Morcillo, S.P. Radical-Promoted C−C Bond Cleavage: A Deconstructive Approach for Selective Functionalization. *Angew. Chem. Int. Ed.* **2019**, *58*, 14044–14054. [CrossRef]
68. Bolleddula, J.; Chowdhury, S.K. Carbon–Carbon Bond Cleavage and Formation Reactions in Drug Metabolism and the Role of Metabolic Enzymes. *Drug Metabol. Rev.* **2015**, *47*, 534–557. [CrossRef]
69. Leonard, D.K.; Li, W.; Junge, K.; Beller, M. Improved Bimetallic Cobalt–Manganese Catalysts for Selective Oxidative Cleavage of Morpholine Derivatives. *ACS Catal.* **2019**, *9*, 11125–11129. [CrossRef]
70. Dong, C.; Huang, L.; Guan, Z.; Huang, C.; He, Y. Visible-Light-Mediated Aerobic Oxidative C(sp^3)–C(sp^3) Bond Cleavage of Morpholine Derivatives Using 4CzIPN as a Photocatalyst. *Adv Synth Catal* **2021**, *363*, 3803–3811. [CrossRef]
71. Singh, A.K.; Chawla, R.; Yadav, L.D.S. Eosin Y Catalyzed Visible Light Mediated Aerobic Photo-Oxidative Cleavage of the C–C Double Bond of Styrenes. *Tetrahedron Lett.* **2015**, *56*, 653–656. [CrossRef]
72. Deng, Y.; Wei, X.J.; Wang, H.; Sun, Y.; Noël, T.; Wang, X. Disulfide-Catalyzed Visible-Light-Mediated Oxidative Cleavage of C=C Bonds and Evidence of an Olefin–Disulfide Charge-Transfer Complex. *Angew. Chem. Int. Ed.* **2017**, *56*, 832–836. [CrossRef]
73. Wan, J.P.; Gao, Y.; Wei, L. Recent Advances in Transition-Metal-Free Oxygenation of Alkene C=C Double Bonds for Carbonyl Generation. *Chem. Asian J.* **2016**, *11*, 2092–2102. [CrossRef]
74. Wise, D.E.; Gogarnoiu, E.S.; Duke, A.D.; Paolillo, J.M.; Vacala, T.L.; Hussain, W.A.; Parasram, M. Photoinduced Oxygen Transfer Using Nitroarenes for the Anaerobic Cleavage of Alkenes. *J. Am. Chem. Soc.* **2022**, *144*, 15437–15442. [CrossRef] [PubMed]
75. Chen, Y.X.; He, J.T.; Wu, M.C.; Liu, Z.L.; Tang, K.; Xia, P.J.; Chen, K.; Xiang, H.Y.; Chen, X.Q.; Yang, H. Photochemical Organocatalytic Aerobic Cleavage of C=C Bonds Enabled by Charge-Transfer Complex Formation. *Org. Lett.* **2022**, *24*, 3920–3925. [CrossRef] [PubMed]
76. Zhang, Y.; Yue, X.; Liang, C.; Zhao, J.; Yu, W.; Zhang, P. Photo-Induced Oxidative Cleavage of C-C Double Bonds of Olefins in Water. *Tetrahedron Lett.* **2021**, *80*, 153321. [CrossRef]
77. Balskus, E.P.; Jacobsen, E.N. α,β-Unsaturated β-Silyl Imide Substrates for Catalytic, Enantioselective Conjugate Additions: A Total Synthesis of (+)-Lactacystin and the Discovery of a New Proteasome Inhibitor. *J. Am. Chem. Soc.* **2006**, *128*, 6810–6812. [CrossRef] [PubMed]
78. Cheng, Z.; Zhu, X.; Kang, E.T.; Neoh, K.G. Modification of Poly(Ether Imide) Membranes via Surface-Initiated Atom Transfer Radical Polymerization. *Macromolecules* **2006**, *39*, 1660–1663. [CrossRef]
79. Sood, A.; Panchagnula, R. Peroral Route: An Opportunity for Protein and Peptide Drug Delivery. *Chem. Rev.* **2001**, *101*, 3275–3304. [CrossRef]
80. Castral, T.C.; Matos, A.P.; Monteiro, J.L.; Araujo, F.M.; Bondancia, T.M.; Batista-Pereira, L.G.; Fernandes, J.B.; Vieira, P.C.; da Silva, M.F.G.F.; Corrêa, A.G. Synthesis of a Combinatorial Library of Amides and Its Evaluation against the Fall Armyworm, Spodoptera Frugiperda. *J. Agric. Food Chem.* **2011**, *59*, 4822–4827. [CrossRef]
81. Forbis, R.M.; Rinehart, K.L. Nybomycin. VII. Preparative Routes to Nybomycin and Deoxynybomycin. *J. Am. Chem. Soc.* **1973**, *95*, 5003–5013. [CrossRef]
82. Srivastava, V.; Singh, P.K.; Singh, P.P. Eosin Y Catalysed Visible-Light Mediated Aerobic Oxidation of Tertiary Amines. *Tetrahedron Lett.* **2019**, *60*, 151041. [CrossRef]
83. Zhang, Y.; Riemer, D.; Schilling, W.; Kollmann, J.; Das, S. Visible-Light-Mediated Efficient Metal-Free Catalyst for α-Oxygenation of Tertiary Amines to Amides. *ACS Catal.* **2018**, *8*, 6659–6664. [CrossRef]
84. Zhang, Y.; Schilling, W.; Riemer, D.; Das, S. Metal-Free Photocatalysts for the Oxidation of Non-Activated Alcohols and the Oxygenation of Tertiary Amines Performed in Air or Oxygen. *Nat. Protoc.* **2020**, *15*, 822–839. [CrossRef] [PubMed]
85. Neerathilingam, N.; Anandhan, R. Metal-Free Photoredox-Catalyzed Direct α-Oxygenation of N,N-Dibenzylanilines to Imides under Visible Light. *RSC Adv.* **2022**, *12*, 8368–8373. [CrossRef] [PubMed]
86. Mandigma, M.J.P.; Žurauskas, J.; MacGregor, C.I.; Edwards, L.J.; Shahin, A.; d'Heureuse, L.; Yip, P.; Birch, D.J.S.; Gruber, T.; Heilmann, J.; et al. An Organophotocatalytic Late-Stage N–CH$_3$ Oxidation of Trialkylamines to N-Formamides with O$_2$ in Continuous Flow. *Chem. Sci.* **2022**, *13*, 1912–1924. [CrossRef] [PubMed]
87. Nakao, Y.; Idei, H.; Kanyiva, K.S.; Hiyama, T. Hydrocarbamoylation of Unsaturated Bonds by Nickel/Lewis-Acid Catalysis. *J. Am. Chem. Soc.* **2009**, *131*, 5070–5071. [CrossRef]
88. Ding, S.; Jiao, N. N,N-Dimethylformamide: A Multipurpose Building Block. *Angew. Chem. Int. Ed.* **2012**, *51*, 9226–9237. [CrossRef]
89. Kohnen-Johannsen, K.; Kayser, O. Tropane Alkaloids: Chemistry, Pharmacology, Biosynthesis and Production. *Molecules* **2019**, *24*, 796. [CrossRef]
90. Nakai, S.; Yatabe, T.; Suzuki, K.; Sasano, Y.; Iwabuchi, Y.; Hasegawa, J.; Mizuno, N.; Yamaguchi, K. Methyl-Selective α-Oxygenation of Tertiary Amines to Formamides by Employing Copper/Moderately Hindered Nitroxyl Radical (DMN-AZADO or 1-Me-AZADO). *Angew. Chem. Int. Ed.* **2019**, *58*, 16651–16659. [CrossRef]

91. Su, J.; Ma, X.; Ou, Z.; Song, Q. Deconstructive Functionalizations of Unstrained Carbon–Nitrogen Cleavage Enabled by Difluorocarbene. *ACS Cent. Sci.* **2020**, *6*, 1819–1826. [CrossRef]
92. Cheng, X.; Yang, B.; Hu, X.; Xu, Q.; Lu, Z. Visible-Light-Promoted Metal-Free Aerobic Oxidation of Primary Amines to Acids and Lactones. *Chem. Eur. J.* **2016**, *22*, 17566–17570. [CrossRef]
93. Eicher, T.; Hauptmann, S.; Speicher, A. *The Chemistry of Heterocycles: Structure, Reactions Synthesis and Applications*, 2nd ed.; Wiley-VCH: Weinheim, Germany, 2003; pp. 1–556. [CrossRef]
94. Zhang, C.; Li, S.; Bureš, F.; Lee, R.; Ye, X.; Jiang, Z. Visible Light Photocatalytic Aerobic Oxygenation of Indoles and pH as a Chemoselective Switch. *ACS Catal.* **2016**, *6*, 6853–6860. [CrossRef]
95. Schilling, W.; Zhang, Y.; Riemer, D.; Das, S. Visible-Light-Mediated Dearomatisation of Indoles and Pyrroles to Pharmaceuticals and Pesticides. *Chem. Eur. J.* **2020**, *26*, 390–395. [CrossRef] [PubMed]
96. Kushwaha, A.K.; Maury, S.K.; Kumari, S.; Kamal, A.; Singh, H.K.; Kumar, D.; Singh, S. Visible-Light-Initiated Oxidative Coupling of Indole and Active Methylene Compounds Using Eosin Y as a Photocatalyst. *Synthesis* **2022**, *54*, 5099–5109. [CrossRef]
97. Zhao, D.; Wang, R. Recent Developments in Metal Catalyzed Asymmetric Addition of Phosphorus Nucleophiles. *Chem. Soc. Rev.* **2012**, *41*, 2095–2108. [CrossRef] [PubMed]
98. Clarion, L.; Jacquard, C.; Sainte-Catherine, O.; Loiseau, S.; Filippini, D.; Hirlemann, M.-H.; Volle, J.N.; Virieux, D.; Lecouvey, M.; Pirat, J.L.; et al. Oxaphosphinanes: New Therapeutic Perspectives for Glioblastoma. *J. Med. Chem.* **2012**, *55*, 2196–2211. [CrossRef] [PubMed]
99. Hibner-Kulicka, P.; Joule, J.A.; Skalik, J.; Bałczewski, P. Recent Studies of the Synthesis, Functionalization, Optoelectronic Properties and Applications of Dibenzophospholes. *RSC Adv.* **2017**, *7*, 9194–9236. [CrossRef]
100. Zhang, Y.; Ye, C.; Li, S.; Ding, A.; Gu, G.; Guo, H. Eosin Y-Catalyzed Photooxidation of Triarylphosphines under Visible Light Irradiation and Aerobic Conditions. *RSC Adv.* **2017**, *7*, 13240–13243. [CrossRef]
101. Ding, A.; Li, S.; Chen, Y.; Jin, R.; Ye, C.; Hu, J.; Guo, H. Visible Light-Induced 4-Phenylthioxanthone-Catalyzed Aerobic Oxidation of Triarylphosphines. *Tetrahedron Lett.* **2018**, *59*, 3880–3883. [CrossRef]
102. Kondo, M.; Agou, T. Catalytic Aerobic Photooxidation of Triarylphosphines Using Dibenzo-Fused 1,4-Azaborines. *Chem. Commun.* **2022**, *58*, 5001–5004. [CrossRef]
103. Denmark, S.E. The Interplay of Invention, Discovery, Development, and Application in Organic Synthetic Methodology: A Case Study. *J. Org. Chem.* **2009**, *74*, 2915–2927. [CrossRef]
104. Mewald, M.; Schiffner, J.A.; Oestreich, M. A New Direction in C-H Alkenylation: Silanol as a Helping Hand. *Angew. Chem. Int. Ed.* **2012**, *51*, 1763–1765. [CrossRef]
105. Tran, N.T.; Wilson, S.O.; Franz, A.K. Cooperative Hydrogen-Bonding Effects in Silanediol Catalysis. *Org. Lett.* **2012**, *14*, 186–189. [CrossRef] [PubMed]
106. Kim, J.K.; Sieburth, S. McN. Synthesis and Properties of a Sterically Unencumbered δ-Silanediol Amino Acid. *J. Org. Chem.* **2012**, *77*, 2901–2906. [CrossRef]
107. Franz, A.K.; Wilson, S.O. Organosilicon Molecules with Medicinal Applications. *J. Med. Chem.* **2013**, *56*, 388–405. [CrossRef]
108. Mitsudome, T.; Noujima, A.; Mizugaki, T.; Jitsukawa, K.; Kaneda, K. Supported gold nanoparticlecatalyst for the selective oxidation of silanes to silanols in water. *Chem. Commun.* **2009**, *35*, 5302–5304. [CrossRef]
109. Jeon, M.; Han, J.; Park, J. Catalytic Synthesis of Silanols from Hydrosilanes and Applications. *ACS Catal.* **2012**, *2*, 1539–1549. [CrossRef]
110. Adam, W.; Mello, R.; Curci, R. O-Atom Insertion into Si-H Bonds by Dioxiranes: A Stereospecific and Direct Conversion of Silanes into Silanols. *Angew. Chem. Int. Ed. Engl.* **1990**, *29*, 890–891. [CrossRef]
111. Li, J.; Xu, D.; Shi, G.; Liu, X.; Zhang, J.; Fan, B. Oxidation of Silanes to Silanols with Oxygen via Photoredox Catalysis. *ChemistrySelect* **2021**, *6*, 8345–8348. [CrossRef]
112. Arzumanyan, A.V.; Goncharova, I.K.; Novikov, R.A.; Milenin, S.A.; Boldyrev, K.L.; Solyev, P.N.; Tkachev, Y.V.; Volodin, A.D.; Smol'yakov, A.F.; Korlyukov, A.A.; et al. Aerobic Co or Cu/NHPI-Catalyzed Oxidation of Hydride Siloxanes: Synthesis of Siloxanols. *Green Chem.* **2018**, *20*, 1467–1471. [CrossRef]
113. Li, H.; Chen, L.; Duan, P.; Zhang, W. Highly Active and Selective Photocatalytic Oxidation of Organosilanes to Silanols. *ACS Sust. Chem. Eng.* **2022**, *10*, 4642–4649. [CrossRef]
114. Wang, J.; Li, B.; Liu, L.C.; Jiang, C.; He, T.; He, W. Metal-Free Visible-Light-Mediated Aerobic Oxidation of Silanes to Silanols. *Sci. China Chem.* **2018**, *61*, 1594–1599. [CrossRef]
115. He, P.; Xhang, F.; Si, X.; Jiang, W.; Shen, Q.; Li, Z.; Zhu, Z.; Tang, S.; Gui, Q.W. Visible-light-induced aerobic oxidation of tertiary silanes to silanols using molecular oxygen as an oxidant. *Synthesis* **2022**. [CrossRef]
116. Zeng, Q.; Gao, S.; Chelashaw, A. Advances in Titanium-Catalyzed Synthesis of Chiral Sulfoxide Drugs. *Mini-Reviews in Org. Chem.* **2013**, *10*, 198–206. [CrossRef]
117. Xu, N.; Zhu, J.; Wu, Y.Q.; Zhang, Y.; Xia, J.Y.; Zhao, Q.; Lin, G.Q.; Yu, H.L.; Xu, J.H. Enzymatic Preparation of the Chiral (S)-Sulfoxide Drug Esomeprazole at Pilot-Scale Levels. *Org. Process Res. Dev.* **2020**, *24*, 1124–1130. [CrossRef]
118. Peng, T.; Cheng, X.; Chen, Y.; Yang, J. Sulfoxide Reductases and Applications in Biocatalytic Preparation of Chiral Sulfoxides: A Mini-Review. *Front. Chem.* **2021**, *9*, 714899. [CrossRef]
119. Yoshida, Y.; Otsuka, S.; Nogi, K.; Yorimitsu, H. Palladium-catalyzed amination of aryl sulfoxides. *Org. Lett.* **2018**, *20*, 1134–1137. [CrossRef]

120. Yang, P.; Xu, W.; Wang, R.; Zhang, M.; Xie, C.; Zeng, X.; Wang, M. Potassium *Tert*-Butoxide-Mediated Condensation Cascade Reaction: Transition Metal-Free Synthesis of Multisubstituted Aryl Indoles and Benzofurans. *Org. Lett.* **2019**, *21*, 3658–3662. [CrossRef]
121. Venier, C.G.; Squires, T.G.; Chen, Y.Y.; Smith, B.F. Peroxytrifluoroacetic acid oxidation of sulfides to sulfoxides and sulfones. *J. Org. Chem.* **1982**, *47*, 3773–3774. [CrossRef]
122. Colonna, S.; Gaggero, N. Enantioselective oxidation of sulphides by dioxiranes in the presence of bovine serum albumin. *Tetrahedron Lett.* **1989**, *30*, 6233–6236. [CrossRef]
123. Kropp, P.J.; Breton, G.W.; Fields, J.D.; Tung, J.C.; Loomis, B.R. Surface-Mediated Reactions. 8. Oxidation of Sulfides and Sulfoxides with *tert*-Butyl Hydroperoxide and Oxone. *J. Am. Chem. Soc.* **2000**, *122*, 4280–4285. [CrossRef]
124. Li, Q.; Lan, X.; An, G.; Ricardez-Sandoval, L.; Wang, Z.; Bai, G. Visible-Light-Responsive Anthraquinone Functionalized Covalent Organic Frameworks for Metal-Free Selective Oxidation of Sulfides: Effects of Morphology and Structure. *ACS Catal.* **2020**, *10*, 6664–6675. [CrossRef]
125. Chen, Y.; Hu, J.; Ding, A. Synthesis of an anthraquinone-containing polymeric photosensitizer and its application in aerobic photooxidation of thioethers. *RSC Adv.* **2020**, *10*, 10661–10665. [CrossRef] [PubMed]
126. Jiang, D.; Chen, M.; Deng, Y.; Hu, W.; Su, A.; Yang, B.; Mao, F.; Zhang, C.; Liu, Y.; Fu, Z. 9,10-Dihydroanthracene auto-photooxidation efficiently triggered photo-catalytic oxidation of organic compounds by molecular oxygen under visible light. *Mol. Catal.* **2020**, *494*, 111127. [CrossRef]
127. Dang, C.; Zhu, L.; Guo, H.; Xia, H.; Zhao, J.; Dick, B. Flavin Dibromide as an Efficient Sensitizer for Photooxidation of Sulfides. *ACS Sustainable Chem. Eng.* **2018**, *6*, 15254–15263. [CrossRef]
128. Guo, H.; Xia, H.; Ma, X.; Chen, K.; Dang, C.; Zhao, J.; Dick, B. Efficient Photooxidation of Sulfides with Amidated Alloxazines as Heavy-Atom-Free Photosensitizers. *ACS Omega* **2020**, *5*, 10586–10595. [CrossRef] [PubMed]
129. Li, W.; Xie, Z.; Jing, X. BODIPY photocatalyzed oxidation of thioanisole under visible light. *Catal. Commun.* **2011**, *16*, 94–97. [CrossRef]
130. Li, W.; Li, L.; Xiao, H.; Qi, R.; Huang, Y.; Xie, Z.; Jing, X.; Zhang, H. Iodo-BODIPY: A visible-light-driven, highly efficient and photostable metal-free organic photocatalyst. *RSC Adv.* **2013**, *3*, 13417–13421. [CrossRef]
131. Gao, Y.; Xu, H.; Zhang, Y.; Tang, C.; Fan, W. Visible-light photocatalytic aerobic oxidation of sulfides to sulfoxides with a perylene diimide photocatalyst. *Org. Biomol. Chem.* **2019**, *17*, 7144–7149. [CrossRef]
132. Ye, C.; Zhang, Y.; Ding, A.; Hu, Y.; Guo, H. Visible Light Sensitizer-Catalyzed Highly Selective Photo Oxidation from Thioethers into Sulfoxides under Aerobic Condition. *Sci. Rep.* **2018**, *8*, 2205. [CrossRef]
133. Zhao, B.; Hammond, G.B.; Xu, B. Modulation of Photochemical Oxidation of Thioethers to Sulfoxides or Sulfones Using an Aromatic Ketone as the Photocatalyst. *Tetrahedron Lett.* **2021**, *82*, 153376. [CrossRef]
134. Forchetta, M.; Sabuzi, F.; Stella, L.; Conte, V.; Galloni, P. KuQuinone as a Highly Stable and Reusable Organic Photocatalyst in Selective Oxidation of Thioethers to Sulfoxides. *J. Org. Chem.* **2022**, *87*, 14016–14025. [CrossRef]

Disclaimer/Publisher's Note: The statements, opinions and data contained in all publications are solely those of the individual author(s) and contributor(s) and not of MDPI and/or the editor(s). MDPI and/or the editor(s) disclaim responsibility for any injury to people or property resulting from any ideas, methods, instructions or products referred to in the content.

Article

Thermally Stable and Highly Efficient *N,N,N*-Cobalt Olefin Polymerization Catalysts Affixed with *N*-2,4-Bis(Dibenzosuberyl)-6-Fluorophenyl Groups

Muhammad Zada [1,2], Desalegn Demise Sage [1], Qiuyue Zhang [1], Yanping Ma [1], Gregory A. Solan [1,3,*], Yang Sun [1] and Wen-Hua Sun [1,*]

[1] Key Laboratory of Engineering Plastics, Beijing National Laboratory for Molecular Sciences, Institute of Chemistry, Chinese Academy of Sciences, Beijing 100190, China
[2] Department of Chemistry, Government Postgraduate College Khar, Bajaur 18650, Pakistan
[3] Department of Chemistry, University of Leicester, University Road, Leicester LE1 7RH, UK
* Correspondence: gas8@leicester.ac.uk (G.A.S.); whsun@iccas.ac.cn (W.-H.S.); Tel.: +44(0)-116-2522096 (G.A.S.); +86-10-62557955 (W.-H.S.)

Citation: Zada, M.; Sage, D.D.; Zhang, Q.; Ma, Y.; Solan, G.A.; Sun, Y.; Sun, W.-H. Thermally Stable and Highly Efficient *N,N,N*-Cobalt Olefin Polymerization Catalysts Affixed with *N*-2,4-Bis(dibenzosuberyl)-6-Fluorophenyl Groups. *Catalysts* **2022**, *12*, 1569. https://doi.org/10.3390/catal12121569

Academic Editor: Carmine Capacchione

Received: 14 November 2022
Accepted: 30 November 2022
Published: 2 December 2022

Publisher's Note: MDPI stays neutral with regard to jurisdictional claims in published maps and institutional affiliations.

Copyright: © 2022 by the authors. Licensee MDPI, Basel, Switzerland. This article is an open access article distributed under the terms and conditions of the Creative Commons Attribution (CC BY) license (https:// creativecommons.org/licenses/by/ 4.0/).

Abstract: The cobalt(II) chloride *N,N,N*-pincer complexes, [2-{(2,4-($C_{15}H_{13}$)$_2$-6-FC_6H_2)N=CMe}-6-(ArN=CMe)C_5H_3N]CoCl$_2$ (Ar = 2,6-Me$_2C_6H_3$) (**Co1**), 2,6-Et$_2C_6H_3$ (**Co2**), 2,6-*i*-Pr$_2C_6H_3$ (**Co3**), 2,4,6-Me$_3C_6H_2$ (**Co4**), 2,6-Et$_2$-4-MeC$_6H_2$ (**Co5**), and [2,6-{(2,4-($C_{15}H_{13}$)$_2$-6-FC_6H_2)N=CMe}$_2C_5H_3$N]CoCl$_2$ (**Co6**), each containing at least one *N*-2,4-bis(dibenzosuberyl)-6-fluorophenyl group, were synthesized in good yield from their corresponding unsymmetrical (**L1–L5**) and symmetrical bis(imino)pyridines (**L6**). The molecular structures of **Co1** and **Co2** spotlighted their distorted square pyramidal geometries (τ_5 value range: 0.23–0.29) and variations in steric hindrance offered by the dissimilar *N*-aryl groups. On activation with either MAO or MMAO, **Co1–Co6** all displayed high activities for ethylene polymerization, with levels falling in the order: **Co1** > **Co4** > **Co5** > **Co2** > **Co3** > **Co6**. Indeed, the least sterically hindered 2,6-dimethyl **Co1** in combination with MAO exhibited a very high activity of 1.15×10^7 g PE mol^{-1} (Co) h^{-1} at the operating temperature of 70 °C, which dropped by only 15% at 80 °C and 43% at 90 °C. Vinyl-terminated polyethylenes of high linearity and narrow dispersity were generated by all catalysts, with the most sterically hindered, **Co3** and **Co6**, producing the highest molecular weight polymers [M_w range: 30.26–33.90 kg mol^{-1} (**Co3**) and 42.90–43.92 kg mol^{-1} (**Co6**)]. In comparison with structurally related cobalt catalysts, it was evident that the presence of the *N*-2,4-bis(dibenzosuberyl)-6-fluorophenyl groups had a limited effect on catalytic activity but a marked effect on thermal stability.

Keywords: cobalt; ethylene polymerization; dibenzosuberyl; *ortho*-fluoride; thermal stability; linear polyethylene; high activity

1. Introduction

The deployment of late transition metal (e.g., Fe, Co, Ni, Pd) complexes as catalysts in ethylene polymerization has been extensively investigated since the groundbreaking discoveries in the mid-to-late 1990's [1–4]. Much of the interest in the area stems from their ease of preparation and their ability to generate various types of industrially important polyethylenes, ranging from highly linear to highly branched with differing levels of end-group unsaturation [3–6].

Regarding iron and cobalt catalysts, most incorporate the tridentate 2,6-bis(arylimino) pyridine ligand frame (**A**, Figure 1). Indeed, for this class of catalyst, numerous studies have been dedicated to exploring the relationships between structure and productivity, as well as the properties of the resulting polymers [1,2]. As a consequence, factors such as the choice of metal center and the substitution pattern displayed by the *N,N,N*-ligand have been shown to be of crucial importance. Furthermore, such variations to the chelating ligand structure can also impact the thermal stability of the catalyst, with poor thermal stability being a characteristic that has, in part, limited their industrial application [1,2,6–8]. In most cases, these structural modifications have been concerned with the N-aryl groups and, in particular, the influence of substituents with differing steric and electronic properties [1,2,5,8–22]. As a more recent modification, the fusion of carbocyclic rings to the central pyridine in **A** has seen improvements in both catalytic performance and thermal stability [23–35].

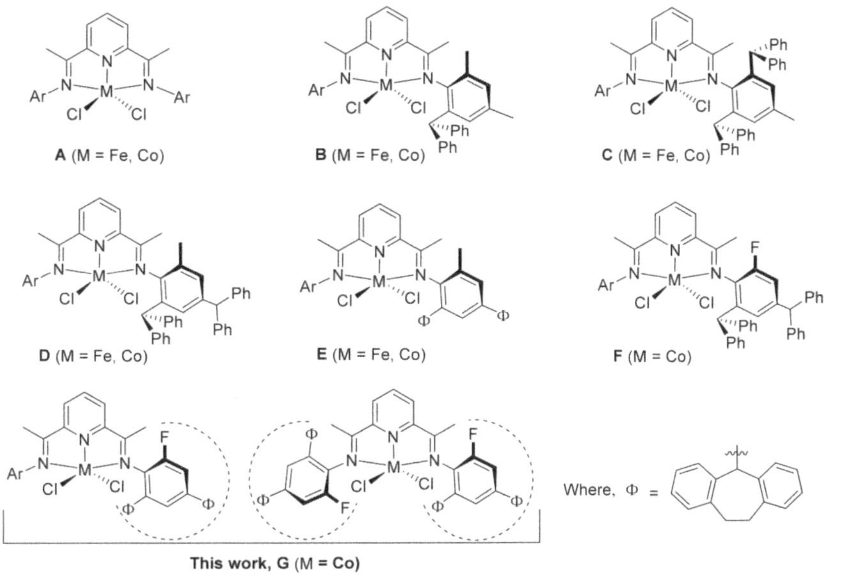

Figure 1. Structural modifications to the ligand framework in 2,6-bis(arylimino)pyridine-iron(II) and -cobalt(II) chloride complexes, **A–F**, including the target of this work, **G**.

As an on-going theme in our research, we have been interested in developing families of unsymmetrical bis(imino)pyridine–iron and –cobalt complexes that contain one fixed N-aryl group and the other variable. For example, precatalysts of type **B** (Figure 1) incorporate a N-2-benzhydryl-4,6-dimethylphenyl group as the fixed aryl group, while the other aryl group offers a means to fine-tune performance through steric and electronic variation. Indeed, iron and cobalt examples of **B** exhibit very good catalytic activity [36], while **C** (Figure 1), bearing two sterically hindered *ortho*-benzhydryl groups along with an electron-donating *para*-methyl group, not only display high catalytic activity but also enhanced thermal stability (**C**, Figure 1) [37–42]. Moreover, analogues of **C** with the *para*-methyl group replaced by electron-withdrawing substituents have seen further improvements, particularly for iron [39–47]. Incorporation of fluoride into the benzhydryl periphery has also led to catalysts displaying high activity and good thermal stability [43,44]. Elsewhere, the positioning of benzhydryl groups within the N-aryl ring has been the subject of a number of reports. For instance, cobalt-containing $\mathbf{D_{Co}}$ (Figure 1), affixed with a N-2,4-dibenzhydryl-6-methylphenyl group, exhibited exceptional catalytic activity (up to 18.1×10^6 g of PE mol^{-1} (Co) h^{-1}), while the thermal stability was somewhat reduced [48,49].

As a development of our work using benzhydryl as a substituent on the N-aryl ring, we have also been interested in employing dibenzosuberyl groups, in which the two phenyl groups in benzhydryl have been tethered by an ethyl linker [50–53]. For example, $\mathbf{E_{Co}}$ (Figure 1), the dibenzosuberyl equivalent of $\mathbf{D_{Co}}$, displayed comparatively higher thermal stability, while the polyethylene was of similarly high molecular weight [52,53]. As a more recent finding, the introduction of fluoride in conjunction with two benzhydryl groups on the N-aryl ring has seen the disclosure of cobalt-containing $\mathbf{F_{Co}}$ (Figure 1). Indeed, $\mathbf{F_{Co}}$

exhibited very high catalytic activity and superior thermal stability [54]. Similarly, the bis(4,4'-difluorobenzhydryl) counterpart of **F** showed promising performance in terms of catalytic productivity and thermostability [55,56].

In order to further refine our understanding of the factors that influence catalytic performance and temperature resilience in this type of polymerization catalyst, we targeted in this work a series of bis(imino)pyridine-cobalt precatalysts appended with N-2,4-bis(dibenzosuberyl)-6-fluorophenyl groups (**G**, Figure 1). We anticipated that the electronic effect of the 6-fluoride in close proximity to the central metal and the steric effect imposed by the dibenzosuberyl groups, along with the anticipated acidity of its methine protons, might have some impact on catalytic performance and particularly on thermal stability. With this in mind, we disclose five distinct examples of unsymmetrical cobalt precatalysts differing in the steric and electronic properties of the second N-aryl group and one symmetrical example bearing two N-2,4-bis(dibenzosuberyl)-6-fluorophenyl groups. These new complexes were subjected to a comprehensive polymerization evaluation that probed the type and amount of co-catalyst, reaction temperature, run time, and ethylene pressure. The results of this study are then compared to previously reported cobalt examples of **B–F** (Figure 1). In addition, the preparative details for the six novel 2,6-bis(imino)pyridines and their cobalt precatalysts are reported in detail.

2. Results and Discussion

The unsymmetrical 2,6-bis(arylimino)pyridines, 2-{(2,4-($C_{15}H_{13}$)$_2$-6-FC_6H_2)N=CMe}-6-(ArN=CMe)C_5H_3N (Ar = 2,6-$Me_2C_6H_3$) (**L1**), 2,6-$Et_2C_6H_3$ (**L2**), 2,6-i-$Pr_2C_6H_3$ (**L3**), 2,4,6-$Me_3C_6H_2$ (**L4**), and 2,6-Et_2-4-MeC_6H_2 (**L5**), were obtained in two steps in satisfactory overall yield (Scheme 1). Firstly, the condensation of 2,6-diacetylpyridine with one molar equivalent of N-2,4-bis(dibenzosuberyl)-6-fluoroaniline afforded the imine-ketone 2-{(2,4-($C_{15}H_{13}$)$_2$-6-FC_6H_2)N=CMe}-6-(O=CMe)C_5H_3N as the main product, along with some 2,6-{(2,4-($C_{15}H_{13}$)$_2$-6-FC_6H_2)N=CMe}$_2C_5H_3$N (**L6**) as the by-product. Treatment of the imine-ketone intermediate with the corresponding aniline, 2,6-R^1_2-4-$R^2C_6H_2NH_2$, under acid catalyzed conditions, gave **L1–L5**. The N-2,4-bis(dibenzosuberyl)-6-fluoroaniline was not commercially available and was prepared using a reported procedure [57–59].

Further reaction of **L1–L6** with anhydrous cobalt(II) chloride in a mixture of ethanol and dichloromethane generated [2-{(2,4-($C_{15}H_{13}$)$_2$-6-FC_6H_2)N=CMe}-6-(ArN=CMe)C_5H_3N]$CoCl_2$ (Ar = 2,6-$Me_2C_6H_3$) (**Co1**, 2,6-$Et_2C_6H_3$ (**Co2**), 2,6-i-$Pr_2C_6H_3$ (**Co3**), 2,4,6-$Me_3C_6H_2$ (**Co4**), 2,6-Et_2-4-MeC_6H_2 (**Co5**) and [2,6-{(2,4-($C_{15}H_{13}$)$_2$-6-FC_6H_2)N=CMe}$_2C_5H_3$N]$CoCl_2$ (**Co6**) in good yields (Scheme 1). All of the organic compounds were characterized by IR, ^{1}H, ^{13}C, and ^{19}F NMR spectroscopy and elemental analysis, while **Co1–Co6** were analyzed by elemental analysis, ^{19}F NMR, and IR spectroscopy, and **Co1** and **Co2** were analyzed by single crystal X-ray diffraction.

Single crystals of **Co1** and **Co2** employed for the X-ray studies were grown by slow diffusion of n-heptane into a dichloromethane solution of the corresponding complex. Perspective views of both are shown in Figures 2 and 3, while selected bond lengths and angles are provided in Table 1. Crystallographic parameters are given in the supporting information (Table S1 from Supplementary Materials). The structures of **Co1** and **Co2** were similar and will be described as a pair. Each comprised a single cobalt center bound by a N,N,N-chelating bis(arylimino)pyridine and two chloride ligands to complete a geometry that can be best described as distorted square pyramidal. More accurately, this distortion can be quantified using the geometric tau value (τ_5). This parameter can be defined from the equation $\tau_5 = (\beta-\alpha)/60$, where β is the largest angle and α is the second largest angle in the coordination sphere. A tau value of zero is indicative of a perfect square pyramid and a value of one for a perfect trigonal bipyramid [49–57,60–64]. For **Co1** and **Co2**, the τ_5 values were 0.23 and 0.29, respectively, reflecting some degree of distortion from a perfect square pyramid. In each structure, the three nitrogen atoms N1, N2, and N3, along with Cl1, formed the basal plane, while Cl2 occupied the apical site and the cobalt center itself sat above the basal plane, by approximately 0.147 Å and 0.137 Å, respectively. With regard to the cobalt-nitrogen bond lengths, the central Co(1)–N(1) distance [2.035(2) Å **Co1**, 2.0512(12) Å **Co2**] was noticeably shorter than Co(1)–N(2) [2.238(2) Å **Co1**, 2.1810(12) Å **Co2**] and Co(1)–N(3) [2.260(2) Å **Co1**, 2.1784(11) Å **Co2**], indicating more effective coordination of the $N_{pyridine}$ with the cobalt atom. Similar findings have been reported in the literature [45,46,48–56]. For **Co1**, the exterior Co-N_{imine} bond lengths also showed some minor variation, with Co1-N3 slightly longer than Co1-N2, although this was not reproduced in **Co2**. The N(2)–C(8) [1.281(3)–1.2819(19) Å] and N(3)–C(2) [1.271(3)–1.2851(18) Å] bond lengths in both complexes were typical of C=N double bonds. In addition, the N-aryl planes were inclined towards the perpendicular with respect to the N,N,N-chelating plane, with some modest variation in the dihedral angles (76.3°, 78.3° **Co1**; 77.7°, 83.3° **Co2**). Similar inclinations have been observed in other unsymmetrical bis(arylimino)pyridine complexes [36–46,50–56]. As a variation between the two structures, the *ortho*-dibenzosuberyl group in **Co1** was positioned on the same side as apical Cl2, while in **Co2**, the *ortho*-fluoride adopted this position. Within each *ortho*-/*para*-substituted dibenzosuberyl group, the central seven-membered ring was puckered on account of the three sp^3-hybridized carbon atoms. There were no intermolecular contacts of note.

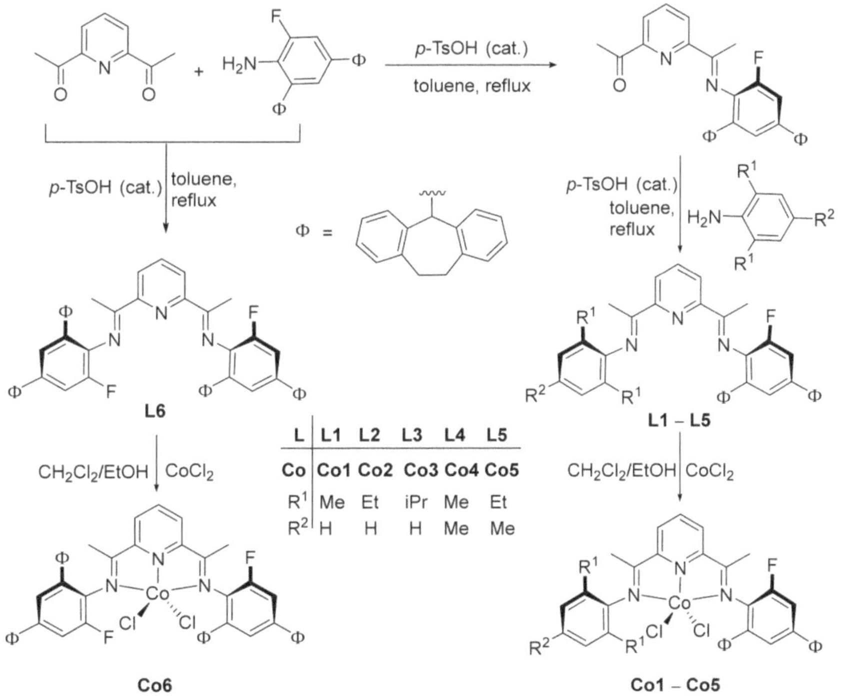

Scheme 1. Synthetic route to the bis(arylimino)pyridine derivatives (**L1–L6**) and their cobalt(II) chloride complexes (**Co1–Co6**).

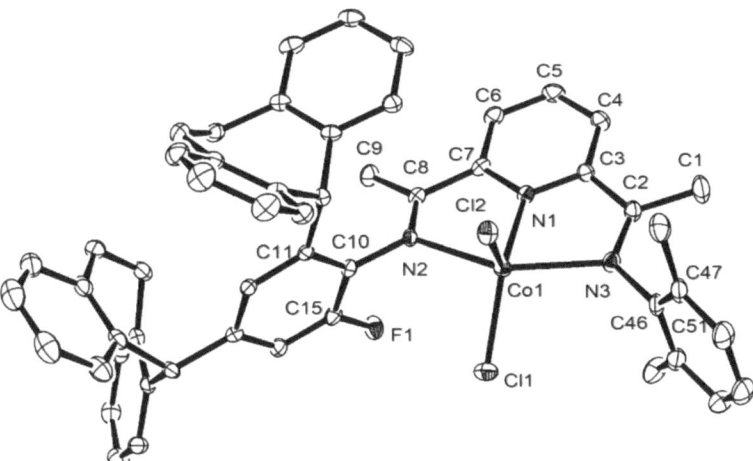

Figure 2. ORTEP representation of **Co1** with the thermal ellipsoids set at 30% probability level. All hydrogen atoms have been omitted for clarity.

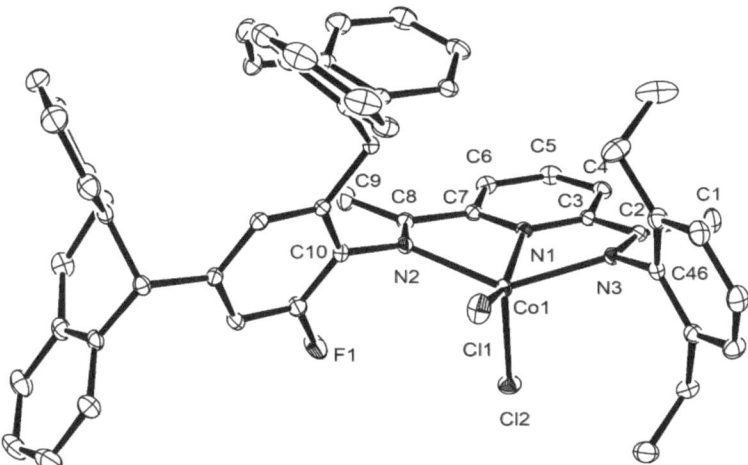

Figure 3. ORTEP representation of **Co2** with the thermal ellipsoids set at the 30% probability level. All hydrogen atoms have been omitted for clarity.

^{19}F NMR spectroscopy was performed on **L1–L6** and **Co1–Co6** in order to probe the effect of the fluorine environment and metal complexation on chemical shift. As expected, the spectra of the free ligands showed single peaks for the *ortho*-fluoride with chemical shifts of approximately δ −128.7 ppm (Figure S1 from Supplementary Materials), which compared to a range of between δ −146.9 and −129.3 ppm for **Co1–Co5** (see Figure 4 for representative spectra). Evidently, complexation with the paramagnetic Co(II) ion resulted in an upfield shift and a broadening of the fluorine resonances. Interestingly, the spectrum of symmetrical **Co6** showed major and minor peaks at δ −146.9 and −145.8 ppm, respectively (see Figure S2 from Supplementary Materials), which suggested the existence of two isomers in solution [55,56].

Table 1. Selected bond lengths (Å) and angles (°) for **Co1** and **Co2**.

	Co1	Co2
Bond Lengths (Å)		
Co(1)–N(1)	2.035 (2)	2.0512 (12)
Co(1)–N(2)	2.238 (2)	2.1810 (12)
Co(1)–N(3)	2.260 (2)	2.1784 (11)
Co(1)–Cl(1)	2.2370 (8)	2.2424 (4)
Co(1)–Cl(2)	2.2673 (8)	2.2960 (4)
N(2)–C(10)	1.437 (3)	1.4292 (18)
N(3)–C(46)	1.449 (3)	1.488 (2)
N(2)–C(8)	1.281 (3)	1.2819 (19)
N(3)–C(2)	1.271 (3)	1.2851 (18)
C(15)–F(1)	1.350 (3)	1.3591 (17)
Bond Angles (°)		
N(1)–Co(1)–N(2)	74.87 (8)	73.56 (4)
N(1)–Co(1)–N(3)	74.78 (8)	73.97 (4)
N(2)–Co(1)–N(3)	149.16 (8)	138.63 (4)
N(1)–Co(1)–Cl(1)	135.22 (6)	156.00 (4)
N(2)–Co(1)–Cl(1)	95.29 (6)	99.83 (3)
N(3)–Co(1)–Cl(1)	102.35 (6)	99.46 (3)
N(2)–Co(1)–Cl(2)	101.27 (6)	101.57 (3)
N(3)–Co(1)–Cl(2)	94.49 (6)	103.81 (3)
Cl(1)–Co(1)–Cl(2)	113.91 (3)	113.233 (18)
N(1)–Co(1)–Cl(2)	110.86 (6)	90.76 (3)

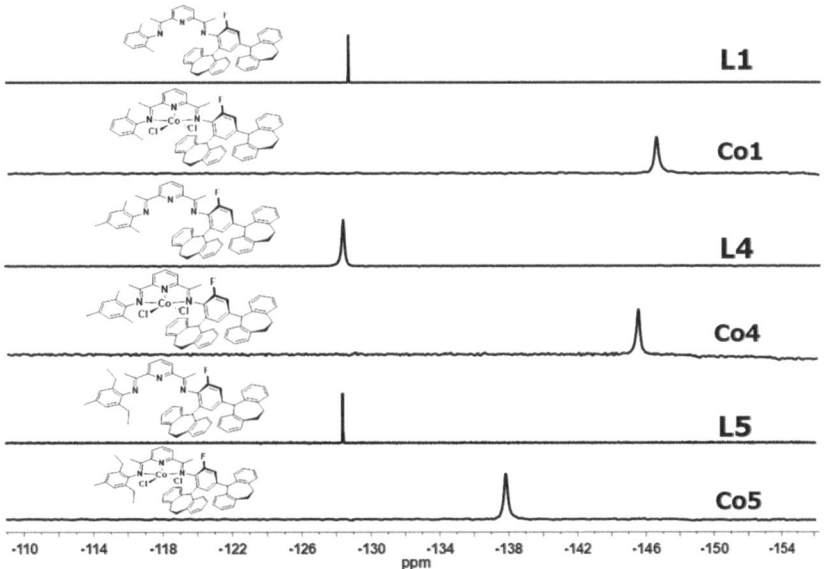

Figure 4. ^{19}F NMR spectra of **L1**, **L4**, and **L5**, along with those for their corresponding complexes **Co1**, **Co4**, and **Co5**; recorded in CDCl$_3$ at ambient temperature.

In the FT-IR spectra of **Co1–Co6**, the stretching vibrations for the C=N$_{imine}$ bonds appeared in the range of 1623–1632 cm^{-1}, which compared to 1641–1651 cm^{-1} for the free bis(imino)pyridines, **L1–L6**. This reduction in wavenumber was in accordance with the effective coordination of the imine nitrogen donor atoms to the metal center [36–39,48–57]. The elemental analyses of **Co1–Co6** were consistent with the proposed LCoCl$_2$ composition.

3. Ethylene Polymerization Studies

Previous work in the area indicated that aluminoxanes, such as MAO (methylaluminoxane) or MMAO (modified methylaluminoxane), are effective co-catalysts for generating the active bis(imino)pyridine-cobalt catalyst [13,14,36–42,48–57]. Accordingly, both MAO and MMAO were used during the catalytic evaluation of **Co1–Co6**, and their catalytic performance was compared. To identify an effective set of reaction conditions for the polymerizations, **Co1** was initially selected and parameters such as temperature, Al:Co molar ratio, run time, and ethylene pressure were systematically investigated with the ethylene pressure kept at 10 atm. All runs were performed in toluene as the solvent.

3.1. Optimization of Polymerization Conditions Using Co1/MAO

With **Co1**/MAO initially employed as the test catalyst system, the influence of run temperature was first investigated with the Al:Co molar ratio fixed at 2000:1 and the run time set at 30 min (runs 1–7, Table 2).

Table 2. Optimization of the polymerization conditions using **Co1**/MAO [a].

Run	T (°C)	t (min)	Al:Co	PE (g)	Activity [b]	M_w [c]	M_w/M_n [c]	T_m (°C) [d]
1	30	30	2000	1.82	2.43	16.80	2.43	130.4
2	40	30	2000	2.14	2.85	14.72	2.67	129.5
3	50	30	2000	2.83	3.77	11.15	2.63	128.9
4	60	30	2000	4.54	6.05	9.26	2.63	128.5
5	70	30	2000	8.61	11.48	7.40	2.42	127.7
6	80	30	2000	7.36	9.81	6.74	2.37	127.1
7	90	30	2000	4.92	6.56	5.44	2.27	126.1
8	70	30	1500	6.85	9.13	7.52	2.47	127.6
9	70	30	1750	8.02	10.70	7.75	2.37	128.2
10	70	30	2250	7.43	9.91	7.27	2.41	127.8
11	70	30	2500	6.37	8.49	7.07	2.29	127.8
12	70	30	3000	5.60	7.47	6.33	2.16	128.4
13	70	05	2000	2.57	20.56	6.73	2.27	127.3
14	70	15	2000	4.61	12.29	7.22	2.27	128.0
15	70	45	2000	9.02	8.02	7.53	2.46	127.6
16	70	60	2000	10.51	7.01	8.23	2.34	127.6
17 [e]	70	30	2000	3.87	5.16	6.27	2.55	127.2
18 [f]	70	30	2000	0.71	0.95	2.92	2.40	123.2

[a] General conditions: 1.5 μmol of **Co1**, 100 mL toluene, 10 atm C_2H_4, [b] 10^6 g of PE mol^{-1} (Co) h^{-1}, [c] M_w: kg mol^{-1}, determined by GPC, [d] determined by DSC, [e] 5 atm of C_2H_4, [f] 1 atm of C_2H_4.

When the temperature was increased from 30 to 70 °C, the catalytic activity steadily increased, reaching a peak level of 11.48 × 10^6 g of PE mol^{-1} of Co h^{-1} at the higher temperature (run 5, Table 2). Above 70 °C, some loss in activity was observed, with levels decreasing to 9.81 × 10^6 g of PE mol^{-1} of Co h^{-1} at 80 °C and then to 6.56 × 10^6 g of PE mol^{-1} of Co h^{-1} at 90 °C. Clearly, the active species generated using **Co1**/MAO displayed good thermal stability with only a 15% reduction in performance at 80 °C. It would seem probable that this latter dip in effectiveness was due to partial deactivation of the active species and/or lower solubility of the ethylene monomer at elevated temperature [36–46,49,52–57]. In contrast, the molecular weight of the polyethylene displayed a gradual decrease when the temperature increased from 30 to 90 °C, with M_w values dropping from 16.80 to 5.44 kg mol^{-1} (Figure 5). This reduction in molecular weight was ascribed to increased rates of chain transfer reactions as compared to chain propagation at higher run temperatures [38–46,49,52–55].

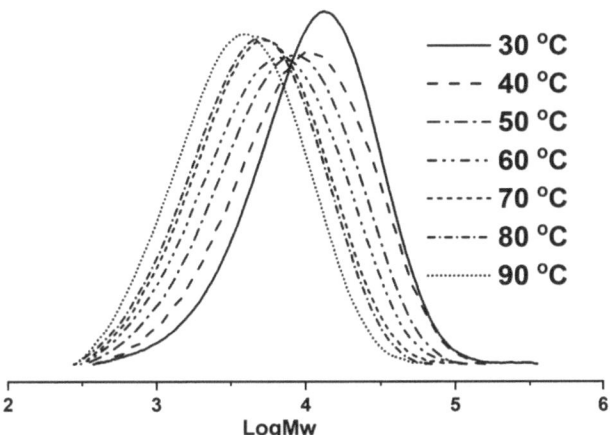

Figure 5. GPC curves showing logM_w of the polyethylenes produced using **Co1**/MAO as a function of the reaction temperature (runs 1–7, Table 2).

Next, the amount of MAO employed was systematically varied by altering the Al:Co molar ratio for **Co1**/MAO from 1500:1 to 3000:1 with the run temperature retained at 70 °C (runs 5, 8–12, Table 2). At 2000:1, the highest catalytic activity of 11.48 × 10^6 g of PE mol^{-1} of Co h^{-1} was achieved (run 5, Table 2), which then declined at higher ratios reaching a low point of 7.47 × 10^6 g of PE mol^{-1} of Co h^{-1} at 3000:1 (run 12, Table 2). In terms of the molecular weight of the polyethylene, the highest value of 7.75 kg mol^{-1} was attained at 1750:1 (run 9, Table 2) and then progressively decreased as the ratio was raised from 2000:1 to 3000:1 (runs 10–12, Table 2). The corresponding GPC curves are given in Figure S3 (Supplementary Materials). It would seem likely that this drop in M_w was due to increased rates in chain transfer and termination reactions in the presence of larger amounts of co-catalyst [48,50,52–55,64–67].

With the Al:Co molar ratio retained at 2000:1 and the temperature set at 70 °C, the time/activity profile of **Co1**/MAO was investigated at selected intervals between 5 and 60 min (runs 5, 13–16, Table 2). Examination of the results indicated an inverse relationship between catalytic activity and run time, with a peak performance observed after 5 min (20.56 × 10^6 g of PE mol^{-1} of Co h^{-1}), which then proceeded to drop-off and reached its lowest value after 60 min (7.01 × 10^6 g of PE mol^{-1} of Co h^{-1}). These results further highlighted the short induction period required to produce the active species (run 13, Table 2) and, moreover, its appreciable lifetime even over longer run times. Meanwhile, the molecular weights of the polyethylenes increased over a prolonged reaction time and remained narrowly dispersed in all cases. The GPC curves of the resulting polyethylenes are collected in Figure S4 (Supplementary Materials).

To explore the influence of ethylene pressure on **Co1**/MAO, the polymerization runs were additionally performed at 5 and 1 atm (runs 5, 17, and 18, Table 2). The results showed a clear correlation between pressure and catalytic activity, as well as the molecular weight of the polyethylene, with higher pressures leading to improved catalytic performance and higher molecular weight polymers. In particular, the catalytic activity displayed at 5 atm of ethylene was almost half that displayed at 10 atm, while the molecular weight of the polyethylene lowered to 6.27 kg mol^{-1} (run 17, Table 2). In comparison, at 1 atm, the catalytic activity dramatically decreased to 0.95 × 10^6 g of PE mol^{-1} of Co h^{-1} while the polymer molecular weight declined to 2.92 kg mol^{-1} (run 18, Table 2). These observations suggested that a high pressure of ethylene was favorable for coordination and insertion, which was likely related to the increased solubility of the ethylene monomer. The GPC curves as a function of ethylene pressure are given in Figure 6.

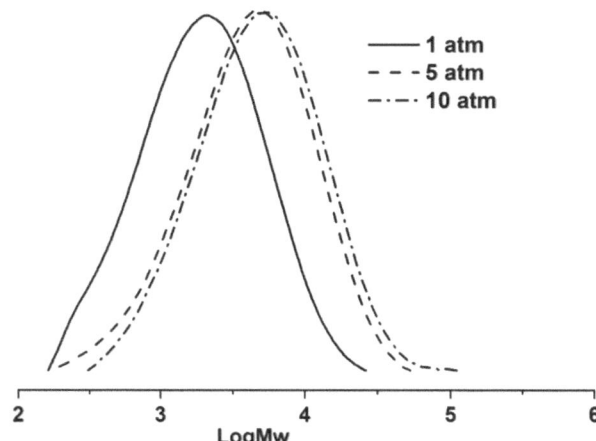

Figure 6. GPC curves showing logM_w of the polyethylenes produced using **Co1**/MAO as a function of the ethylene pressure (runs 5, 17, and 18, Table 2).

As a general observation for all the different polymerization runs performed using **Co1**/MAO (runs 1–18, Table 2), the polyethylenes showed a narrow dispersity (M_w/M_n range: 2.16 to 2.67; runs 1–18, Table 2), suggesting well-controlled polymerizations as the result of single-site active species.

3.2. Optimization of Polymerization Conditions Using Co1/MMAO

With MAO now replaced with MMAO, a similar strategy was employed to optimize the performance of **Co1**/MMAO. The complete set of results are collected in Table 3.

Table 3. Optimization of the polymerization conditions using **Co1**/MMAO [a].

Run	T (°C)	t (min)	Al:Co	PE (g)	Activity [b]	M_w [c]	M_w/M_n [c]	T_m (°C) [d]
1	30	30	2000	1.43	1.91	20.14	2.42	131.2
2	40	30	2000	1.87	2.49	14.36	2.47	130.2
3	50	30	2000	2.54	3.39	11.67	2.54	129.5
4	60	30	2000	3.28	4.37	8.05	2.45	128.2
5	70	30	2000	4.86	6.48	7.28	2.38	128.0
6	80	30	2000	3.72	4.96	6.42	2.25	127.2
7	90	30	2000	1.85	2.47	5.53	2.33	126.4
8	70	30	1500	3.12	4.16	7.60	2.34	127.9
9	70	30	1750	3.97	5.29	7.45	2.31	128.1
10	70	30	2250	5.63	7.51	7.09	2.51	127.9
11	70	30	2500	3.85	5.13	6.64	1.94	128.2
12	70	30	3000	2.74	3.65	3.67	2.45	128.7
13	70	05	2250	2.14	17.12	6.68	2.06	128.0
14	70	15	2250	3.48	9.28	6.90	2.31	127.6
15	70	45	2250	5.92	5.26	7.09	2.10	127.8
16	70	60	2250	6.41	4.27	7.91	2.12	128.0
17 [e]	70	30	2250	2.15	2.87	5.51	2.34	127.7
18 [f]	70	30	2250	0.53	0.71	3.58	1.92	124.1

[a] General conditions: 1.5 µmol of **Co1**, 100 mL toluene, 10 atm C_2H_4, [b] 10^6 g of PE mol^{-1} (Co) h^{-1}, [c] M_w: kg mol^{-1}, determined by GPC, [d] determined by DSC, [e] 5 atm of C_2H_4, [f] 1 atm of C_2H_4.

By running the polymerizations at temperatures between 30 and 90 °C (runs 1–7, Table 3), a peak activity of 6.48 × 10^6 g of PE mol^{-1} of Co h^{-1} was again observed at 70 °C (run 5, Table 3). However, this activity level was not as high as that observed for **Co1**/MAO at the same temperature. Nonetheless, the optimal temperature of 70 °C again highlighted the high thermal stability of this class of catalyst [40–46,49,52–57]. At temperatures in excess of 70 °C, the catalytic activity decreased to 4.96×10^6 g of PE mol^{-1} of Co h^{-1} at 80 °C and then to 2.47×10^6 g of PE mol^{-1} of Co h^{-1} at 90 °C (runs 6 and 7, Table 3), highlighting once again the appreciable resilience of the active species to high operating temperature. The molecular weights of the polyethylenes showed a similar trend to that observed with MAO, with the values decreasing from 20.14 to 5.53 kg mol^{-1} when the temperature was raised from 30 to 90 °C (Figure 7). The generated polyethylenes displayed a reasonably narrow dispersity across the temperature range (runs 1–7, Table 3).

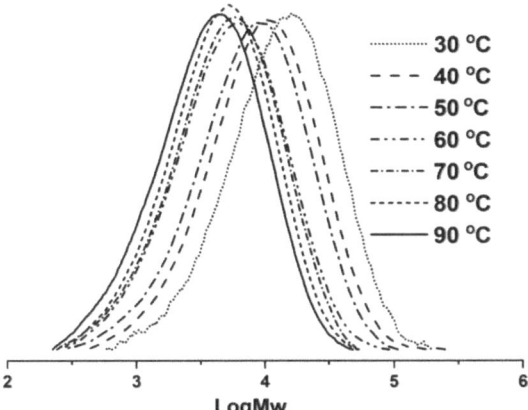

Figure 7. GPC curves showing logM_w of the polyethylenes produced using **Co1**/MMAO as a function of the reaction temperature (runs 1–7, Table 3).

The response of **Co1**/MMAO to variations in Al:Co molar ratio was then explored by altering the ratio from 1500:1 to 3000:1 (runs 5, 8–12, Table 3) with the temperature set at 70 °C. The results revealed a gradual increase in catalytic activity, reaching a maximum of 7.51×10^6 g of PE mol^{-1} of Co h^{-1} at 2250:1 (run 10, Table 4). Further increases in the amount of co-catalyst led to a reduction in activity, with the lowest value of 3.65×10^6 g of PE mol^{-1} of Co h^{-1} obtained at 3000:1 (run 12, Table 3). On the other hand, the molecular weights of the polyethylenes steadily decreased from 7.60 kg mol^{-1} to 3.65 kg mol^{-1} (runs 5, 8–12, Table 3) as the ratio increased (Figure S5 from Supplementary Materials). This finding aligned with increased rates of both chain transfer and termination reactions at higher concentration of co-catalyst [48,50,52–55,64–67].

Table 4. Screening of **Co1**–**Co6** with either MAO or MMAO under optimized conditions [a].

Run	Precat.	Co-Cat.	PE (g)	Activity [b]	M_w [c]	M_w/M_n [c]	T_m (°C) [d]
1	Co1	MAO	8.61	11.48	7.40	2.42	127.7
2	Co2	MAO	7.03	9.37	13.06	2.54	129.8
3	Co3	MAO	3.91	5.21	30.26	2.47	132.9
4	Co4	MAO	7.54	10.05	8.10	2.29	128.4
5	Co5	MAO	6.88	9.17	13.38	2.53	129.9
6	Co6	MAO	2.43	3.24	42.90	2.28	132.8
7	Co1	MMAO	5.63	7.51	7.10	2.51	127.9
8	Co2	MMAO	3.42	4.56	13.40	2.35	130.0
9	Co3	MMAO	2.90	3.87	33.90	1.88	132.7
10	Co4	MMAO	5.16	6.88	8.07	2.61	128.0
11	Co5	MMAO	3.28	4.37	13.93	2.32	130.0
12	Co6	MMAO	2.06	2.74	43.92	2.03	132.4

[a] General conditions: 1.5 µmol of cobalt precatalyst, 100 mL toluene, 10 atm C$_2$H$_4$, 70 °C, 30 min, Al:Co ratio of 2000:1 (MAO) and 2250:1 (MMAO), [b] 10^6 g PE mol^{-1} (Co) h^{-1}, [c] kg mol^{-1} determined by GPC, [d] determined by DSC.

The impact of run time on **Co1**/MMAO was then explored between 5 and 60 min (runs 10, 13–16, Table 3) with the temperature maintained at 70 °C and the Al:Co molar ratio set at 2250:1. Similar to the results observed with MAO, the peak level of 17.12×10^6 g of PE $\mathrm{mol^{-1}}$ of Co $\mathrm{h^{-1}}$ was found after 5 min (run 13, Table 3). Over a prolonged reaction time, a gradual decrease in catalytic activity was observed, reaching a value of 4.27×10^6 g of PE $\mathrm{mol^{-1}}$ of Co $\mathrm{h^{-1}}$ after 60 min (runs 16, Table 3). Nonetheless, the levels of activity after longer run times remained appreciable, highlighting the good lifetime of the active species. In terms of the polyethylenes, a gradual increase in molecular weight was observed over time (Figure S6 from Supplementary Materials).

To explore the influence of ethylene pressure on **Co1**/MMAO, we then investigated this parameter using the most effective conditions identified in terms of run temperature and Al:Co molar ratio (runs 10, 17, and 16, Table 3). As with the MAO study, the highest catalytic activity was achieved at the highest pressure of ethylene. At 5 atm, the catalytic activity dropped to 2.87×10^6 g of PE $\mathrm{mol^{-1}}$ of Co $\mathrm{h^{-1}}$, while at 1 atm the catalytic activity lowered to 0.71×10^6 g of PE $\mathrm{mol^{-1}}$ of Co $\mathrm{h^{-1}}$. Similarly, the molecular weight of the polyethylene dropped to 5.51 kg $\mathrm{mol^{-1}}$ at 5 atm (run 17, Table 3) and then fell further to 3.58 kg $\mathrm{mol^{-1}}$ at 1 atm (run 18, Table 3). It would seem probable that less effective mass transfer of the ethylene monomer at lower pressure accounted for the observations described above. The GPC curves are given in Figure 8. As a final remark, under all polymerization conditions employed using **Co1**/MMAO, the dispersity of the polyethylenes (M_w/M_n) was in the range 1.92 to 2.54 (runs 1–18, Table 3), indicating the single-site nature of the active species.

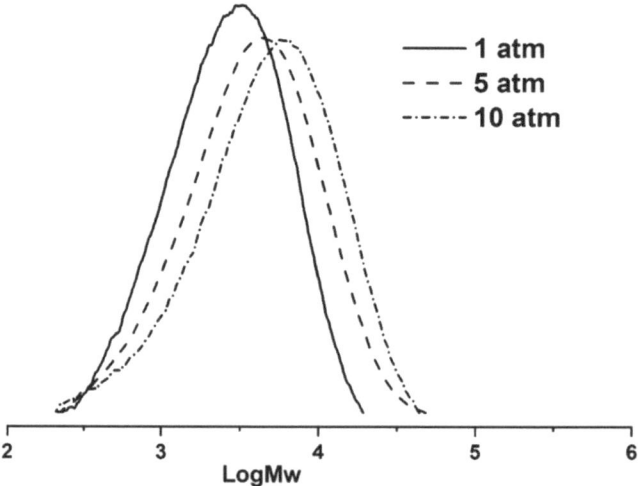

Figure 8. GPC curves showing log M_w of the polyethylenes produced using **Co1**/MMAO as a function of the ethylene pressure (entries 10, 17, and 18, Table 3).

3.3. Screening of **Co1**–**Co6** with MAO and MMAO

Finally, all the remaining cobalt complexes, **Co2**–**Co6**, were investigated as precatalysts for ethylene polymerization using the corresponding set of optimized reaction conditions identified with either **Co1**/MAO (reaction temperature = 70 °C, Al:Co molar ratio = 2000:1, run time = 30 min, P_{C2H4} = 10 atm) or **Co1**/MMAO (reaction temperature = 70 °C, Al:Co molar ratio = 2250:1, run time = 30 min, P_{C2H4} = 10 atm). The results are displayed in Table 4.

By utilizing MAO as co-catalyst, all cobalt precatalysts displayed high catalytic activity for ethylene polymerization (range: 3.24–11.48×10^6 g of PE $\mathrm{mol^{-1}}$ of Co $\mathrm{h^{-1}}$) with the relative order following: **Co1** > **Co4** > **Co5** > **Co2** > **Co3** > **Co6** (runs 1–6, Table 4). At the top end of the range were **Co1** and **Co4**, which both contain the least sterically bulky *ortho*-methyl (R^1) substituents (runs 1, 4, Table 4). Conversely, **Co3**, bearing the bulkier *ortho*-isopropyl substituents, exhibited lower activity, while symmetrical **Co6**, incorporating two *N*-2,4-bis(dibenzosuberyl)-6-fluorophenyl groups, was the least active of the series (runs 3, 6, Table 4). On the other hand, **Co3** and **Co6** generated the highest molecular weight polyethylenes of 30.26 and 42.90 kg mol^{-1}, respectively. Variations in the molecular weight of the polyethylene as a function of precatalyst are shown in Figure 9. These findings suggested that bulkier substituents inhibited chain transfer and protected the active species [42–46,52–57].

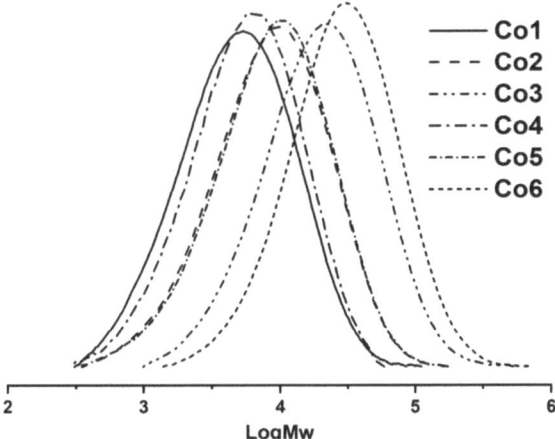

Figure 9. GPC curves showing log M_w of the polyethylenes produced using **Co1–Co6** with MAO as activator in each case (runs 1–6, Table 4).

Similarly, using MMAO, the catalytic performance of **Co2–Co6** was investigated using the optimum conditions established with **Co1**/MMAO (runs 7–12, Table 4). At first glance, it was evident that the activities for all six cobalt precatalysts mirrored the order observed with MAO: **Co1** > **Co4** > **Co5** > **Co2** > **Co3** > **Co6**. However, these systems generally displayed lower catalytic activities (range: 2.74–7.51 × 10^6 g of PE mol^{-1} of Co h^{-1}), which could plausibly be attributed to the different activation processes involved with each co-catalyst. Nevertheless, steric influences imparted by the *N*-aryl substituents once again played a key role on performance, with the least hindered 2,6-dimethyl **Co1** and **Co4** falling at the top end of the catalytic range (runs 7, 10, Table 4). Likewise, the highest molecular weight polyethylene was generated using the bulkiest precatalysts, **Co6** (M_w = 43.93 kg mol^{-1}) and **Co3** (M_w = 33.90 kg mol^{-1}) (runs 9, 12, Table 4). The GPC curves of the polyethylenes generated using **Co1–Co6**/MMAO are shown in Figure 10. As was noted in the initial studies using **Co1**/MAO or **Co1**/MMAO, the polymers produced using **Co1–Co6** were narrowly dispersed, with M_w/M_n values falling in the range of 1.88 to 2.61, highlighting the good control and single-site nature of the active species.

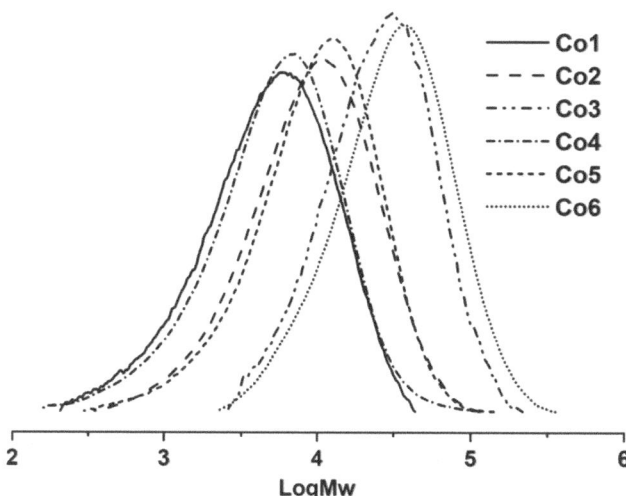

Figure 10. GPC curves showing logM_w of the polyethylenes produced using **Co1–Co6** with MMAO as activator in each case (runs 7–12, Table 4).

3.4. Microstructural Features of the Polyethylenes

As is apparent from Tables 2–4, all of the polyethylene samples produced in this work displayed melting points (T_m) in the range 126.1–132.9 °C, values that are typical of linear polyethylenes. Indeed, bis(imino)pyridine-cobalt catalysts have a history for forming polymers with very little branching [1–6,60,64–71]. To investigate the structural composition of these polyethylenes in more detail, selected samples generated using the more active catalysts that were activated using MAO and MMAO, were investigated using ^{1}H and ^{13}C NMR spectroscopy. To allow suitable solubility, these samples were dissolved in deuterated $C_2D_2Cl_4$ at high temperature and their spectra were recorded at 100 °C.

Firstly, a sample obtained using **Co1**/MAO at the optimal temperature of 70 °C (M_w = 7.40 kg mol^{-1}, run 5, Table 2) was recorded and analyzed using approaches detailed in the literature [36,45–48,52–59,68–71]. The ^{1}H NMR spectrum displayed characteristic peaks for a vinyl-terminated polyethylene displaying high linearity (Figure 11). In particular, a high intensity singlet at δ 1.30 ppm for the –(CH$_2$)$_n$– repeat unit was observed along with two less intense downfield multiplets at δ 5.91 and δ 4.98 ppm in a 1:2 intensity ratio, which was assigned to the vinyl end group (–CH=CH$_2$). In addition, a weak signal for the protons on the carbon adjacent to vinyl group (H$_c$) were visible more upfield at δ 2.12 ppm, while the protons belonging to methyl end group (H$_f$) were detected most upfield at δ 0.98 ppm. On the other hand, signals for H$_e$ and H$_f$ could not be observed and were presumably masked by the signal for –(CH$_2$)$_n$– repeat unit, while H$_d$ could be just identified at δ 1.54 ppm.

In the ^{13}C NMR spectrum of this polyethylene sample (Figure 12), the signal for the –(CH$_2$)$_n$– repeat unit took the form of an intense resonance at δ 30.00 ppm. On either side of this peak, weaker signals for the methylene carbons, C$_c$, C$_d$, and C$_f$, could be detected, with the methyl chain end observed most upfield at δ 14.2 ppm. Conversely, the vinylic carbons, C$_b$ and C$_a$ were found most downfield at δ 139.61 and 114.39 ppm, whereas the signal for the carbon atom C$_c$ adjacent to this vinyl end group appeared at δ 33.98 ppm. Notably, the integral ratio of the methyl and vinyl end (Cg:Ca) in ^{13}C NMR spectrum was found to be approximately 1:1, which provided support for a vinyl-terminated polymer. Overall, this observation of a vinyl end group in both the ^{1}H NMR and ^{13}C NMR spectra implied that β-hydrogen elimination represented the termination pathway in this polymerization process [50–53,70,71].

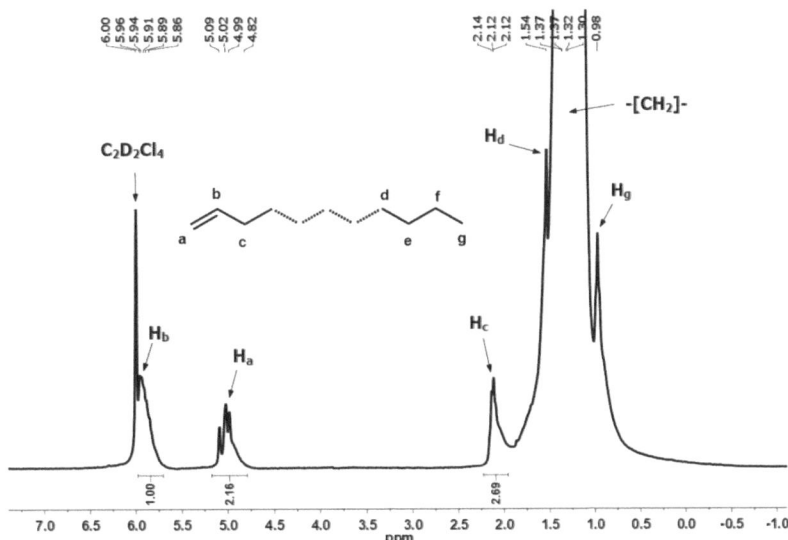

Figure 11. ^{1}H NMR spectrum of the polyethylene generated using **Co1**/MAO (run 5, Table 2); recorded in $C_2D_2Cl_4$ at 100 °C.

A similar analysis was performed on the ^{1}H and ^{13}C NMR spectra of the polyethylene afforded using **Co1**/MMAO at 70 °C (M_w = 7.09 kg mol^{-1}, run 10, Table 3). As was noted above, signals characteristic of a vinyl-terminated polyethylene with high linearity were evident (Figures S7 and S8 from Supplementary Materials) [36,45–48,52–59,68–71]. Clearly, β-hydrogen elimination again accounted for the key termination pathway when using MMAO as co-catalyst.

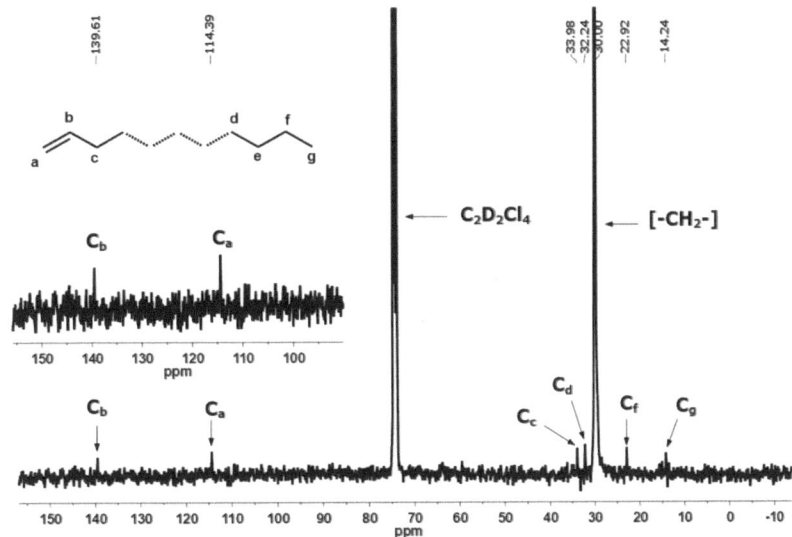

Figure 12. ^{13}C NMR spectrum of the polyethylene generated using **Co1**/MAO (run 5, Table 2); recorded in $C_2D_2Cl_4$ at 100 °C.

3.5. Comparative Study of the Current Catalyst System with Previously Reported Examples

To facilitate a comparison of the current class of precatalysts with related unsymmetrical cobalt(II) chloride precatalysts [36,37,48,52,54], Figure 13 presents various performance data for previously reported **B–F** alongside that for **G**. All the tests were carried out under optimal reaction conditions at $P_{C_2H_4}$ = 10 atm using MAO as the co-catalyst.

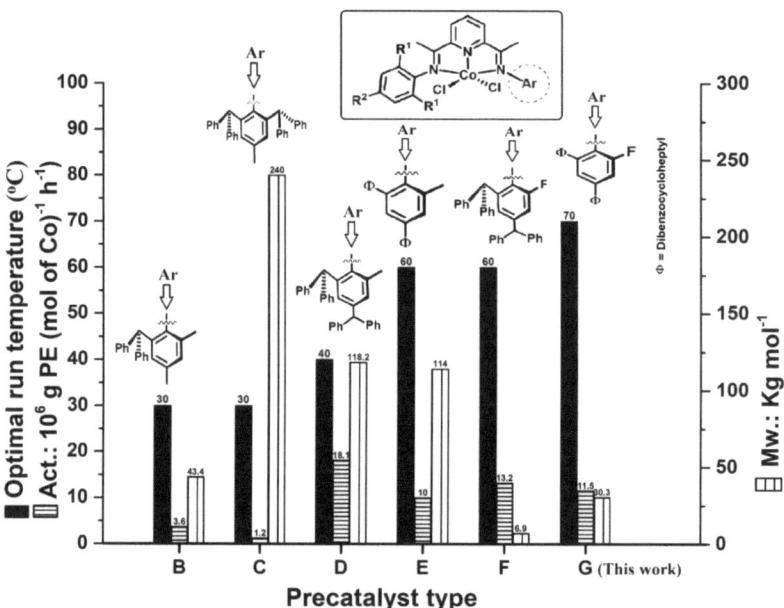

Figure 13. Comparison of catalytic activity, optimal run temperature, and polymer molecular weight data for **G** (this work) with that obtained using **B–F**. All runs were performed under optimized conditions at 10 atm of C_2H_4 using MAO as the activator.

Several notable findings emerged from inspection of the figure. Firstly, the highest optimal run temperature was displayed by **G** (70 °C); for **B–F** this temperature fell anywhere between 30 and 60 °C. This observation lent support for the beneficial effect of the introduction of a *N*-2,4-bis(dibenzosuberyl)-6-fluorophenyl group on the thermal stability of the cobalt catalyst. In terms of catalytic activity, **G** (11.5 × 10^6 g of PE mol^{-1} (Co) h^{-1}) was the third most active system, with only **D** [48] and **F** exceeding it under optimal operating conditions. Nonetheless, **G** was more active than **B** [36], **C** [37], and **E** [52], as well as several other reported precatalysts bearing *N*-aryl substituents appended with electron withdrawing/donating groups in combination with benzhydryl groups [39–42,45]. Evidently, the presence of the *N*-2,4-bis(dibenzosuberyl)-6-fluorophenyl group did not significantly enhance activity. In terms of the molecular weight of the polyethylene, that generated with **G** (30.3 kg/mol) was higher than that with **F** [54], although less than that produced using **B–E**. However, it would seem most likely that this relatively low molecular weight polymer was due to the higher run temperature leading to a higher rate of chain transfer compared to **B–E**.

4. Experimental
4.1. General Considerations

Manipulations of all air and/or moisture sensitive compounds were undertaken using standard Schlenk techniques under inert nitrogen atmosphere conditions. All solvents prior to use were distilled under nitrogen. The aluminum-alkyls, methylaluminoxane

(MAO, 1.46 M in toluene), and modified-methylaluminoxane (MMAO, 1.93 M in *n*-heptane) were purchased from Akzo Nobel Corp. (Nanjing, China), while the high-purity ethylene was bought from Beijing Yanshan Petrochemical Co. (Beijing, China). All other reagents were purchased from either Aldrich (Beijing, China) or Acros (Beijing, China). The *N*-2,4-bis(dibenzosuberyl)-6-fluoroaniline was prepared according to a described procedure [52,53,58,59]. The ^{1}H, ^{13}C, and ^{19}F NMR spectra of all bis(imino)pyridines and their precursors were recorded on a Bruker DMX 400M Hz NMR (Bruker, Karlsruhe, Germany) at ambient temperature using TMS as an internal standard. The FT-IR spectra were recorded on the Perkin Elmer Spectrum 2000 FT-IR spectrometer (Shanghai, China). The elemental analyses were carried out on a Flash EA 1112 microanalyzer (Thermo Electron SPA, Beijing, China). The molecular weight (M_w) and dispersity (M_w/M_n) of the polyethylenes were measured using a PL-GPC 220 instrument (PL, Shropshire, UK) operating at 150 °C and using 1,2,4-trichlorobenzene as the eluting solvent. The melting points of the polyethylenes were measured using differential scanning calorimetry (Q2000 DSC; TA Instruments, New Castle, DE, USA) under a nitrogen atmosphere. Typically, a PE sample of 5.0 mg was heated to 160 °C at a heating rate of 20 °C min^{-1} and maintained for 3 min at the same temperature to remove its thermal history. This sample was then cooled to −20 °C at a rate of 20 °C min^{-1}. The ^{1}H and ^{13}C NMR spectra of the polyethylenes were recorded in deuterated 1,1,2,2-tetrachloroethane at 100 °C. Sample preparation typically involved taking a known amount of polyethylene (90–100 mg) and dissolving it in deuterated 1,1,2,2-tetrachloroethane (2.0 mL) at high temperature before an aliquot was transferred to a 5 mm standard glass NMR tube. Inverse-gated ^{13}C NMR spectra were recorded on a Bruker DMX 300 MHz spectrometer (Bruker) with the number of scans anywhere between 1798 and 1935. The conditions used for the spectral analysis were as follows: a spectral width of 22675.7 kHz, acquisition time of 0.7225 s, relaxation delay of 5.0 s, and pulse width of 15.5 µs.

4.2. Synthesis of 2-{(2,4-($C_{15}H_{13}$)$_2$-6-FC_6H_2)N=CMe}-6-(O=CMe)C_5H_3N

To a round-bottom flask containing 2,6-diacetylpyridine (3.30 g, 20.2 mmol) in toluene (100 mL) was added 2,4-bis(dibenzosuberyl)-6-fluoroaniline (10.02 g, 20.2 mmol), before a Dean-Stark trap was fitted. The reaction mixture was stirred under reflux for 10 min and then a catalytic amount of *p*-toluenesulfonic acid (ca. 15 mol%) was slowly added. After a further 6 h under reflux, the reaction mixture was allowed to cool to room temperature, at which point all volatiles were evaporated under reduced pressure. The resulting residue was then purified by column chromatography (on basic alumina) using an eluent composed of petroleum ether and diethyl ether (25:2) to afford the title compound as a pale yellow solid (4.36 g, 34%). Mp: 203–205 °C. ^{1}H NMR (400 MHz, CDCl$_3$, TMS): δ 8.52 (d, *J* = 7.9 Hz, Py–H, 1H), 8.13 (d, *J* = 7.7 Hz, Py–H, 1H), 7.97 (t, *J* = 7.8 Hz, Py–H, 1H), 7.26–6.85 (m, Ar–H, 14H), 6.58 (s, Ar–H$_m$, 1H), 6.32 (d, *J* = 11.2 Hz, Ar–H, 2H), 6.12 (s, Ar–H$_m$, 1H), 5.16 (s, –CH–, 1H), 4.98 (s, –CH–, 1H), 3.02–2.86 (m, –CH$_2$–, 3H), 2.81 (s, –CH$_3$, 3H), 2.80–2.72 (m, –CH$_2$–, 5H), 1.63 (s, –CH$_3$, 3H). ^{13}C NMR (100 MHz, CDCl$_3$, TMS): δ 200.2, 170.7, 155.1, 152.3, 140.8, 140.7, 139.8, 137.1, 135.0, 134.0, 131.3, 130.6, 127.3, 126.7, 126.1, 125.7, 125.0, 122.5, 112.6, 112.4, 57.7, 56.2, 32.3, 25.7, 16.2. TMS FT-IR (KBr cm^{-1}): 3051 (w), 3060 (w), 3012 (w), 2926 (w), 2836 (w), 2829 (w), 1704 (υ(C=O), s), 1649 (υ(C=N), s), 1584 (w), 1528 (w), 1491 (m), 1463 (w), 1429 (w), 1364 (s), 1309 (m), 1234 (m), 1150 (w), 1120 (w), 1078 (w), 1042 (w), 1014 (w), 990 (w), 940 (w), 892 (w), 828 (w), 816 (m), 790 (w), 760 (s), 734 (m), 701 (m), 668 (w).

4.3. Synthesis of 2-{(2,4-($C_{15}H_{13}$)$_2$-6-FC_6H_2)N=CMe}-6-(ArN=CMe)C_5H_3N (**L1–L5**)

4.3.1. Ar = 2,6-Me$_2$C$_6$H$_3$ (**L1**)

A mixture of the imine-ketone 2-{(2,4-($C_{15}H_{13}$)$_2$-6-FC_6H_2)N=CMe}-6-(O=CMe)C_5H_3N (1.01 g, 1.58 mmol) and 2,6-dimethylaniline (0.25 g, 2.06 mmol), along with a catalytic amount of *p*-toluenesulfonic acid (ca. 15%) in toluene, were loaded into a round-bottom flask equipped with a Dean-Stark trap. The reaction mixture was stirred under reflux for

6 h and then allowed to cool to room temperature before all volatiles were removed under reduced pressure. The resulting residue was purified by column chromatography (using basic alumina) using a mixture of petroleum ether and diethyl ether (25:2) as the eluent to give **L1** as a yellow powder (0.68 g, 8%). Mp: 254–256 °C. ^{1}H NMR (400 MHz, CDCl$_3$, TMS): δ 8.47 (d, J = 7.6 Hz, Py–H, 1H), 8.42 (d, J = 8.0 Hz, Py–H, 1H), 7.94 (t, J = 7.8 Hz, Py–H, 1H), 7.22–7.06 (m, Ar–H, 19H), 6.52 (d, J = 11.2 Hz, Aryl–H, 1H), 6.36 (s, aryl–H$_m$, 1H), 5.13 (s, –CH–, 1H), 5.09 (s, –CH–, 1H), 3.01–2.93 (m, –CH$_2$–, 3H), 2.83–2.79 (m, –CH$_2$–, 1H), 2.70–2.62 (m, –CH$_2$–, 3H), 2.52–2.48 (m, –CH$_2$–, 1H), 2.18 (s, –CH$_3$, 3H), 2.08 (s, –CH$_3$, 6H), 1.67 (s, –CH$_3$, 3H). ^{13}C NMR (100 MHz, CDCl$_3$, TMS): δ 171.3, 167.3, 155.0, 154.7, 151.8, 149.4, 148.8, 140.6, 140.5, 140.1, 140.0, 139.8, 139.1, 136.7, 135.0, 134.4, 134.3, 131.4, 131.2, 131.0, 130.6, 130.1, 127.9, 127.3, 127.2, 127.1, 126.9, 126.3, 126.1, 126.0, 125.7, 125.5, 123.0, 122.7, 122.2, 112.6, 112.4, 57.7, 57.4, 56.2, 32.3, 31.3, 18.0, 16.4. ^{19}F NMR (470 MHz, CDCl$_3$): δ –128.68. FT-IR (KBr cm^{-1}): 3060 (w), 3022 (w), 2930 (w), 2889 (w), 2827 (w), 1645 (υ(C=N), s), 1581 (w), 1493 (w), 1452 (m), 1360 (s), 1301 (w), 1247 (w), 1202 (m), 1130 (m), 1101(w), 1050 (w), 1021 (w), 994 (w), 961 (w), 934 (w), 872 (w), 8061 (m), 771 (s), 718 (m), 687 (w). Anal. Calcd. for C$_{53}$H$_{46}$N$_3$F·0.5H$_2$O (752.98): C, 84.54; H, 6.29; N, 5.58. Found: C, 84.60; H, 6.29; N, 5.37.

4.3.2. Ar = 2,6-Et$_2$C$_6$H$_3$ (**L2**)

Using a similar procedure as described for the synthesis of **L1**, **L2** was prepared as a yellow powder (0.34 g, 26%). Mp: 212–214 °C. ^{1}H NMR (400 MHz, CDCl$_3$, TMS): ^{1}H NMR (400 MHz, CDCl$_3$, TMS): δ 8.45 (d, J = 8.0 Hz, Py–H, 1H), 8.42 (d, J = 8.0 Hz, Py–H, 1H), 7.94 (t, J = 7.8 Hz, Py–H, 1H), 7.22–7.09 (m, Ar–H, 17H), 7.20 (Ar–H$_m$, 1H), 6.87 (d, J = 11.2 Hz, Ar–H, 2H), 6.36 (s, Aryl–H$_m$, 1H), 5.12 (s, –CH–, 1H), 5.08 (s, –CH–, 1H), 3.00–2.95 (m, –CH$_2$–, 3H), 2.70–2.56 (m, –CH$_2$–, 4H), 2.74–2.34 (m, –CH$_2$–, 5H), 2.19 (s, –CH$_3$, 3H), 1.61 (s, –CH$_3$, 3H), 1.17 (t, J = 7.4 Hz, –CH$_3$, 6H). ^{13}C NMR (100 MHz, CDCl$_3$, TMS): δ 171.3, 167.0, 155.0, 154.7, 151.8, 149.4, 147.8, 140.6, 140.5, 139.8, 136.7, 135.0, 134.4, 134.3, 131.5, 131.2, 127.3, 126.7, 126.1, 126.0, 125.8, 123.3, 122.7, 122.2, 112.6, 112.4, 57.7, 56.2, 32.3, 24.6, 16.8, 16.4, 16.3, 13.8. ^{19}F NMR (470 MHz, CDCl$_3$): δ –128.64. FT-IR (KBr cm^{-1}): 3060 (w), 3010 (w), 2956 (m), 2922 (w), 2854 (w), 1650 (υ(C=N), s), 1580 (w), 1554 (w), 1510 (w), 1481(w), 1443 (m), 1427 (m), 1382 (w), 1357 (m), 1319 (w), 1304 (w), 1241 (w), 1217 (w), 1188 (w), 1116 (m), 1081 (w), 1049 (w), 1016 (w), 961 (w), 924 (w), 886 (w), 840 (w), 772 (w), 762 (m), 736 (s), 712 (w), 662 (w). Anal. Calcd. for C$_{55}$H$_{50}$N$_3$F·0.5H$_2$O (781.03): C, 84.58; H, 6.58; N, 5.38. Found: C, 84.77; H, 6.59; N, 5.04.

4.3.3. Ar = 2,6-i-Pr$_2$C$_6$H$_3$ (**L3**)

Using a similar procedure as described for the synthesis of **L1**, **L3** was prepared as a yellow powder (0.38 g, 28%). Mp: 243–245 °C. ^{1}H NMR (400 MHz, CDCl$_3$, TMS): δ 8.45 (d, J = 7.6 Hz, Py–H, 1H), 8.41 (d, J = 7.6 Hz, Py–H, 1H), 7.95 (t, J = 8.0 Hz, Py–H, 1H), 7.22–7.00 (m, Py–H, 17H), 6.88 (d, J = 7.2 Hz, Ar–H, 1H), 6.46 (s, Ar–H$_m$, 1H), 6.25 (d, J = 11.2 Hz, Ar–H, 1H), 6.36 (s, Ar–H$_m$, 1H), 5.12 (s, –CH–, 1H), 5.09 (s, –CH– 1H), 3.01–2.95 (m, –CH$_2$–, 3H), 2.80–2.70 (m, –CH$_2$–, 5H), 2.52–2.36 (m, –CH–, 2H), 2.21 (s, –CH$_3$, 3H), 1.68 (s, –CH$_3$, 3H), 1.19 (t, J = 6.4 Hz, –CH$_3$, 12H). ^{13}C NMR (100 MHz, CDCl$_3$, TMS): δ 171.3, 167.1, 155.0, 154.7, 146.7, 140.5, 139.8, 136.6, 135.5, 135.0, 134.4, 131.4, 127.2, 126.6, 126.1, 125.6, 123.6, 123.0, 122.6, 122.2, 112.5, 112.3, 57.6, 56.1, 32.2, 28.3, 23.2, 22.9, 17.1, 16.4. ^{19}F NMR (470 MHz, CDCl$_3$): δ –128.69. FT-IR (KBr cm^{-1}): 3058 (w), 3008 (w), 2957(m), 2918 (w), 2862 (w), 1651 (υ(C=N), s), 1588 (w), 1564 (w), 1512 (w), 1491(w), 1448 (m), 1429 (w), 1392 (w), 1359 (m), 1321(w), 1301 (w), 1239 (w), 1219 (w), 1191 (w), 1156 (w), 1114 (m), 1078 (w), 1046 (w), 1013 (w), 957 (w), 928 (w), 895 (w), 844 (w), 817 (w), 778 (w), 765 (m), 751 (s), 741 (s), 717 (w), 682 (w). Anal. Calcd. for C$_{57}$H$_{54}$N$_3$F·0.5H$_2$O (809.09): C, 84.62; H, 6.85; N, 5.19. Found: C, 84.50; H, 6.88; N, 5.51.

4.3.4. Ar = 2,4,6-Me$_3$C$_6$H$_2$ (**L4**)

Using a similar procedure as described for the synthesis of **L1**, **L4** was prepared as a yellow powder (0.42 g, 32%). Mp: 217–219 °C. ^{1}H NMR (400 MHz, CDCl$_3$, TMS): δ 8.46 (d, J = 7.6 Hz, Py–H, 1H), 8.42 (d, J = 8.0 Hz, Py–H, 1H), 7.94 (t, J = 7.8 Hz, Py–H, 1H), 7.22–6.91 (m, Ar–H, 16H), 6.87 (d, J = 7.6 Hz, Ar–H, 1H), 6.96 (s, Ar–H$_m$, 1H), 6.52 (d, J = 11.2 Hz, Ar–H, 1H), 6.36 (s, Ar–H$_m$, 1H), 5.12 (s, –CH–, 1H), 5.09 (s, –CH–, 1H), 3.01–2.96 (m, –CH$_2$–, 3H), 2.83–2.47 (m, –CH$_2$–, 5H), 2.31 (s, –CH$_3$, 3H), 2.18 (s, –CH$_3$, 3H), 2.04 (s, –CH$_3$, 6H), 1.67 (s, –CH$_3$, 3H). ^{13}C NMR (100 MHz, CDCl$_3$, TMS): δ 171.5, 167.5, 155.1, 154.6, 151.8, 149.4, 147.8, 146.3, 140.6, 140.5, 139.8, 136.6, 135.0, 134.4, 134.3, 132.2, 131.4, 131.2, 130.6, 128.6, 127.3, 127.1, 126.7, 126.1, 125.7, 125.3, 122.7, 122.2, 112.6, 112.4, 57.7, 56.2, 32.3, 31.3, 20.8, 17.9, 16.4. ^{19}F NMR (470 MHz, CDCl$_3$): δ –128.67. FT-IR (KBr cm^{-1}): 3060 (w), 3014 (w), 2963 (w), 2894 (w), 2860 (w), 1644 (υ(C=N), s), 1576 (w), 1542 (w), 1491 (s), 1447 (w), 1428 (w), 1386 (w), 1351 (m), 1293 (w), 1256 (w), 1221 (s), 1192 (w), 1124 (m), 1078 (w), 1045 (w), 1011 (w), 981 (w), 938 (w), 890 (w), 848 (s), 809 (w), 755 (s), 710 (w), 682 (w). Anal. Calcd. for C$_{54}$H$_{48}$N$_3$F·0.5H$_2$O (767.00): C, 84.56; H, 6.44; N, 5.48. Found: C, 84.47; H, 6.54; N, 5.16.

4.3.5. Ar = 2,6-Et$_2$-4-MeC$_6$H$_2$ (**L5**)

Using a similar procedure as described for the synthesis of **L1**, **L5** was prepared as a yellow powder (0.75 g, 54%). Mp: 234–236 °C. ^{1}H NMR (400 MHz, CDCl$_3$, TMS): δ 8.44 (d, J = 8.0 Hz, Py–H, 1H), 8.41 (d, J = 7.6 Hz, Py–H, 1H), 7.93 (t, J = 7.8 Hz, Py–H, 1H), 7.20–6.94 (m, Ar–H, 16H), 6.87 (s, Ar–H$_m$, 1H), 6.51 (d, J = 10.8 Hz, Ar–H, 2H), 6.36 (s, Ar–H$_m$, 1H), 5.11 (s, –CH–, 1H), 5.08 (s, –CH–, 1H), 3.01–2.95 (m, –CH$_2$–, 3H), 2.69–2.49 (m, –CH$_2$–, 5H), 2.43–2.40 (m, –CH$_2$–, 4H), 2.38 (s, –CH$_3$, 3H), 2.18 (s, –CH$_3$, 3H), 1.66 (s, –CH$_3$, 3H), 1.15 (t, J = 7.2 Hz, –CH$_3$, 6H). ^{13}C NMR (100 MHz, CDCl$_3$, TMS): δ 171.3, 167.2, 155.2, 154.5, 145.3, 140.5, 139.8, 136.6, 135.0, 134.3, 132.4, 131.5, 131.1, 130.8, 127.3, 126.7, 126.6, 126.1, 125.7, 122.6, 122.2, 112.6, 112.4, 57.7, 56.2, 32.2, 24.6, 21.0, 16.8, 16.4, 13.9. ^{19}F NMR (470 MHz, CDCl$_3$): δ –128.68. FT-IR (KBr cm^{-1}): 3026 (w), 2966 (w), 2941 (w), 2870 (w), 2829 (w) 1648 (υ(C=N), s), 1579 (w), 1526 (w), 1490 (w), 1461 (s), 1418 (w), 1411 (w), 1356 (m), 1329 (w), 1305 (w), 1283 (w), 1240 (w), 1211 (s), 1180 (w), 1138 (w), 1114 (m), 1068 (w), 1034 (w), 992 (w), 938 (w), 908 (w), 892 (w), 854 (s), 826 (m), 784 (m), 756 (s), 714 (w), 667 (w). Anal. Calcd. for C$_{56}$H$_{52}$N$_3$F·1.5H$_2$O (813.07): C, 82.73; H, 6.82; N, 5.17. Found: C, 82.50; H, 6.83; N, 4.94.

4.4. Synthesis of 2,6-{(2,4-(C$_{15}$H$_{13}$)$_2$-6-FC$_6$H$_2$)N=CMe}$_2$C$_5$H$_3$N (L6)

During the chromatographic purification process used to isolate the imine-ketone 2-{(2,4-(C$_{15}$H$_{13}$)$_2$-6-FC$_6$H$_2$)N=CMe}-6-(O=CMe)C$_5$H$_3$N (see above), a second fraction was removed from the column that was identified as bis(imino)pyridine **L6** and isolated as a pale yellow solid (3.24 g, 14%). Mp: 204–206 °C. ^{1}H NMR (400 MHz, CDCl$_3$, TMS): δ 8.44 (d, J = 8.0 Hz, Py–H, 2H), 7.98 (t, J = 8.00 Hz, Py–H, 1H), 7.21–7.01 (m, Ar–H, 26H), 6.80 (d, J = 4.0 Hz, Ar–H, 6H), 6.51 (Ar–H$_m$, 2H), 6.34 (s, Ar–H$_m$, 2H), 5.08 (s, –CH–, 2H), 3.00–2.94 (m, –CH$_2$–, 8H), 2.72–2.68 (m, –CH$_2$–, 5H), 2.45–2.21 (m, –CH$_2$–, 3H), 1.55 (s, –CH$_3$, 6H). ^{13}C NMR (100 MHz, CDCl$_3$, TMS): δ 171.4, 154.5, 151.8, 149.4, 140.6, 140.5, 140.1, 140.0, 139.8, 136.4, 135.1, 134.4, 134.2, 131.4, 131.2, 131.0, 130.6, 130.1, 129.1, 128.3, 127.3, 127.1, 126.7, 126.3, 126.1, 126.0, 125.7, 122.7, 112.6, 112.3, 57.7, 56.2, 32.3, 31.3, 16.4, 16.3. ^{19}F NMR (470 MHz, CDCl$_3$): δ –128.66. FT-IR (KBr cm^{-1}): 3026 (w), 2966 (w), 2941 (w), 2870 (w), 2829 (w) 1650 (υ(C=N), s), 1576 (w), 1536 (w), 1472 (w), 1452 (m), 1412 (w), 1356 (s), 1320 (w), 1301 (w), 1244 (w), 1228 (w), 1201 (m), 1176 (w), 1125 (m), 1056 (w), 1014 (w), 982 (w), 918 (w), 872 (w), 834 (m), 764 (m), 750 (s), 714 (w), 670 (w). Anal. Calcd. for C$_{81}$H$_{65}$N$_3$F$_2$·0.5H$_2$O (1126.52): C, 86.29; H, 5.90; N, 3.73. Found: C, 86.40; H, 5.68; N, 3.80.

4.5. Synthesis of [2-{(2,4-(C$_{15}$H$_{13}$)$_2$-6-FC$_6$H$_2$)N=CMe}-6-(ArN=CMe)C$_5$H$_3$N]CoCl$_2$ (Co1–Co6)

4.5.1. Ar = 2,6-Me$_2$C$_6$H$_3$ (**Co1**)

To a Schlenk tube containing **L1** (0.20 g, 0.27 mmol) and anhydrous CoCl$_2$ (0.035 g, 0.27 mmol) was added dichloromethane (5 mL) and ethanol (5 mL). The reaction mixture was stirred at room temperature for 12 h. All volatiles were then removed under reduced pressure and an excess of diethyl ether added to induce precipitation. The precipitate was filtered, washed with more diethyl ether (3 × 15 mL), and dried to afford **Co1** as a green powder (0.15 g, 62%). FT-IR (KBr cm^{-1}): 3057 (w), 3016 (w), 2926 (w), 2883 (w), 2828 (w), 1625 (υ(C=N), m), 1587 (s), 1572 (m), 1492 (m), 1471 (m), 1444 (m), 1421 (s), 1370 (m), 1310 (w), 1261 (s), 1215 (m), 1184 (w), 1162 (w), 1120 (w), 1045 (w), 1027 (w), 995 (w), 946 (w), 922 (w), 876 (w), 841 (w), 814 (m), 785 (w), 765 (s), 706 (m), 683 (w). ^{19}F NMR (470 MHz, CDCl$_3$): δ −146.68. Anal. Calcd. for $C_{53}H_{46}Cl_2CoN_3F\cdot EtOH$ (919.87): C, 71.81; H, 5.70; N, 4.57. Found: C, 75.23; H, 5.73; N, 4.86.

4.5.2. Ar = 2,6-Et$_2$C$_6$H$_3$ (**Co2**)

Using the same procedure and molar ratios as described for **Co1**, **Co2** was isolated as a green powder (0.14 g, 58%). FT-IR (KBr cm^{-1}): 3064 (w), 3016 (w), 2965 (w), 2930 (w), 2878 (w), 2834 (w), 1623 (υ(C=N), m), 1588 (s),1571 (w), 1492 (m), 1470 (m), 1445 (s), 1420 (w), 1371 (m), 1312 (w), 1261 (s), 1214 (m), 1204 (w), 1169 (w), 1102 (w), 1061 (w), 1047 (w), 1028 (w), 999 (w), 949 (w), 916 (w), 879 (w), 841 (w), 809 (m), 791 (w), 769 (s),755 (s), 742 (m), 716 (w). ^{19}F NMR (470 MHz, CDCl$_3$): δ −141.96. Anal. Calcd, for $C_{55}H_{50}Cl_2CoN_3F$ (901.86): C, 73.25; H, 5.59; N, 4.66. Found: C, 72.99; H, 5.50; N, 4.67.

4.5.3. Ar = 2,6-i-Pr$_2$C$_6$H$_3$ (**Co3**)

Using the same procedure and molar ratios as described for **Co1**, **Co3** was isolated as a green powder (0.13 g, 54%). FT-IR (KBr cm^{-1}): 3060 (w), 3016 (w), 2963 (w), 2928 (w), 2867 (w), 1623 (υ(C=N), m), 1591 (m), 1568 (m), 1561 (w), 1496 (s), 1465 (w), 1441 (w), 1418 (s), 1384 (w), 1370 (s), 1314 (m), 1201 (w), 1184 (w), 1162 (w), 1101 (w), 1061 (w), 1046 (w), 1026 (w), 991 (w), 978 (w), 938 (w), 878 (w), 843 (w), 813 (w), 797 (w), 785 (w), 769 (s), 760 (m), 742 (w). ^{19}F NMR (470 MHz, CDCl$_3$): δ −129.32. Anal. Calcd, for $C_{57}H_{54}Cl_2CoN_3F\cdot EtOH$ (975.98): C, 72.61; H, 6.20; N, 4.31. Found: C, 72.60; H, 5.86; N, 4.43.

4.5.4. Ar = 2,4,6-Me$_3$C$_6$H$_2$ (**Co4**)

Using the same procedure and molar ratios as described for **Co1**, **Co4** was isolated as a green powder (0.16 g, 70%). FT-IR (KBr cm^{-1}): 3058 (w), 3011 (w), 2959 (w), 2921 (w), 2864 (w), 1632 (υ(C=N), w), 1589 (υ(C=N), s), 1577 (w), 1546 (w), 1471 (m), 1456 (s), 1423 (s), 1399 (w), 1369 (m), 1320 (w), 1291 (w), 1259 (s), 1219 (s), 1189 (w), 1160 (w), 1122 (w), 1100 (w), 1081 (w), 1030 (m), 1000 (w), 977 (w), 942 (w), 883 (w), 850 (m), 811 (m), 769 (s), 760 (s), 747 (w). ^{19}F NMR (470 MHz, CDCl$_3$): δ −145.14. Anal. Calcd. for $C_{54}H_{48}Cl_2CoN_3F$ (887.83): C, 73.05; H, 5.45; N, 4.73. Found: C, 72.67; H, 5.23; N, 4.73.

4.5.5. Ar = 2,6-Et$_2$-4-MeC$_6$H$_2$ (**Co5**)

Using the same procedure and molar ratios as described for **Co1**, **Co5** was isolated as a green powder (0.14 g, 56%). FT-IR (KBr cm^{-1}): 3061 (w), 2961 (w), 2928 (w), 2893 (w), 2830 (w), 1626 (υ(C=N), m), 1588 (m),1572 (w), 1515 (w), 1494 (m), 1458 (m), 1448 (w), 1422 (s), 1371 (s), 1340 (w), 1318 (w), 1261 (s), 1189 (w), 1158 (w), 1120 (w), 1098 (w), 1058 (w), 1030 (m), 984 (w), 946 (w), 921 (w), 881 (w), 861 (w), 841 (w), 815 (m), 769 (s), 758 (s), 745 (w). ^{19}F NMR (470 MHz, CDCl$_3$): δ −137.97. Anal. Calcd. for $C_{57}H_{55}Cl_2CoN_3$ (911.91): C, 73.44; H, 5.72; N, 4.59. Found: C, 73.19; H, 5.85; N, 4.50.

4.6. Synthesis of [2,6-{(2,4-(C$_{15}$H$_{13}$)$_2$-6-FC$_6$H$_2$)N=CMe}$_2$C$_5$H$_3$N]CoCl$_2$ (Co6)

Using the same procedure as described for **L1**, but using **L6** (0.15 g, 0.135 mmol) and CoCl$_2$ (0.0175 g, 0.135 mmol), **Co6** was isolated as a pale green powder (0.11 g, 73%). FT-IR (KBr cm^{-1}): 3059 (w), 2956 (w), 2930 (w), 2891 (w), 2833 (w), 1618 (υ(C=N), m), 1581 (m), 1570 (w), 1520 (w), 1488 (m), 1454 (w), 1440 (w), 1418 (s), 1374 (m), 1338 (w), 1316 (w), 1264 (s), 1187 (w), 1160 (w), 1124 (w), 1092 (w), 1061 (w), 1028 (m), 980 (w), 941 (w), 918 (w), 887 (w), 852 (w), 833 (m), 812 (w), 766 (s), 754 (s), 720 (w). ^{19}F NMR (470 MHz, CDCl$_3$): δ

−145.83, −146.89. Anal. Calcd. for $C_{81}H_{65}Cl_2CoN_3F_2 \cdot 0.5EtOH$ (1271.30): C, 77.47; H, 5.39; N, 3.31. Found: C, 77.09; H, 5.58; N, 3.48.

4.7. Procedures for the Ethylene Polymerization Runs at 1, 5, and 10 atm

4.7.1. Ethylene Polymerization at 5 and 10 atm Pressure

These runs were performed in a stainless-steel autoclave (250 mL) equipped with a mechanical stirrer and an ethylene pressure and temperature control system. In a typical procedure, the autoclave was evacuated and refilled with nitrogen gas; this process was repeated three times. After the final evacuation, ethylene was introduced to provide an ethylenic atmosphere inside the autoclave, at which point a solution of the corresponding precatalyst (1.5 µmol) in toluene (25 mL) was injected, followed by more toluene (25 mL). Then, the required amount of a co-catalyst (MAO, or MMAO) was loaded, followed by the addition of more toluene (50 mL). Finally, the autoclave was pressurized with ethylene (5 or 10 atm) and stirring commenced at a rate of 400 rpm. Upon completion of the reaction, the stirring was stopped, the reactor was cooled to room temperature, and the pressure slowly released. The contents of the autoclave were quenched with hydrochloric acid (10%) in ethanol and the polymer further washed with ethanol. Finally, the polymer was filtered, dried under reduced pressure at 40 °C, and weighed.

4.7.2. Ethylene Polymerization at 1 atm Pressure

The polymerizations undertaken at 1 atm of C_2H_4 were carried out in a Schlenk vessel. Under an atmosphere of ethylene (ca. 1 atm), the cobalt precatalyst (1.5 µmol) was added to the vessel, followed by toluene (30 mL) and then the required amount of co-catalyst was introduced by syringe. The resulting solution was stirred at 30 °C under an ethylene atmosphere (1 atm). After 30 min, the pressure was slowly released and the solution was quenched with 10% hydrochloric acid in ethanol. The polymer was washed with ethanol, dried under reduced pressure at 40 °C, and then weighed.

5. Conclusions

Five types of unsymmetrical 2,6-bis(arylimino)pyridine cobalt(II) complex (**Co1**–**Co5**), each incorporating one *N*-2,4-bis(dibenzosuberyl)-6-fluorophenyl group together with one sterically and electronically variable *N*-aryl group were successfully prepared. The symmetrical comparator **Co6**, containing two *N*-2,4-bis(dibenzosuberyl)-6-fluorophenyl groups, was also prepared. All compounds, including the free ligands (**L1**–**L6**), were characterized by various spectroscopic techniques, including single crystal X-ray diffraction for **Co1** and **Co2**. Upon activation with MAO or MMAO, **Co1**–**Co6** displayed high activities for ethylene polymerization, with levels reaching as high as 1.15×10^7 g PE mol^{-1} (Co) h^{-1} at 70 °C. Notably, this peak level of performance was observed using the least sterically protected precatalyst, **Co1**, in combination with MAO. Conversely, the most sterically hindered precatalysts, **Co3** and **Co6**, were capable of generating the highest molecular weight polyethylenes in the range of 30.26–33.90 kg mol^{-1} (**Co3**) and 42.90–43.92 kg mol^{-1} (**Co6**). As a key point, the catalytic activity remained significant at temperatures in excess of 70 °C [9.81×10^6 g PE mol^{-1} (Co) h^{-1} at 80 °C and 6.56×10^6 g PE mol^{-1} (Co) h^{-1} at 90 °C]. Evidently, the introduction of a *N*-2,4-bis(dibenzosuberyl)-6-fluorophenyl group contributes to the catalyst's good thermal stability, highlighting a structural feature that could be integrated into industrially relevant catalysts for ethylene polymerization.

Supplementary Materials: The following supporting information can be downloaded at: https://www.mdpi.com/article/10.3390/catal12121569/s1, X-ray diffraction analysis; Table S1: Details of the crystal data and structure refinement parameters for **Co1** and **Co2**. References [72,73] are cited in the Supplementary Materials; Figure S1: ^{19}F NMR spectra of **L1**–**L6**; Figure S2: ^{19}F NMR spectra of **L2**, **L3** and **L6** along with those for **Co2**, **Co3** and **Co6**; Figure S3: GPC curves showing log M_w for the polyethylene produced using **Co1**/MAO as a function of Al:Co molar ratio (runs 5, 8–12, Table 2); Figure S4: GPC curves showing log M_w for the polyethylene produced using **Co1**/MAO as a function of reaction time (runs 5, 13–16, Table 2); Figure S5: GPC curves showing log M_w for the polyethylene produced using **Co1**/MMAO as a function of Al:Co molar ratio (runs 5, 8–12, Table 3); Figure S6: GPC curves showing log M_w for the polyethylene produced using **Co1**/MMAO as a function of reaction time runs 10, 13–16, Table 3); Figure S7: ^{1}H NMR spectrum of the polyethylene produced using **Co1**/MMAO (run 10, Table 3); recorded in $C_2D_2Cl_4$ at 100 °C; Figure S8: ^{13}C NMR spectrum of the polyethylene produced using **Co1**/MMAO (run 10, Table 3; recorded in deuterated $C_2D_2Cl_4$ at 100 °C.

Author Contributions: Conceptualization, W.-H.S. and G.A.S.; methodology, M.Z.; software, Y.S.; validation, M.Z., W.-H.S. and Y.M.; formal analysis, M.Z.; investigation, M.Z.; resources, Q.Z.; data curation, D.D.S.; writing—original draft preparation, M.Z.; writing—review and editing, G.A.S.; visualization, W.-H.S.; supervision, W.-H.S.; project administration, W.-H.S. All authors have read and agreed to the published version of the manuscript.

Funding: This work was supported by the National Natural Science Foundation of China (No. 21871275).

Data Availability Statement: All the samples of the organic compounds, cobalt complexes, and data supporting the study are available from the authors.

Acknowledgments: G.A.S. thanks the Chinese Academy of Sciences for a President's International Fellowship for Visiting Scientists. M.Z. is thankful for the CAS-TWAS President's Fellowship for International Students.

Conflicts of Interest: There are no conflict of interest to declare.

References

1. Small, B.L.; Brookhart, M.; Bennett, A.M.A. Highly Active Iron and Cobalt Catalysts for the Polymerization of Ethylene. *J. Am. Chem. Soc.* **1998**, *120*, 4049–4050. [CrossRef]
2. Britovsek, G.J.P.; Gibson, V.C.; Kimberley, B.S.; Maddox, P.J.; McTavish, S.J.; Solan, G.A.; White, A.J.P.; Williams, D.J. Novel olefin polymerization catalysts based on iron and cobalt. *Chem. Commun.* **1998**, *7*, 849–850. [CrossRef]
3. Johnson, L.K.; Killian, C.M.; Brookhart, M. New Pd(II)- and Ni(II)-Based Catalysts for Polymerization of Ethylene and α-Olefins. *J. Am. Chem. Soc.* **1995**, *117*, 6414–6415. [CrossRef]
4. Killian, C.M.; Tempel, D.J.; Johnson, L.K.; Brookhart, M. Living Polymerization of r-Olefins Using NiII-α-Diimine Catalysts. Synthesis of New Block Polymers Based on α-Olefins. *J. Am. Chem. Soc.* **1996**, *118*, 11664–11665. [CrossRef]
5. Wang, Z.; Liu, Q.; Solan, G.A.; Sun, W.-H. Recent advances in Ni-mediated ethylene chain growth: N_{imine}-donor ligand effects on catalytic activity, thermal stability and oligo-/polymer structure. *Coord. Chem. Rev.* **2017**, *350*, 68–83. [CrossRef]
6. Edgecombe, B.D.; Stein, J.A.; Frechet, J.M.J. The Role of Polymer Architecture in Strengthening Polymer-Polymer Interfaces: A Comparison of Graft, Block, and Random Copolymers Containing Hydrogen-Bonding Moieties. *Macromolecules* **1998**, *31*, 1292–1304. [CrossRef]
7. Harth, E.M.; Hecht, S.; Helms, B.; Malmstrom, E.E.; Frechet, J.M.J.; Hawker, C.J. The Effect of Macromolecular Architecture in Nanomaterials: A Comparison of Site Isolation in Porphyrin Core Dendrimers and Their Isomeric Linear Analogues. *J. Am. Chem. Soc.* **2002**, *124*, 3926–3938. [CrossRef]
8. Gibson, V.C.; Solan, G.A. Iron-Based and Cobalt-Based Olefin Polymerisation Catalysts. *Top. Organomet. Chem.* **2009**, *2*, 107–158.
9. Gibson, V.C.; Redshaw, C.; Solan, G.A. Bis(imino)pyridines: Surprisingly reactive ligands and a gateway to new families of catalysts. *Chem. Rev.* **2007**, *107*, 1745–1776. [CrossRef]
10. Gibson, V.C.; Solan, G.A. *Catalysis without Precious Metals*; Bullock, R.M., Ed.; Wiley-VCH: Weinheim, Germany, 2010; pp. 111–141.
11. Zhang, W.; Sun, W.-H.; Redshaw, C. Tailoring iron complexes for ethylene oligomerization and/or polymerization. *Dalton Trans.* **2013**, *42*, 8988–8997. [CrossRef]
12. Ma, J.; Feng, C.; Wang, S.; Zhao, K.-Q.; Sun, W.-H.; Redshaw, C.; Solan, G.A. Bi- and tri-dentate imino-based iron and cobalt pre-catalysts for ethylene oligo-/polymerization. *Inorg. Chem. Front.* **2014**, *1*, 14–34. [CrossRef]
13. Flisak, Z.; Sun, W.-H. Progression of Diiminopyridines: From Single Application to Catalytic Versatility. *ACS Catal.* **2015**, *5*, 4713–4724. [CrossRef]
14. Britovsek, G.J.P.; Gibson, V.C.; Kimberley, B.S.; Mastroianni, S.; Redshaw, C.; Solan, G.A.; White, A.J.P.; Williams, D.J. Bis(imino)pyridyl iron and cobalt complexes: The effect of nitrogen substituents on ethylene oligomerisation and polymerisation. *J. Chem. Soc. Dalton Trans.* **2001**, *6*, 1639–1644. [CrossRef]
15. Yue, E.; Zeng, Y.; Zhang, W.; Sun, Y.; Cao, X.-P.; Sun, W.-H. Highly linear polyethylenes using the 2-(1-(2,4-dibenzhydrylnaphthy limino)ethyl)-6-(1-(arylimino)ethyl)pyridylcobalt chlorides: Synthesis, characterization and ethylene polymerization. *Sci. China Chem.* **2016**, *59*, 1291–1300. [CrossRef]
16. Lu, Z.; Liao, Y.; Fan, W.; Dai, S. Efficient suppression of the chain transfer reaction in ethylene coordination polymerization with dibenzosuberyl substituents. *Polym. Chem.* **2022**, *13*, 4090–4099. [CrossRef]
17. Bariashir, C.; Wang, Z.; Du, S.; Solan, G.A.; Huang, C.; Liang, T.; Sun, W.-H. Cycloheptyl-Fused NNN-Ligands as Electronically Modifiable Supports for M(II) (M = Co, Fe) Chloride Precatalysts; Probing Performance in Ethylene Oligo-/Polymerization. *J. Polym. Sci. Part A Polym. Chem.* **2017**, *55*, 3980–3989. [CrossRef]

18. McTavish, S.; Britovsek, G.J.P.; Smit, T.M.; Gibson, V.C.; White, A.J.P.; Williams, D.J. Iron-based ethylene polymerization catalysts supported by bis(imino)pyridine ligands: Derivatization via deprotonation/alkylation at the ketimine methyl position. *J. Mol. Catal. A Chem.* **2007**, *261*, 293–300. [CrossRef]
19. Smit, T.M.; Tomov, A.K.; Britovsek, G.J.P.; Gibson, V.C.; White, A.J.P.; Williams, D.J. The effect of imine-carbon substituents in bis(imino)pyridine-based ethylene polymerisation catalysts across the transition series. *Catal. Sci. Technol.* **2012**, *2*, 643–655. [CrossRef]
20. Sun, W.H.; Hao, P.; Li, G.; Zhang, S.; Wang, W.Q.; Yi, J.; Asma, M.; Tang, N. Synthesis and characterization of iron and cobalt dichloride bearing 2-quinoxalinyl-6-iminopyridines and their catalytic behavior toward ethylene reactivity. *J. Organomet. Chem.* **2007**, *692*, 4506–4518. [CrossRef]
21. Gao, R.; Li, Y.; Wang, F.; Sun, W.-H.; Bochmann, M. 2-Benzoxazolyl-6-[1-(arylimino)ethyl]pyridyliron(II) Chlorides as Ethylene Oligomerization Catalysts. *Eur. J. Inorg. Chem.* **2009**, *2009*, 4149–4156. [CrossRef]
22. Sun, W.-H.; Hao, P.; Zhang, S.; Shi, Q.; Zuo, W.; Tang, X. 2-(Benzimidazolyl)-6-(1-(arylimino)ethyl)pyridyl Complexes as Catalysts for Ethylene Oligomerization and Polymerization. *Organometallics* **2007**, *26*, 2720–2734. [CrossRef]
23. Wang, K.; Wedeking, K.; Zuo, W.; Zhang, D.; Sun, W.-H. Iron(II) and cobalt(II) complexes bearing N-((pyridin-2-yl)methylene)-quinolin-8-amine derivatives: Synthesis and application to ethylene oligomerization. *J. Organomet. Chem.* **2008**, *693*, 1073–1080. [CrossRef]
24. Wang, Z.; Solan, G.A.; Zhang, W.; Sun, W.-H. Carbocyclic-fused NNN-pincer ligands as ring-strain adjustable supports for iron and cobalt catalysts in ethylene oligo-/polymerization. *Coord. Chem. Rev.* **2018**, *363*, 92–108. [CrossRef]
25. Pelletier, J.D.A.; Champouret, Y.D.M.; Cadarso, J.; Clowes, L.; Gañete, M.; Singh, K.; Thanarajasingham, V.; Solan, G.A. Electronically variable imino-phenanthrolinyl-cobalt complexes; synthesis, structures and ethylene oligomerisation studies. *J. Organomet. Chem.* **2006**, *691*, 4114–4123. [CrossRef]
26. Jie, S.; Zhang, S.; Wedeking, K.; Zhang, W.; Ma, H.; Lu, X.; Deng, Y.; Sun, W.-H. Cobalt(II) complexes bearing 2-imino-1,10-phenanthroline ligands: Synthesis, characterization and ethylene oligomerization. *C. R. Chim.* **2006**, *9*, 1500–1509. [CrossRef]
27. Jie, S.; Zhang, S.; Sun, W.H.; Kuang, X.; Liu, T.; Guo, J. Iron(II) complexes ligated by 2-imino-1,10-phenanthrolines: Preparation and catalytic behavior toward ethylene oligomerization. *J. Mol. Catal. A Chem.* **2007**, *269*, 85–96. [CrossRef]
28. Han, M.; Oleynik, I.I.; Liu, M.; Ma, Y.; Oleynik, I.V.; Solan, G.A.; Liang, T.; Sun, W.-H. Ring size enlargement in an ortho-cycloalkyl-substituted bis(imino)pyridine-cobalt ethylene polymerization catalyst and its impact on performance and polymer properties. *Appl. Organomet. Chem.* **2021**, *36*, e6529. [CrossRef]
29. Han, M.; Oleynik, I.I.; Ma, Y.; Oleynik, I.V.; Solan, G.A.; Hao, X.; Sun, W.-H. Modulating Thermostability and Productivity of Benzhydryl-Substituted Bis(imino)pyridine-Iron C_2H_4 Polymerization Catalysts through ortho-C_nH_{2n-1} 1 (n = 5, 6, 8, 12) Ring Size Adjustment. *Eur. J. Inorg. Chem.* **2022**, *2022*, e202200224. [CrossRef]
30. Xiao, L.; Gao, R.; Zhang, M.; Li, Y.; Cao, X.; Sun, W.-H. 2-(1H-2-Benzimidazolyl)-6-(1-(arylimino)ethyl)pyridyl Iron(II) and Cobalt(II) Dichlorides: Syntheses, Characterizations, and Catalytic Behaviors toward Ethylene Reactivity. *Organometallics* **2009**, *28*, 2225–2233. [CrossRef]
31. Appukuttan, V.K.; Liu, Y.; Son, B.C.; Ha, C.-S.; Suh, H.; Kim, I. Iron and Cobalt Complexes of 2,3,7,8-Tetrahydroacridine-4,5(1H,6H)-diimine Sterically Modulated by Substituted Aryl Rings for the Selective Oligomerization to Polymerization of Ethylene. *Organometallics* **2011**, *30*, 2285–2294. [CrossRef]
32. Zhang, W.; Chai, W.; Sun, W.-H.; Hu, X.; Redshaw, C.; Hao, X. 2-(1-(Arylimino)ethyl)-8-arylimino-5,6,7-trihydroquinoline Iron(II) Chloride Complexes: Synthesis, Characterization, and Ethylene Polymerization Behavior. *Organometallics* **2012**, *31*, 5039–5048. [CrossRef]
33. Sun, W.-H.; Kong, S.; Chai, W.; Shiono, T.; Redshaw, C.; Hu, X.; Guo, C.; Hao, X. 2-(1-(Arylimino)ethyl)-8-arylimino-5,6,7-trihydroquinolylcobalt dichloride: Synthesis and polyethylene wax formation. *Appl. Catal. A* **2012**, *447*, 67–73. [CrossRef]
34. Huang, F.; Xing, Q.; Liang, T.; Flisak, Z.; Ye, B.; Hu, X.; Yang, W.; Sun, W.H. 2-(1-Arylimimoethyl)-9-arylimino-5,6,7,8 tetrahydrocycloheptapyridyl iron(II) dichloride: Synthesis, characterization, and the highly active and tunable active species in ethylene polymerization. *Dalton Trans.* **2014**, *43*, 16818–16829. [CrossRef]
35. Huang, F.; Zhang, W.; Yue, E.; Liang, T.; Hu, X.; Sun, W.-H. Controlling the molecular weights of polyethylene waxes using the highly active precatalysts of 2-(1-aryliminoethyl)-9-arylimino-5,6,7,8tetrahydrocycloheptapyridylcobalt chlorides: Synthesis, characterization, and catalytic behaviour. *Dalton Trans.* **2016**, *45*, 657–666. [CrossRef]
36. Wang, S.; Li, B.; Liang, T.; Redshaw, C.; Li, Y.; Sun, W.-H. Synthesis, characterization and catalytic behavior toward ethylene of 2-[1-(4,6-dimethyl-2-benzhydrylphenylimino)ethyl]-6-[1-(arylimino)ethyl]-pyridylmetal (iron or cobalt) chlorides. *Dalton Trans.* **2013**, *42*, 9188–9197. [CrossRef]
37. Yu, J.; Huang, W.; Wang, L.; Redshaw, C.; Sun, W.H. 2-[1-(2,6-Dibenzhydryl-4-methylphenylimino)ethyl]-6-[1-(arylimino) ethyl]pyridylcobalt(II) dichlorides: Synthesis, characterization and ethylene polymerization behaviour. *Dalton Trans.* **2011**, *40*, 10209–10214. [CrossRef]
38. Yu, J.; Liu, H.; Zhang, W.; Hao, X.; Sun, W.H. Access to highly active and thermally stable iron precatalysts using bulky 2-[1-(2,6-dibenzhydryl-4-methylphenylimino)ethyl]-6-[1-(arylimino)ethyl]pyridine ligands. *Chem. Commun.* **2011**, *47*, 3257–3259. [CrossRef]

39. He, F.; Zhao, W.; Cao, X.-P.; Liang, T.; Redshaw, C.; Sun, W.-H. 2-[1-(2,6-dibenzhydryl-4-chlorophenylimino)ethyl]-6-[1-aryliminoethyl]pyridyl cobalt dichlorides: Synthesis, characterization and ethylene polymerization behaviou. *J. Organomet. Chem.* **2012**, *713*, 209–216. [CrossRef]
40. Cao, X.; He, F.; Zhao, W.; Cai, Z.; Hao, X.; Shiono, T.; Redshaw, C.; Sun, W.-H. 2-[1-(2,6-Dibenzhydryl-4-chlorophenylimino)ethyl]-6-[1-(arylimino)ethyl]pyridyliron(II) dichlorides: Synthesis, characterization and ethylene polymerization behaviour. *Polymer* **2012**, *53*, 1870–1880. [CrossRef]
41. Sun, W.-H.; Zhao, W.; Yu, J.; Zhang, W.; Hao, X.; Redshaw, C. Enhancing the Activity and Thermal Stability of Iron Precatalysts Using 2-(1-{2,6-bis[bis(4fluorophenyl)methyl]-4-methylphenylimino}ethyl)-6-[1-(arylimino)ethyl]pyridines. *Macromol. Chem. Phys.* **2012**, *213*, 1266–1273. [CrossRef]
42. Zhao, W.; Yue, E.; Wang, X.; Yang, W.; Chen, Y.; Hao, X.; Cao, X.; Sun, W.H. Activity and Stability Spontaneously Enhanced Toward Ethylene Polymerization by Employing 2-(1-(2,4-Dibenzhydrylnaphthylimino)Ethyl)-6-(1-(Arylimino)Ethyl)Pyridyliron(II) Dichlorides. *J. Polym. Sci. Part A Polym. Chem.* **2017**, *55*, 988–996. [CrossRef]
43. Wang, S.; Zhao, W.; Hao, X.; Li, B.; Redshaw, C.; Li, Y.; Sun, W.-H. 2-(1-{2,6-Bis[bis(4-fluorophenyl)methyl]-4-methylphenylimino}ethyl)-6-[1(arylimino)ethyl]pyridylcobalt dichlorides: Synthesis, characterization and ethylene polymerization behaviour. *J. Organomet. Chem.* **2013**, *731*, 78–84. [CrossRef]
44. Zhang, W.; Wang, S.; Du, S.; Guo, C.-Y.; Hao, X.; Sun, W.-H. 2-(1-(2,4-Bis((di(4-fluorophenyl)methyl-6methylphenylimino)ethyl)-6-(1-(arylimino)ethyl)pyridylmetal (iron or cobalt) Complexes: Synthesis, Characterization, and Ethylene Polymerization Behavior. *Macromol. Chem. Phys.* **2014**, *215*, 1797–1809. [CrossRef]
45. Mahmood, Q.; Ma, Y.; Hao, X.; Sun, W.-H. Substantially enhancing the catalytic performance of bis(imino)pyridylcobalt chloride pre-catalysts adorned with benzhydryl and nitro groups for ethylene polymerization. *Appl. Organomet. Chem.* **2019**, *33*, e4857. [CrossRef]
46. Mahmood, Q.; Zeng, Y.; Wang, X.; Sun, Y.; Sun, W.-H. Advancing polyethylene properties by incorporating NO$_2$ moiety in 1,2-bis(arylimino)acenaphthylnickel precatalysts: Synthesis, characterization and ethylene polymerization. *Dalton Trans.* **2017**, *46*, 6934–6947. [CrossRef]
47. Mitchell, N.E.; Anderson, W.C., Jr.; Long, B.K. Mitigating Chain-Transfer and Enhancing the Thermal Stability of Co-Based Olefin Polymerization Catalysts through Sterically Demanding Ligands. *J. Polym. Sci., Part A Polym. Chem.* **2017**, *55*, 3990–3995. [CrossRef]
48. Lai, J.; Zhao, W.; Yang, W.; Redshaw, C.; Liang, T.; Liu, Y.; Sun, W.-H. 2-[1-(2,4-Dibenzhydryl-6-methylphenylimino)ethyl]-6-[1-(arylimino)ethyl]pyridylcobalt(II) dichlorides: Synthesis and ethylene polymerization behaviour. *Polym. Chem.* **2012**, *3*, 787–793. [CrossRef]
49. Zhao, W.; Yu, J.; Song, S.; Yang, W.; Liu, H.; Hao, X.; Redshaw, C.; Sun, W.-H. Controlling the ethylene polymerization parameters in iron pre-catalysts of the type 2-[1-(2,4-dibenzhydryl-6-methylphenylimino)ethyl]-6-[1-(arylimino)ethyl]pyridyliron dichloride. *Polymer* **2012**, *53*, 130–137. [CrossRef]
50. Guo, L.; Zada, M.; Zhang, W.; Vignesh, A.; Zhu, D.; Ma, Y.; Liang, T.; Sun, W.-H. Highly linear polyethylenes tailored by 2,6-bis[1-(p-dibenzocycloheptylarylimino)ethyl]pyridylcobalt dichlorides. *Dalton Trans.* **2019**, *48*, 5604–5613. [CrossRef]
51. Guo, L.; Zhang, W.; Cao, F.; Jian, Y.; Zhang, R.; Ma, Y.; Solan, G.A.; Sun, W.-H. Remote dibenzocycloheptyl substitution on a bis(arylimino)pyridyl-iron ethylene polymerization catalyst; enhanced thermal stability and unexpected effects on polymer properties. *Polym. Chem.* **2021**, *12*, 4214–4225. [CrossRef]
52. Zada, M.; Guo, L.; Ma, Y.; Zhang, W.; Flisak, Z.; Sun, Y.; Sun, W.-H. Activity and Thermal Stability of Cobalt(II)-Based Olefin Polymerization Catalysts Adorned with Sterically Hindered Dibenzocycloheptyl Groups. *Molecules* **2019**, *24*, 2007. [CrossRef] [PubMed]
53. Zada, M.; Vignesh, A.; Guo, L.; Suo, H.; Ma, Y.; Liu, H.; Sun, W.-H. NNN type iron(II) complexes consisting sterically hindered dibenzocycloheptyl groups: Synthesis and catalytic activity towards ethylene polymerization. *Mol. Catal.* **2020**, *492*, 110981. [CrossRef]
54. Bariashir, C.; Zhang, R.; Vignesh, A.; Ma, Y.; Liang, T.; Sun, W.-H. Enhancing Ethylene Polymerization of NNN-Cobalt(II) Precatalysts Adorned with a Fluoro-substituent. *ACS Omega* **2021**, *6*, 4448–4460. [CrossRef] [PubMed]
55. Zheng, Q.; Li, Z.; Han, M.; Xiang, J.; Solan, G.A.; Liang, T.; Sun, W.-H. Fluorinated cobalt catalysts and their use in forming narrowly dispersed polyethylene waxes of high linearity and incorporating vinyl functionality. *Catal. Sci. Technol.* **2021**, *11*, 656–670. [CrossRef]
56. Zheng, Q.; Zuo, Z.; Ma, Y.; Liang, T.; Yang, X.; Sun, W.-H. Fluorinated 2,6-bis(arylimino)pyridyl iron complexes targeting bimodal dispersive polyethylene: Probing chain termination pathways via combined experimental and DFT. *Dalton Trans.* **2022**, *51*, 8290–8302. [CrossRef]
57. Meiries, S.; Speck, K.; Cordes, D.B.; Slawin, A.M.Z.; Nolan, S.P. [Pd(IPr*OMe)(acac)Cl]: Tuning the N-Heterocyclic Carbene in Catalytic C−N Bond Formation. *Organometallics* **2012**, *32*, 330–339. [CrossRef]
58. Zada, M.; Guo, L.; Zhang, R.; Zhang, W.; Ma, Y.; Solan, G.A.; Sun, Y.; Sun, W.-H. Moderately branched ultra-high molecular weight polyethylene by using N,N′-nickel catalysts adorned with sterically hindered dibenzocycloheptyl groups. *Appl. Organomet. Chem.* **2019**, *33*, e4749. [CrossRef]

59. Zada, M.; Vignesh, A.; Guo, L.; Zhang, R.; Zhang, W.; Ma, Y.; Sun, Y.; Sun, W.-H. Sterically and Electronically Modified Aryliminopyridyl-Nickel Bromide Precatalysts for an Access of Branched Polyethylene with Vinyl/Vinylene End Groups. *ACS Omega* **2020**, *5*, 10610–10625. [CrossRef]
60. Britovsek, G.J.P.; Bruce, M.; Gibson, V.C.; Kimberley, B.S.; Maddox, P.J.; Mastroianni, S.; McTavish, S.J.; Redshaw, C.; Solan, G.A.; Stromberg, S.; et al. Iron and Cobalt Ethylene Polymerization Catalysts Bearing 2,6-Bis(Imino)Pyridyl Ligands: Synthesis, Structures, and Polymerization Studies. *J. Am. Chem. Soc.* **1999**, *121*, 8728–8740. [CrossRef]
61. Britovsek, G.J.P.; Gibson, V.C.; Spitzmesser, S.K.; Tellmann, K.P.; White, A.J.P.; Williams, D.J. Cationic 2,6-bis(imino)pyridine iron and cobalt complexes: Synthesis, structures, ethylene polymerisation and ethylene/polar monomer co-polymerisation studies. *J. Chem. Soc. Dalton Trans.* **2002**, *6*, 1159–1171. [CrossRef]
62. Cantalupo, S.A.; Ferreira, H.E.; Bataineh, E.; King, A.J.; Petersen, M.V.; Wojtasiewicz, T.; DiPasquale, A.G.; Rheingold, A.L.; Doerrer, L.H. Synthesis with Structural and Electronic Characterization of Homoleptic Fe(II)- and Fe(III)-Fluorinated Phenolate Complexes. *Inorg. Chem.* **2011**, *50*, 6584–6596. [CrossRef] [PubMed]
63. Yuan, J.; Shi, W.-B.; Kou, H.-Z. Syntheses, crystal structures and magnetism of azide-bridged five-coordinate binuclear nickel(II) and cobalt(II) complexes. *Transit. Met. Chem.* **2015**, *40*, 807–811. [CrossRef]
64. Britovsek, G.J.P.; Gibson, V.C.; Hoarau, O.D.; Spitzmesser, S.K.; White, A.J.P.; Williams, D.J. Iron and Cobalt Ethylene Polymerization Catalysts: Variations on the Central Donor. *Inorg. Chem.* **2003**, *42*, 3454–3465. [CrossRef] [PubMed]
65. Tomov, A.K.; Gibson, V.C.; Britovsek, G.J.P.; Long, R.J.; Meurs, M.V.; Jones, D.J.; Tellmann, K.P.; Chirinos, J.J. Distinguishing Chain Growth Mechanisms in Metal-catalyzed Olefin Oligomerization and Polymerization Systems: C_2H_4/C_2D_4 Cooligomerization/Polymerization Experiments Using Chromium, Iron, and Cobalt Catalysts. *Organometallics* **2009**, *28*, 7033–7040. [CrossRef]
66. Barbaro, P.; Bianchini, C.; Giambastiani, G.; Rios, I.G.; Meli, A.; Oberhauser, W.; Segarra, A.M.; Sorace, L.; Toti, A. Synthesis of New Polydentate Nitrogen Ligands and Their Use in Ethylene Polymerization in Conjunction with Iron(II) and Cobalt(II) Bis-halides and Methylaluminoxane. *Organometallics* **2007**, *26*, 4639–4651. [CrossRef]
67. Liu, M.; Jiang, S.; Ma, Y.; Solan, G.A.; Sun, Y.; Sun, W.-H. CF_3O-Functionalized Bis(arylimino)pyridine−Cobalt Ethylene Polymerization Catalysts: Harnessing Solvent Effects on Performance and Polymer Properties. *Organometallics* **2022**, *41*, 3237–3248. [CrossRef]
68. Pooter, M.D.; Smith, P.B.; Dohrer, K.K.; Bennett, K.F.; Meadows, M.D.; Smith, C.G.; Schouwenaars, H.P.; Geerards, R.A. Determination of the Composition of Common linear low Density Polyethylene Copolymers by ^{13}C-NMR Spectroscopy. *J. Polym. Sci.* **1991**, *42*, 399–408. [CrossRef]
69. Galland, G.B.; Quijada, R.; Rojas, R.; Bazan, G.; Komon, Z.J.A. NMR Study of Branched Polyethylenes Obtained with Combined Fe and Zr Catalysts. *Macromolecules* **2002**, *35*, 339–345. [CrossRef]
70. Hansen, E.W.; Blom, R.B.; Bade, O.M. N.m.r. characterization of polyethylene with emphasis on internal consistency of peak intensities and estimation of uncertainties in derived branch distribution numbers. *Polymer* **1997**, *38*, 4295–4304. [CrossRef]
71. Semikolenova, N.V.; Sun, W.-H.; Soshnikov, I.E.; Matsko, M.A.; Kolesova, O.V.; Zakharov, V.A.; Bryliakov, K.P. Origin of "Multisite-like" Ethylene Polymerization Behavior of the Single-Site Nonsymmetrical Bis(imino)pyridine Iron(II) Complex in the Presence of Modified Methylaluminoxane. *ACS Catal.* **2017**, *7*, 2868–2877. [CrossRef]
72. Sheldrick, G.M. Crystal structure refinement with SHELXL. *Acta Crystallogr. C Struct. Chem.* **2015**, *71*, 3–8. [CrossRef] [PubMed]
73. Sheldrick, G.M. SHELXT—Integrated space-group and crystal structure determination. *Acta Crystallogr. A Found. Adv.* **2015**, *71*, 3–8. [CrossRef] [PubMed]

Article

Improving Catalytic Activity towards the Direct Synthesis of H_2O_2 through Cu Incorporation into AuPd Catalysts

Alexandra Barnes [1,†], Richard J. Lewis [1,*,†], David J. Morgan [1,2], Thomas E. Davies [1] and Graham J. Hutchings [1,*]

[1] Max Planck-Cardiff Centre on the Fundamentals of Heterogeneous Catalysis FUNCAT, Cardiff Catalysis Institute, School of Chemistry, Cardiff University, Main Building, Park Place, Cardiff CF10 3AT, UK
[2] Harwell XPS, Research Complex at Harwell (RCaH), Didcot OX11 0FA, UK
[*] Correspondence: lewisr27@cardiff.ac.uk (R.J.L.); hutch@cardiff.ac.uk (G.J.H.)
[†] These authors contributed equally to this work.

Abstract: With a focus on catalysts prepared by an excess-chloride wet impregnation procedure and supported on the zeolite ZSM-5(30), the introduction of low concentrations of tertiary base metals, in particular Cu, into supported AuPd nanoparticles can be observed to enhance catalytic activity towards the direct synthesis of H_2O_2. Indeed the optimal catalyst formulation (1%AuPd$_{(0.975)}$Cu$_{(0.025)}$/ZSM-5) is able to achieve rates of H_2O_2 synthesis (115 mol$_{H_2O_2}$ kg$_{cat}^{-1}$h^{-1}) approximately 1.7 times that of the bi-metallic analogue (69 mol$_{H_2O_2}$ kg$_{cat}^{-1}$h^{-1}) and rival that previously reported over comparable materials which use Pt as a dopant. Notably, the introduction of Cu at higher loadings results in an inhibition of performance. Detailed analysis by CO-DRFITS and XPS reveals that the improved performance observed over the optimal catalyst can be attributed to the electronic modification of the Pd species and the formation of domains of a mixed Pd^{2+}/Pd^0 oxidation state as well as structural changed within the nanoalloy.

Keywords: hydrogen peroxide; gold; palladium; copper; trimetallic; green chemistry

Citation: Barnes, A.; Lewis, R.J.; Morgan, D.J.; Davies, T.E.; Hutchings, G.J. Improving Catalytic Activity towards the Direct Synthesis of H_2O_2 through Cu Incorporation into AuPd Catalysts. *Catalysts* **2022**, *12*, 1396. https://doi.org/10.3390/catal12111396

Academic Editor: Carl Redshaw

Received: 18 October 2022
Accepted: 2 November 2022
Published: 9 November 2022

Publisher's Note: MDPI stays neutral with regard to jurisdictional claims in published maps and institutional affiliations.

Copyright: © 2022 by the authors. Licensee MDPI, Basel, Switzerland. This article is an open access article distributed under the terms and conditions of the Creative Commons Attribution (CC BY) license (https://creativecommons.org/licenses/by/4.0/).

1. Introduction

The direct synthesis of H_2O_2 from molecular H_2 and O_2 (Scheme 1) represents an attractive alternative to the current means of large-scale production of this environmentally benign, powerful oxidant, the anthraquinone oxidation (AO) process. Indeed, the direct approach would allow for the production of appropriate concentrations of H_2O_2 at the point of final use, avoiding the substantial economic and environmental drawbacks associated with the industrial route. Due to production costs, H_2O_2 production via the AO process is typically centralised, with H_2O_2 transported at concentrations in excess of that required by the end-user [1]. The subsequent dilution of the oxidant prior to use effectively wastes a significant amount of energy associated with the initial purification and concentration steps [2]. Additionally, H_2O_2 is relatively unstable, decomposing readily to H_2O in the presence of mild temperatures or weak bases and, as such, requires acid stabilizers to prolong its shelf-life [3], which results in complex product streams and can deleteriously affect the reactor lifetime [4]. These cumulative drawbacks associated with off-site H_2O_2 production pass on significant costs to the end user, which would be greatly reduced or removed altogether via a direct synthesis approach to H_2O_2 production. In particular, the direct route may find the greatest application for oxidative transformations where the synthesised H_2O_2 is readily utilised for chemical valorisation or generated in situ [5–7].

A Langmuir–Hinshelwood mechanism involving the successive hydrogenation of molecular O_2 has been often proposed for the direct synthesis reaction [8,9]. However, in recent years Flaherty and co-workers have advanced an alternative, non-Langmuirian mechanism [10], which involves a water-mediated proton-electron transfer and have further reported the role of protic solvents in the formation of surface-bound intermediates that shuttle both the protons and electrons to active sites [11]. Indeed, detailed theoretical studies have further demonstrated that energy barriers associated with a solvent-mediated protonation of adsorbed O_2 are not prohibitive and, indeed, are as low as O_2 hydrogenation [12].

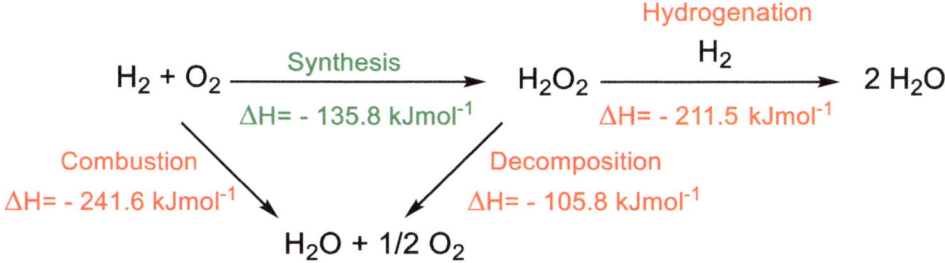

Scheme 1. Reaction pathways associated with the direct synthesis of H_2O_2 from H_2 and O_2.

Pd-based catalysts have been well studied for their application in the direct synthesis reaction [13–15] since the first patent was granted to Henkel and Weber in 1914 [16]. However, a major challenge associated with catalytic selectivity has prevented the development of an industrial-scale direct synthesis process [17,18]. This can be understood as the formation of H_2O is thermodynamically more favourable than that of H_2O_2, with H_2O formation driven via combustion or through the subsequent degradation of H_2O_2 (via decomposition and hydrogenation pathways).

Measures to improve catalytic selectivity have often focused on the introduction of secondary metals into supported Pd catalysts, with AuPd systems being perhaps the most extensively studied [19–21]. In recent years, significant attention has been placed on the alloying of Pd with a range of earth-abundant metals [22–27]. Further investigations have focused on the introduction of dopant levels of precious metals, such as Pt, into supported Pd [28–33] and AuPd [34–37] catalysts. The resulting improved catalytic activity towards H_2O_2 production is often attributed to a combination of the electronic promotion of Pd and the isolation of contiguous Pd domains, widely considered to be key in promoting the cleavage of O–O bonds (in *O_2, *H_2O_2, or *OOH) and the resultant formation of H_2O [38–40].

The use of zeolitic and zeo-type materials as catalyst supports for use in the direct synthesis of H_2O_2 has received significant attention, with such catalysts typically offering improved activity and selectivity compared to oxide-supported analogues [5,36,41–43]. This has often been attributed to the improved dispersion of metal species and the acidic nature of the support materials, with supports of low isoelectric points (i.e., the pH at which the surface has zero net charges and an indication of catalyst acidity/basicity) and are reported to offer enhanced catalytic performance compared to those with a high isoelectric point [20].

Recently, we have demonstrated that significant improvements in catalytic performances can be achieved through the introduction of low concentrations of base metals into AuPd nanoalloys when supported on TiO_2 (P25) [44]. Indeed, the catalysts developed within this earlier work offered comparable H_2O_2 synthesis rates to those previously observed through Pt incorporation while avoiding the additional cost associated with the precious metal dopant [44]. With these earlier studies in mind, we now investigate the efficacy of base metal-incorporated AuPd catalysts supported on the zeolite ZSM-5(30) towards the direct synthesis of H_2O_2.

2. Experimental Section

2.1. Catalyst Preparation

Prior to the co-impregnation of metal salts, the NH_4-ZSM-5 material ($SiO_2:Al_2O_3$ = 30:1, Alfa Aesar) was calcined (flowing air, 550 °C, 3 h, 20 °C min^{-1}).

A series of bi- and tri-metallic 1%AuPdX/ZSM-5 (X = Cu, Ni, Zn) catalysts were prepared by an excess chloride wet co-impregnation procedure, based on the methodology previously reported in the literature, which has been shown to result in the enhanced dispersion of precious metals, in particular Au, when compared to conventional wet co-impregnation methodologies [45]. The procedure to produce 1%AuPd$_{(0.975)}$Cu$_{(0.025)}$/ZSM-5 (1 g) is outlined below where the total metal loading was 1 wt.%, the combined weight loading of Au and Pd was 0.975 wt.%, and that of Cu was 0.025 wt.%, and in all cases the Au:Pd ratio was 1:1 (mol/mol). A similar methodology to that outlined below was utilised for all catalysts investigated, with the exact quantities of metal precursor used to synthesise the key catalysts used within this work reported in Table S1.

Aqueous solutions of $HAuCl_4 \cdot 3H_2O$ (0.322 mL, 12.25 mgmL^{-1}, Strem Chemicals), $PdCl_2$ (0.356 mL, 6 mgmL^{-1}, 0.58 M HCl, Sigma Aldrich, Burlington, MA, USA), and $CuCl_2$ (106 µL, 2.36 mgmL^{-1}, Sigma Aldrich) were mixed in a 50 mL round bottom flask and heated to 60 °C with stirring (1000 rpm) in a thermostatically controlled oil bath, with the total volume fixed to 16 mL using H_2O (HPLC grade). Upon reaching 65 °C, ZSM-5 (0.99 g, $SiO_2:Al_2O_3$ = 30:1, Alfa Aesar) was added over the course of 10 min with constant stirring. The resultant slurry was stirred at 60 °C for a further 15 min, and following this, the temperature was raised to 95 °C for 16 h to allow for the complete evaporation of water. The resulting solid was ground prior to heat treatment in a reductive atmosphere (5%H_2/Ar, 400 °C, 4 h, 10 °Cmin^{-1}).

2.2. Catalyst Testing

Note 1. *Reaction conditions used within this study operate under the flammability limits of gaseous mixtures of H_2 and O_2.*

Note 2. *The conditions used within this work for H_2O_2 synthesis and degradation have previously been investigated, with the presence of CO_2 as a diluent for reactant gases and a methanol co-solvent being identified as key to maintaining high catalytic efficacy towards H_2O_2 production [45]. In regard to the role of the CO_2 diluent, this was found to act as an in-situ promoter of H_2O_2 stability through its dissolution in the reaction solution and the formation of carbonic acid. We have previously reported that the use of the CO_2 diluent has a comparable promotive effect to that observed when acidifying the reaction solution to a pH of 4 using HNO_3 [46].*

2.3. Direct Synthesis of H_2O_2

Hydrogen peroxide synthesis was evaluated using a Parr Instruments stainless steel autoclave with a nominal volume of 100 mL, equipped with a PTFE liner so that the total volume was reduced to 66 mL and a maximum working pressure of 2000 psi. To test each catalyst for H_2O_2 synthesis, the autoclave liner was charged with a catalyst (0.01 g), solvent (methanol (5.6 g, HPLC grade, Fischer Scientific, Waltham, MA, USA), and H_2O (2.9 g, HPLC grade, Fischer Scientific)). The charged autoclave was then purged three times with 5%H_2/CO_2 (100 psi) before filling with 5%H_2/CO_2 to a pressure of 420 psi, followed by the addition of 25%O_2/CO_2 (160 psi), with the pressure of 5%H_2/CO_2 and 25%O_2/CO_2 given as gauge pressures. The reactor was not continually fed with reactant gas. The reaction was conducted at a temperature of 2 °C for 0.5 h with stirring (1200 rpm). The H_2O_2 productivity was determined by titrating aliquots of the final solution after the reaction with acidified $Ce(SO_4)_2$ (0.0085 M) in the presence of a ferroin indicator. Catalyst productivities are reported as $mol_{H_2O_2} kg_{cat}^{-1} h^{-1}$.

The catalytic conversion of H_2 and selectivity towards H_2O_2 were determined using a Varian 3800 GC fitted with TCD and equipped with a Porapak Q column.

H_2 conversion (Equation (1)) and H_2O_2 selectivity (Equation (2)) are defined as follows:

$$H_2 \text{Conversion } (\%) = \frac{\text{mmol}_{H2\ (t(0))} - \text{mmol}_{H2\ (t(1))}}{\text{mmol}_{H2\ (t(0))}} \times 100 \quad (1)$$

$$H_2O_2 \text{ Selectivity } (\%) = \frac{H_2O_2 \text{detected (mmol)}}{H_2 \text{ consumed (mmol)}} \times 100 \quad (2)$$

The total autoclave capacity was determined via water displacement to allow for the accurate determination of H_2 conversion and H_2O_2 selectivity. When equipped with the PTFE liner, the total volume of an unfilled autoclave was determined to be 93 mL, which included all available gaseous space within the autoclave.

2.4. Gas Replacement Experiments for the Direct Synthesis of H_2O_2

An identical procedure to that outlined above for the direct synthesis reaction was followed for a reaction time of 0.5 h. After this, stirring was stopped, and the reactant gas mixture was vented prior to replacement with the standard pressures of 5% H_2/CO_2 (420 psi) and 25% O_2/CO_2 (160 psi). The reaction mixture was then stirred (1200 rpm) for a further 0.5 h. To collect a series of data points, as in the case of Figure 5, it should be noted that individual experiments were carried out, and the reactant mixture was not sampled online.

2.5. Catalyst Reusability in the Direct Synthesis and Degradation of H_2O_2

In order to determine catalyst reusability, a similar procedure to that outlined above for the direct synthesis of H_2O_2 was followed utilising 0.05 g of catalyst. Following the initial test, the catalyst was recovered by filtration and dried (30 °C, 16 h, under vacuum); from the recovered catalyst sample 0.01 g and was used to conduct a standard H_2O_2 synthesis or degradation test.

2.6. Degradation of H_2O_2

Catalytic activity towards H_2O_2 degradation was determined in a similar manner to the direct synthesis activity of a catalyst. The autoclave liner was charged with a solvent (methanol (5.6 g, HPLC grade, Fischer Scientific), H_2O (2.9 g, HPLC grade, Fischer Scientific)), and H_2O_2 (50 wt. % 0.69 g, Sigma Aldrich), with the resultant solvent composition equivalent to a 4 wt. % H_2O_2 solution. From the solution, two 0.05 g aliquots were removed and titrated with acidified $Ce(SO_4)_2$ solution using ferroin as an indicator to determine an accurate concentration of H_2O_2 at the start of the reaction. Subsequently, a catalyst (0.01 g) was added to the reaction media, and the autoclave was purged with 5%H_2/CO_2 (100 psi) prior to being pressurised with 5%H_2/CO_2 (420 psi). The reaction medium was cooled to a temperature of 2 °C prior to stirring (1200 rpm) for 0.5 h. After the reaction was complete, the catalyst was removed from the reaction mixture, and two 0.05 g aliquots were titrated against the acidified $Ce(SO_4)_2$ solution using ferroin as an indicator. The degradation activity is reported as $\text{mol}_{H_2O_2}\text{kg}_{cat}^{-1}\text{h}^{-1}$.

Note 3. *In all cases, the reactor temperature was controlled using a HAAKE K50 bath/circulator and an appropriate coolant. The reactor temperature was maintained at 2 ± 0.2 °C throughout the course of the H_2O_2 synthesis and degradation reaction.*

In all cases, the reactions were run multiple times, over multiple batches of catalysts, with the data being presented as an average of these experiments. The catalytic activity toward the direct synthesis and subsequent degradation of H_2O_2 was found to be consistent to within $\pm 3\%$ on the basis of multiple reactions.

2.7. Characterisation

A Thermo Scientific K-Alpha$^+$ photoelectron spectrometer was used to collect XP spectra utilising a micro-focused monochromatic Al K_α X-ray source operating at 72 W.

Samples were pressed into a copper holder and analysed using the 400 μm spot mode at pass energies of 40 and 150 eV for high-resolution and survey spectra, respectively. Charge compensation was performed using a combination of low-energy electrons and argon ions, which resulted in a C(1s) binding energy of 284.8 eV for the adventitious carbon present in all the samples and all samples also showed a constant Ti(2p$_{3/2}$) of 458.5 eV. All data were processed using CasaXPS v2.3.24 (Casa Software Ltd., Teignmouth, UK) with a Shirley background, Scofield sensitivity factors, and an electron energy dependence of −0.6, as recommended by the manufacturer. Peak fits were performed using a combination of Voigt-type functions and models derived from the bulk reference samples where appropriate.

The bulk structure of the catalysts was determined by powder X-ray diffraction using a (θ-θ) PANalytical X'pert Pro powder diffractometer with a Cu K$_\alpha$ radiation source, operating at between 40 keV and 40 mA. Standard analysis was carried out using a 40 min run with a backfilled sample, between 2θ values of 5 and 75°. Phase identification was carried out using the International Centre for Diffraction Data (ICDD).

Note 4. *X-ray diffractograms of key as-prepared catalysts are reported in Figure S1 (and accompanying text) with no reflections associated with active metals, indicative of the relatively low total loading and high dispersion of the immobilised metals.*

Transmission electron microscopy (TEM) was performed on a JEOL JEM-2100 (Tokyo, Japan) operating at 200 kV. Samples were prepared through their dispersion in ethanol by sonication, and they were deposited on 300 mesh copper grids coated with holey carbon film. Energy dispersive X-ray spectroscopy (XEDS) was performed using an Oxford Instruments (Abingdon, UK) X-Max^N 80 detector, and the data analysed used Aztec software (Abingdon, UK). Aberration corrected scanning transmission electron microscopy (AC-STEM) was performed using a probe-corrected Hitachi (Brisbane, Australia) HF5000 S/TEM, operating at 200 kV. The instrument was equipped with bright field (BF), high angle annular dark field (HAADF), and secondary electron (SE) detectors for high spatial resolution STEM imaging experiments. This microscope was also equipped with a secondary electron detector and dual Oxford Instruments (Abingdon, UK) XEDS detectors (2 × 100 mm^2) with a total collection angle of 2.02 sr.

Total metal leaching from the supported catalyst was quantified via inductively coupled plasma mass spectrometry (ICP-MS). Post-reaction solutions were analysed using an Agilent (Santa Clara, CA, USA) 7900 ICP-MS equipped with an I-AS auto-sampler. All samples were diluted by a factor of 10 using HPLC grade H$_2$O (1%HNO$_3$ and 0.5% HCl matrix). All calibrants were matrix matched and measured against a five-point calibration using certified reference materials purchased from Perkin Elmer and certified internal standards acquired from Agilent.

Fourier-transform infrared spectroscopy (FTIR) was carried out with a Bruker (Hanau, Germany) Tensor 27 spectrometer fitted with a HgCdTe (MCT) detector and was operated with OPUS software (Ettinger, Germany).

Note 5. *FTIR analysis of key as-prepared catalysts is reported in Figure S2 (and accompanying text) and indicates no discernible changes in the structure of the HZSM-5 support upon metal immobilisation and exposure to a reductive heat treatment.*

N$_2$ isotherms were collected on a Micromeritics 3-Flex. Samples (ca. 0.070 g) were degassed (350 °C, 9 h) prior to analysis. Analyses were carried out at 77 K, with P$_0$ measured continuously. Free space was measured post-analysis with He. Data analyses were carried out using the Micromeritics 3-Flex software with the non-local density functional theory (NLDFT), Tarazona model.

Note 6. *The details of the textural properties of key ZSM-5-supported catalysts are reported in Table S2 and Figure S3. The immobilisation of active metals can be seen to lead to a general decrease in both the total surface area and pore volume in comparison to the bare ZSM-5 support. This is ascribed to the deposition of metal nanoparticles inside the zeolitic pore structure.*

3. Results and Discussion

The introduction of small concentrations of precious dopants, in particular Pt, [38,39] into the supported AuPd nanoalloys has been extensively reported to offer improved catalytic activity towards the direct synthesis of H_2O_2, when compared to the bimetallic analogue. We recently demonstrated that comparable enhancements in performance could result from the incorporation of dopant concentrations of base meals into AuPd nanoalloys [44]. In keeping with this earlier work, our initial investigations identified the promotive effect that can result from the introduction of Cu, Ni, and Zn at low concentrations (0.025 wt.%) into a 1%AuPd$_{(1.00)}$/ZSM-5 catalyst (Figure 1). In particular, the introduction of Cu, which is known to be readily incorporated into AuPd alloys [47], was observed to significantly increase activity towards H_2O_2 synthesis, with this metric approximately 1.7 times greater (115 mol$_{H_2O_2}$ kg$_{cat}^{-1}$h^{-1}), than that observed for the bi-metallic 1%AuPd$_{(1.00)}$/ZSM-5 catalyst (69 mol$_{H_2O_2}$ kg$_{cat}^{-1}$h^{-1}). Indeed, the activity of the 1%AuPd$_{(0.975)}$Cu$_{(0.025)}$/ZSM-5 catalyst exceeded that of the analogous formulation supported on TiO_2 (P25) (94 mol$_{H_2O_2}$ kg$_{cat}^{-1}$h^{-1}) [44]. However, a concurrent increase in catalytic activity towards the subsequent degradation of H_2O_2 (via decomposition and hydrogenation pathways) was also observed (529 mol$_{H_2O_2}$ kg$_{cat}^{-1}$h^{-1}). This may be surprising given earlier works which have demonstrated that the addition of Cu, either into AuPd [48] or Pd-only [49] catalysts, can inhibit catalytic activity, with DFT studies indicating that the formation of the intermediate hydroperoxyl species (OOH*) and subsequently H_2O_2 is thermodynamically unfavoured over Cu-containing precious metal surfaces [50]. However, notably, these prior works have focused on the incorporation of Cu at much higher concentrations than that utilised within this study. In comparison, the introduction of Ni (81 mol$_{H_2O_2}$ kg$_{cat}^{-1}$h^{-1}) and Zn (77 mol$_{H_2O_2}$ kg$_{cat}^{-1}$h^{-1}) resulted in only a minor improvement in the catalyst performance compared to the bi-metallic AuPd analogue, although the improved selectivity of the 1%AuPd$_{(0.975)}$Ni$_{(0.025)}$/ZSM-5 catalyst is noteworthy, with H_2O_2 degradation rates (281 mol$_{H_2O_2}$ kg$_{cat}^{-1}$h^{-1}) significantly lower than that observed over the 1%AuPd$_{(1.00)}$/ZSM-5 catalyst (320 mol$_{H_2O_2}$ kg$_{cat}^{-1}$h^{-1}) or, indeed, the other trimetallic formulations. The enhanced performance of the 1%AuPd$_{(0.975)}$Cu$_{(0.025)}$/ZSM-5 catalyst is further evidenced by the comparison of initial reaction rates, where the contribution of competitive H_2O_2 degradation pathways is considered to be negligible (Table S3).

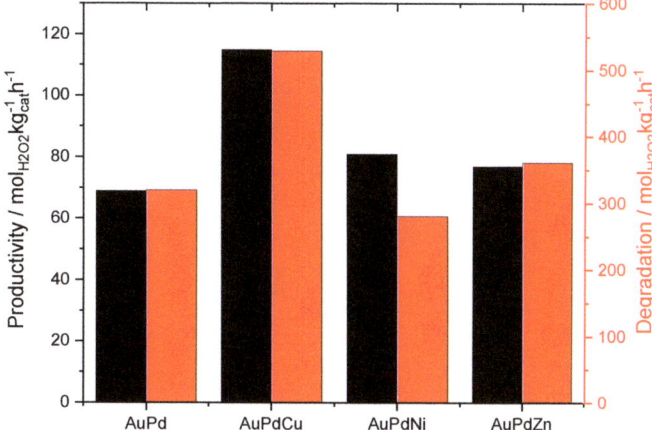

Figure 1. Catalytic activity towards 1%AuPd$_{(0.975)}$X$_{(0.025)}$/ZSM-5, the direct synthesis, and subsequent degradation of H_2O_2. **H_2O_2 direct synthesis reaction conditions:** catalyst (0.01 g), H_2O (2.9 g), MeOH (5.6 g), 5% H_2/CO_2 (420 psi), 25% O_2/CO_2 (160 psi), 0.5 h, 2 °C, and 1200 rpm. **H_2O_2 degradation reaction conditions:** catalyst (0.01 g), H_2O_2 (50 wt.% 0.68 g) H_2O (2.22 g), MeOH (5.6 g), 5% H_2/CO_2 (420 psi), 0.5 h, 2 °C, and 1200 rpm.

An evaluation of the as-prepared 1%AuPd$_{(0.975)}$X$_{(0.025)}$/ZSM-5 catalysts by XPS can be seen in Figure 2 (additional data reported in Table S4). Interestingly, despite exposure to a high-temperature reductive heat treatment (5%H$_2$/Ar, 400 °C, 4 h, 10 °Cmin^{-1}), the 1%AuPd$_{(1.00)}$/ZSM-5 catalyst was found to consist of a relatively high proportion of Pd^{2+}. Such an observation is in keeping with our previous investigations into AuPd systems, where the introduction of Au has been found to modify Pd speciation [20]. Upon the introduction of low quantities of Ni, Cu, and Zn, a significant shift in Pd speciation was observed, towards Pd^{2+}, with the formation of mixed domains of the Pd oxidation state, which is well known to offer improved activity compared to Pd^0 or Pd^{2+} rich analogues [51]. The shift in the Pd oxidation state towards Pd^{2+} upon the introduction of Ni was found to be the greatest, which aligned well with the observed selectivity of the 1%AuPd$_{(0.975)}$Ni$_{(0.025)}$/ZSM-5 catalyst. However, it should be noted that the Pd speciation of the fresh catalyst is likely to be not representative of those under direct synthesis reaction conditions.

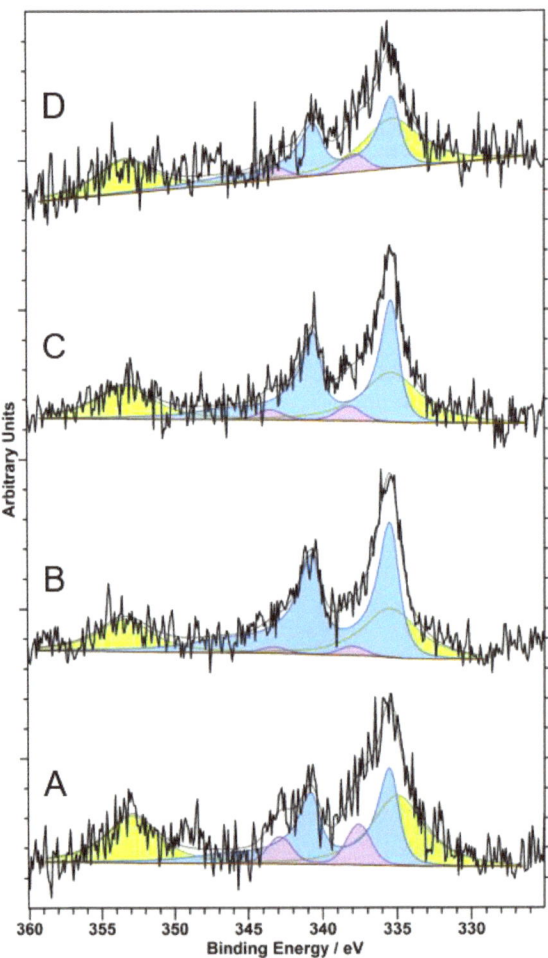

Figure 2. XPS spectra of Pd(3d) regions for the as-prepared 1%AuPd$_{(0.975)}$X$_{(0.025)}$/ZSM-5 catalysts as a function of tertiary metal dopant. (**A**) 1%AuPd$_{(1.00)}$/ZSM-5; (**B**) 1%AuPd$_{(0.975)}$Cu$_{(0.025)}$/ZSM-5; (**C**) 1%AuPd$_{(0.975)}$Ni$_{(0.025)}$/ZSM-5; and (**D**) 1%AuPd$_{(0.975)}$Zn$_{(0.025)}$/ZSM-5. **Key:** Au(4d) (green); Pd^0(blue); Pd^{2+}(purple).

With the improved activity of the 1%AuPd$_{(0.975)}$Cu$_{(0.025)}$/ZSM-5 catalyst towards H$_2$O$_2$ production established, we were subsequently motivated to determine the effect of Cu loading on catalytic activity while maintaining the total metal loading at 1 wt.% (Figure 3A,B, with initial reaction rates reported in Table S5). The introduction of low concentrations of Cu (< 0.025 wt.%) was observed to significantly increase the catalytic activity towards both the direct synthesis and subsequent degradation of H$_2$O$_2$, compared to the bimetallic AuPd parent material. However, both these metrics decreased considerably at higher loadings of Cu (75 and 287 mol$_{H_2O_2}$kg$_{cat}^{-1}$h^{-1}, respectively, for H$_2$O$_2$ direct synthesis and degradation pathways at a Cu loading of 0.037%, which is equivalent to 3.7% of the total metal loading). This is in keeping with previous works, which demonstrated a deleterious effect on performance with the introduction of high loadings of Cu into precious metal nanoparticles [48,50] and suggested a high sensitivity towards tertiary metal content. While the evaluation of catalytic activity towards H$_2$O$_2$ synthesis alone (Figure 3A) may suggest that there is very little difference in the performance over a range of Cu loadings (H$_2$O$_2$ synthesis rates between 111 and 116 mol$_{H_2O_2}$kg$_{cat}^{-1}$h^{-1} observed for Cu loadings of 0.012–0.025 wt.%), the determination of H$_2$ conversion rates and H$_2$O$_2$ selectivity indicates that a substantial reduction in catalytic selectivity towards H$_2$O$_2$ coincides with the introduction of Cu (Figure 3B), which would align with determination of trends in H$_2$O$_2$ degradation activity (Figure 3A). Indeed, these observations imply that the enhanced activity of the Cu-containing catalysts is associated with increased reactivity (i.e., the rate of H$_2$ conversion) rather than H$_2$O$_2$ selectivity. However, it is important to consider that such evaluations are not made at comparable rates of H$_2$ conversion and, notably, the high H$_2$ selectivity of the 1%AuPd$_{(1.00)}$/ZSM-5 catalyst (81%) can be related to the low rates of conversion (10%) observed. With the introduction of Cu at concentrations greater than 0.025 wt.%, a substantial decrease in H$_2$ conversion rates was observed, which correlates well with the observed loss in catalytic activity towards both the direct synthesis and subsequent degradation of H$_2$O$_2$.

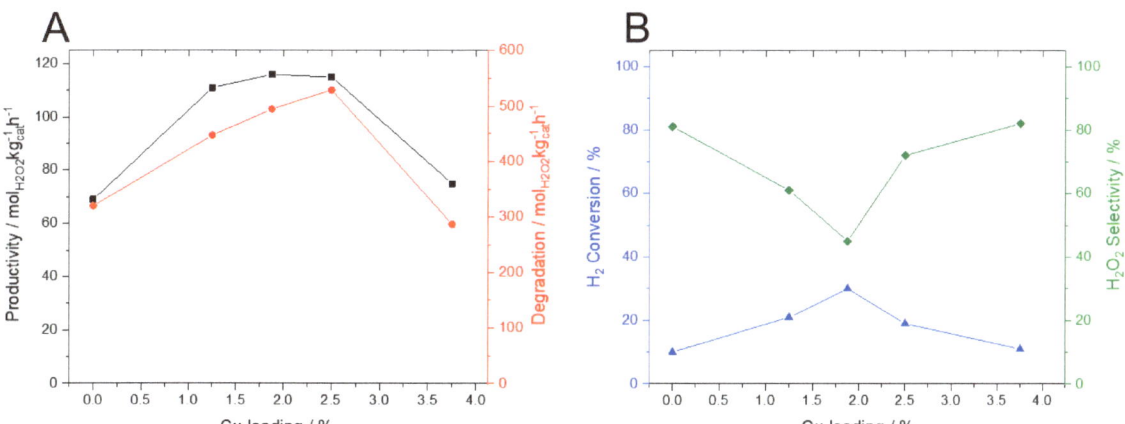

Figure 3. The effect of Cu incorporation into 1%AuPd$_{(1.00)}$/ZSM-5 on catalytic activity. (**A**) Comparison of activity towards the direct synthesis and subsequent degradation of H$_2$O$_2$. (**B**) Evaluation of H$_2$ conversion and selectivity towards H$_2$O$_2$ during the H$_2$O$_2$ synthesis reaction. *Key:* H$_2$O$_2$ synthesis (*black squares*); H$_2$O$_2$ degradation (*red circles*); H$_2$ conversion during H$_2$O$_2$ direct synthesis reaction (*blue triangles*); H$_2$O$_2$ selectivity during H$_2$O$_2$ direct synthesis reaction (*green diamonds*). **H$_2$O$_2$ direct synthesis reaction conditions:** catalyst (0.01 g), H$_2$O (2.9 g), MeOH (5.6 g), 5% H$_2$/CO$_2$ (420 psi), 25% O$_2$/CO$_2$ (160 psi), 0.5 h, 2 °C, and 1200 rpm. **H$_2$O$_2$ degradation reaction conditions:** catalyst (0.01 g), H$_2$O$_2$ (50 wt.% 0.68 g) H$_2$O (2.22 g), MeOH (5.6 g), 5% H$_2$/CO$_2$ (420 psi), 0.5 h, 2 °C, and 1200 rpm.

With our XPS analysis revealing a modification in the Pd oxidation state as a result of the incorporation of dopant metals (Figure 2), we were subsequently motivated to investigate the 1%AuPdCu/ZSM-5 catalytic series via CO-DRIFTS (Figure 4). CO-DRIFTS is a technique that has been extensively utilised to probe the surface of supported precious metal catalysts [51–54]. For each catalyst, spectra were measured in the 1750–2150 cm^{-1} range, which contains the stretching modes associated with the CO adsorbed on Pd and Au surfaces. The DRIFTS spectra of all the catalysts were dominated by Pd–CO bands. The peak observed at approximately 2080 cm^{-1} represents the CO bound in a linear manner to low co-ordination Pd sites (i.e., corner or edge sites), while the broad feature, centred around 1950 cm^{-1}, represents the bi- and tri-dentate bridging modes of CO on Pd [55]. Upon the introduction of small concentrations of Cu into the AuPd nanoalloy, a small blue shift in the band relates to the linearly bonded CO on Pd, which can be observed. This aligns well with previous investigations by Wilson et al. into the formation of AuPd alloys [51]. In particular, such a shift can be attributed to the segregation of Pd at the nanoparticle surface and a corresponding occupation of lower coordination sites. It is, therefore, possible to propose that the introduction of Cu into AuPd nanoalloys at low concentrations results in a similar modification of the nanoparticle composition.

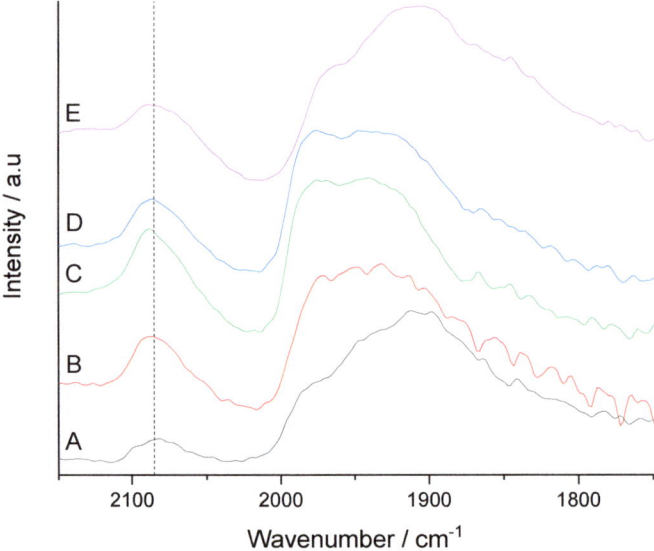

Figure 4. CO-DRIFTS spectra for the as-prepared 1%AuPd$_{(1-X)}$Cu$_{(X)}$/ZSM-5 catalysts. (**A**) 1%AuPd$_{(1.00)}$/ZSM-5; (**B**) 1%AuPd$_{(0.875)}$Cu$_{(0.125)}$/ZSM-5; (**C**) 1%AuPd$_{(0.8125)}$Cu$_{(0.1875)}$/ZSM-5; (**D**) 1%AuPd$_{(0.75)}$Cu$_{(0.25)}$/ZSM-5; and (**E**) 1%AuPd$_{(0.625)}$Cu$_{(0.375)}$/ZSM-5.

With the evident improvement upon the incorporation of Cu into a supported 1%AuPd$_{(1.00)}$/ZSM-5 catalyst, we subsequently set out to further contrast the catalytic performance of the optimal catalyst and the bimetallic AuPd analogue. Time-on-line studies comparing H$_2$O$_2$ synthesis rates are reported in Figure 5A, where a significant difference in catalytic performance is observed. Indeed, the enhanced reactivity of the 1%AuPd$_{(0.975)}$Cu$_{(0.025)}$/ZSM-5 catalyst is clear, achieving H$_2$O$_2$ concentrations (0.35 wt.%) far greater than that of the AuPd analogue (0.21 wt.%) over a 1 h H$_2$O$_2$ synthesis reaction. Further investigation of catalytic performance over several successive H$_2$O$_2$ synthesis reactions can be seen in Figure 5B, where a marked enhancement in H$_2$O$_2$ concentration can be observed over the 1%AuPd$_{(0.975)}$Cu$_{(0.025)}$/ZSM-5 catalyst (0.60 wt.%) compared to that achieved by the 1%AuPd$_{(1.00)}$/ZSM-5 catalyst (0.27 wt.%). Notably, the concentration of H_2O_2 achieved over the AuPdCu catalyst was found to be comparable to that achieved when utilising identical concentrations of Pt as a catalytic promoter for AuPd nanoalloys [44].

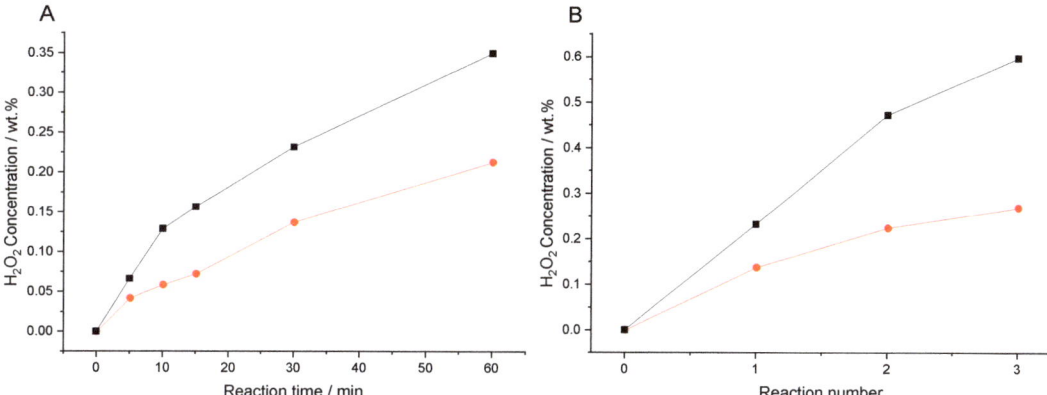

Figure 5. Comparison of catalytic activity towards the direct synthesis of H_2O_2 as (**A**) a function of reaction time and (**B**) over sequential H_2O_2 synthesis reactions. *Key:* 1%AuPd$_{(1.00)}$/ZSM-5 (*red circles*) and 1%AuPd$_{(0.975)}$Cu$_{(0.025)}$/ZSM-5 (*black squares*). **H_2O_2 direct synthesis reaction conditions:** catalyst (0.01 g), H_2O (2.9 g), MeOH (5.6 g), 5% H_2/CO_2 (420 psi), 25% O_2/CO_2 (160 psi), 0.5 h, 2 °C, and 1200 rpm.

Numerous works have elucidated the dependence between catalytic performance towards H_2O_2 synthesis and particle size, with studies by Tian et al., in particular, revealing that particle size in the sub-nanometre range is crucial for achieving optimal catalytic performance, at least in the case of monometallic Pd catalysts [56]. Comparisons of the mean particle size of the as-prepared 1%AuPd$_{(1.00)}$/ZSM-5 and 1%AuPd$_{(0.975)}$Cu$_{(0.025)}$/ZSM-5 catalysts, as determined from the bright field transmission electron micrographs presented in Figure S4, are reported in Table 1 with a negligible variation in particle size observed between the AuPd and optimal AuPdCu catalysts (3.7–4.1 nm). As such, it is possible to conclude that the enhanced catalytic activity towards H_2O_2 synthesis achieved through the introduction of Cu into AuPd nanoparticles is not associated with increased metal dispersion. Rather, it can be considered that the electronic modification of the Pd species is a result of dopant introduction, as indicated by our CO-DRIFTS (Figure 4) and XPS analysis (Figure 2), as well as possible structural changes which were indicated by our CO-DRIFTS analysis (Figure 4)and are responsible for the observed reactivity improvements.

Table 1. Particle size measurements of as-prepared 1%AuPd$_{(1.00)}$/ZSM-5 and 1%AuPdCu$_{(0.025)}$/ZSM-5 catalysts.

Catalyst	Mean Particle Size/nm (Standard Deviation)	Productivity/mol$_{H_2O_2}$ kg$_{cat}^{-1}$ h^{-1} (H_2O_2 Conc./wt.%)
1%AuPd$_{(1.00)}$/ZSM-5	3.7 (1.43)	69 (0.14)
1%AuPd$_{(0.975)}$Cu$_{(0.025)}$/ZSM-5	4.4 (1.93)	115 (0.23)

H_2O_2 direct synthesis reaction conditions: catalyst (0.01 g), H_2O (2.9 g), MeOH (5.6 g), 5% H_2/CO_2 (420 psi), 25% O_2/CO_2 (160 psi), 0.5 h, 2 °C, and 1200 rpm.

For any heterogeneous catalyst operating in a liquid phase reaction, the possibility of catalyst deactivation via the leaching of supported metals and the resultant homogeneous contribution to the observed catalytic activity is a major concern. This is particularly true given the ability of colloidal Pd to catalyse the direct synthesis reaction [57]. It was found that for both the AuPd and AuPdCu catalysts, catalytic activity toward H_2O_2 production decreased upon second use (Table 2). However, the ICP-MS analysis of H_2O_2 reaction solutions (Table 2) and TEM analysis of spent materials (Figure S4) indicated that such a loss in catalyst performance could not be attributed to either the leaching of active metals or nanoparticle agglomeration, with negligible levels of active metals detected through the analysis of H_2O_2 synthesis reaction solutions and the mean particle size determined to be comparable for both the as-prepared and used materials.

Table 2. Catalyst reusability in the direct synthesis and subsequent degradation of H_2O_2.

Catalyst	Productivity/ $mol_{H_2O_2} kg_{cat}^{-1} h^{-1}$		Initial Reaction Rate/ $mmol_{H_2O_2} mmol_{metal}^{-1} h^{-1}$		Degradation/ $mol_{H_2O_2} kg_{cat}^{-1} h^{-1}$		Metal Leached/%		
	Fresh	Used	Fresh	Used	Fresh	Used	Au	Pd	Cu
1%AuPd$_{(1.00)}$/ZSM-5	69	57	37.6	27.0	320	397	0	0.17	-
1%AuPd$_{(0.975)}$Cu$_{(0.025)}$/ZSM-5	115	64	48.5	27.1	529	231	0	0.13	BDL

H_2O_2 **direct synthesis reaction conditions:** catalyst (0.01 g), H_2O (2.9 g), MeOH (5.6 g), 5% H_2/CO_2 (420 psi), 25% O_2/CO_2 (160 psi), 0.5 h, 2 °C, and 1200 rpm. H_2O_2 **degradation reaction conditions:** catalyst (0.01 g), H_2O_2 (50 wt.% 0.68 g) H_2O (2.22 g), MeOH (5.6 g), 5% H_2/CO_2 (420 psi), 0.5 h, 2 °C, and 1200 rpm. BDL: Below detection limits. **Note:** Initial reaction rates were determined after a reaction time of 0.083 h and were calculated based on theoretical metal loading. Metal leaching is presented as a percentage loss of the nominal metal loading of the fresh materials, with the observed levels of leached metal equivalent to approximately 0.002 wt.%.

Similar observations have been recently reported for AuPd-supported catalysts prepared by an identical excess-chloride impregnation procedure, with the loss in catalytic performance upon use found to be associated with a significant loss of surface Cl species, a known promoter of activity in the H_2O_2 synthesis reaction [21]. With these earlier observations in mind, we set out to determine the extent of surface Cl loss (if any) after use in the H_2O_2 direct synthesis reaction via XPS (Figure S5). Interestingly, negligible concentrations of Cl species were observed in either the fresh or used AuPd and AuPdCu catalysts. This is in stark contrast to our earlier investigations into AuPd/TiO$_2$ catalysts prepared via an analogous synthesis technique and possibly highlights the key role of the support in retaining halide species [21]. Regardless, such an observation excludes the possibility of Cl loss as the cause for the observed loss in catalytic performance upon reuse. However, further investigation by XPS (Figures S6 and S7) does reveal a modification in the Au: Pd ratio for both catalysts after use, which may be indicative of nanoalloy restructuring and perhaps, more importantly, a total shift in the Pd speciation towards Pd^0, which can be attributed to the presence of H_2 within the reaction. As such, it is possible to conclude that while the catalytic materials developed within this work represent a promising basis for future study, there is still a need to address stability concerns, particularly around Pd speciation.

4. Conclusions

The introduction of low concentrations of earth-abundant metals (Ni, Cu, Zn) into supported AuPd nanoparticles has been demonstrated to improve catalytic activity towards the direct synthesis of H_2O_2, with the inclusion of Cu in particular, found to offer an enhancement compared to that previously reported upon the use of Pt as a promoter for AuPd nanoalloys. Indeed, the activity of the optimal AuPdCu catalyst is shown to outperform the bimetallic analogue by a factor of 1.7. The underlying cause for the increase in H_2O_2 synthesis activity can be attributed to the electronic modification of the Pd species and changes in the surface composition of the nanoalloys as a result of Cu inclusion, as evidenced by XPS and CO-DRIFTS investigations. While catalytic stability is of concern, with deactivation attributed to in situ reductions in Pd species, it can be considered that these materials represent a promising basis for future exploration in a range of reactions, particularly where the in situ supply of H_2O_2 is required.

Supplementary Materials: The following supporting information can be downloaded at: https://www.mdpi.com/article/10.3390/catal12111396/s1. Table S1. Synthesis details of the precursors used in the preparation of key bi- and tri-metallic 1%AuPd/ZSM-5 catalysts. Table S2. Summary of porosity and surface area of key 1%AuPd$_{(0.0975)}$X$_{(0.025)}$/ZSM-5 catalysts and HZSM-5(30). Table S3. Comparison of initial H_2O_2 synthesis rates over 1%AuPd$_{(0.0975)}$X$_{(0.025)}$/ZSM-5 (X = Cu, Ni, Zn) catalysts, as a function of Cu loading. Table S4. The effect of tertiary metal introduction upon the surface atomic composition of 1%AuPd$_{(0.975)}$X$_{(0.025)}$/ZSM-5 catalysts (X = Cu, Ni, Zn), as determined by XPS. Table S5. Comparison of initial H_2O_2 synthesis rates over 1%AuPdCu/ZSM-5 catalysts, as a function of Cu loading. Figure S1. X-ray diffractograms of 1%AuPd$_{(0.0975)}$X$_{(0.025)}$/ZSM-5 catalysts. (A) ZSM-5, (B) 1%AuPd$_{(1.00)}$/ZSM-5, (C) 1%AuPd$_{(0.0975)}$Cu$_{(0.025)}$/ZSM-5, (D) 1%AuPd$_{(0.0975)}$Ni$_{(0.025)}$/ZSM-5 and (E) 1%AuPd$_{(0.0975)}$Zn$_{(0.025)}$/ZSM-5. Figure S2. Fourier-transform infrared spectroscopy of 1%AuPd$_{(0.975)}$X$_{(0.025)}$/ZSM-5 catalysts. (A) ZSM-5, (B) 1%AuPd$_{(1.00)}$/ZSM-5, (C) 1%AuPd$_{(0.975)}$Cu$_{(0.025)}$/ZSM-5, (D) 1%AuPd$_{(0.975)}$Ni$_{(0.025)}$/ZSM-5 and (E) 1%AuPd$_{(0.975)}$Zn$_{(0.025)}$/ZSM-5. Figure S3. BET analysis plots for key 1%AuPd$_{(0.0975)}$X$_{(0.025)}$/ZSM-5 catalysts and HZSM-5(30). *Key*: ZSM-5(30) (*red triangles*), 1%AuPd$_{(1.00)}$/ZSM-5 (*blue squares*), 1%AuPd$_{(0.975)}$Cu$_{(0.025)}$/ZSM-5 (*green circles*). Note: ZSM-5 support exposed to calcination prior to metal immobilisation (flowing air, 550 °C, 3 h, 20 °Cmin^{-1}). Figure S4. Representative bright field transmission electron micrographs and corresponding particle size histograms of as-prepared (A) 1%AuPd$_{(1.00)}$/ZSM-5 and (C) 1%AuPd$_{(0.975)}$Cu$_{(0.025)}$/ZSM-5 catalysts and analogous (B) 1%AuPd$_{(1.00)}$/ZSM-5 and (D) 1%AuPd$_{(0.975)}$Cu$_{(0.025)}$/ZSM-5 samples after use in the direct synthesis reaction. H_2O_2 direct synthesis reaction conditions: Catalyst (0.01 g), H_2O (2.9 g), MeOH (5.6 g), 5% H_2/CO_2 (420 psi), 25% O_2/CO_2 (160 psi), 0.5 h, 2° C, 1200 rpm. Figure S5. Comparison of surface atomic Cl content of the as-prepared (A) and (B) used 1%AuPd$_{(1.00)}$/ZSM-5 catalyst and the fresh (C) and (D) used 1%AuPd$_{(0.975)}$Cu$_{(0.025)}$/ZSM-5 analogue. Note: No signal is observed for any catalyst formulation within the expected energy window for Cl(2p) (approx. 200 eV). Figure S6. XPS spectra of Pd(3d) regions for the (A) as-prepared and (B) used 1%AuPd$_{(1.00)}$/ZSM-5 catalyst, after use in the H_2O_2 direct synthesis reaction. *Key:* Au(4d) (green), Pd^0(blue), Pd^{2+}(purple), Ca^{2+} (orange). Figure S7. XPS spectra of Pd(3d) regions for the (A) as-prepared and (B) used 1%AuPd$_{(0.975)}$Cu$_{(0.025)}$/ZSM-5 catalyst, after use in the H_2O_2 direct synthesis reaction. *Key:* Au(4d) (green), Pd^0(blue), Pd^{2+}(purple), Ca^{2+} (orange).

Author Contributions: A.B. and R.J.L. conducted catalytic synthesis, testing, and data analysis. A.B., R.J.L., D.J.M., and T.E.D. conducted catalyst characterisation and corresponding data processing. R.J.L. and G.J.H. contributed to the design of the study and provided technical advice and result interpretation. R.J.L. wrote the manuscript and Supplementary Information, with all authors commented on and amended both documents. All authors discussed and contributed to the work. All authors have read and agreed to the published version of the manuscript.

Funding: The authors gratefully acknowledge Cardiff University and the Max Planck Centre for Fundamental Heterogeneous Catalysis (FUNCAT) for financial support.

Data Availability Statement: The data presented in this study are fully available within the manuscript and supporting information.

Acknowledgments: The authors would like to thank the CCI-Electron Microscopy Facility which has been part-funded by the European Regional Development Fund through the Welsh Government, and The Wolfson Foundation. XPS data collection was performed at the EPSRC National Facility for XPS ('HarwellXPS'), operated by Cardiff University and UCL, under contract No. PR16195.

Conflicts of Interest: The authors declare no conflict of interest.

References

1. Lewis, R.J.; Hutchings, G.J. Recent Advances in the Direct Synthesis of H_2O_2. *ChemCatChem* **2019**, *11*, 298–308. [CrossRef]
2. Campos-Martin, J.M.; Blanco-Brieva, G.; Fierro, J.L. Hydrogen peroxide synthesis: An outlook beyond the anthraquinone process. *Angew. Chem. Int. Ed.* **2006**, *45*, 6962–6984. [CrossRef]
3. Blaser, B.; Karl-Heinz, W.; Schiefer, J. (Henkel AG and Co KGaA), Stabilizing Agent for Peroxy-Compounds and Their Solutions. U.S. Patent 3122417A, 3 June 1959.
4. Gao, G.; Tian, Y.; Gong, X.; Pan, Z.; Yong, K.; Zong, B. Advances in the production technology of hydrogen peroxide. *Chin. J. Catal.* **2020**, *41*, 1039–1047. [CrossRef]

5. Lewis, R.J.; Ueura, K.; Liu, X.; Fukuta, Y.; Davies, T.E.; Morgan, D.J.; Chen, L.; Qi, J.; Singleton, J.; Edwards, J.K.; et al. Highly efficient catalytic production of oximes from ketones using in situ generated H_2O_2. *Science* **2022**, *376*, 615–620. [CrossRef] [PubMed]
6. Crombie, C.M.; Lewis, R.J.; Kovačič, D.; Morgan, D.J.; Slater, T.J.A.; Davies, T.E.; Edwards, J.K.; Skjøth-Rasmussen, M.S.; Hutchings, G.J. The Influence of Reaction Conditions on the Oxidation of Cyclohexane via the in-situ Production of H_2O_2. *Catal. Lett.* **2021**, *151*, 164–171. [CrossRef]
7. Jin, Z.; Wang, L.; Zuidema, E.; Mondal, K.; Zhang, M.; Zhang, J.; Wang, C.; Meng, X.; Yang, H.; Mesters, C.; et al. Hydrophobic zeolite modification for in situ peroxide formation in methane oxidation to methanol. *Science* **2020**, *367*, 193–197. [CrossRef]
8. Ford, D.C.; Nilekar, A.U.; Xu, Y.; Mavrikakis, M. Partial and complete reduction of O_2 by hydrogen on transition metal surfaces. *Surf. Sci.* **2010**, *604*, 1565–1575. [CrossRef]
9. Voloshin, Y.; Halder, R.; Lawal, A. Kinetics of hydrogen peroxide synthesis by direct combination of H_2 and O_2 in a microreactor. *Catal. Today* **2007**, *125*, 40–47. [CrossRef]
10. Wilson, N.M.; Flaherty, D.W. Mechanism for the Direct Synthesis of H_2O_2 on Pd Clusters: Heterolytic Reaction Pathways at the Liquid–Solid Interface. *J. Am. Chem. Soc.* **2016**, *138*, 574. [CrossRef]
11. Adams, J.S.; Chemburkar, A.; Priyadarshini, P.; Ricciardulli, T.; Lu, Y.; Maliekkal, V.; Sampath, A.; Winikoff, S.; Karim, A.M.; Neurock, M.; et al. Solvent molecules form surface redox mediators in situ and cocatalyze O_2 reduction on Pd. *Science* **2021**, *371*, 626–632. [CrossRef]
12. Ricciardulli, T.; Gorthy, S.; Adams, J.S.; Thompson, C.; Karim, A.M.; Neurock, M.; Flaherty, D.W. Effect of Pd Coordination and Isolation on the Catalytic Reduction of O_2 to H_2O_2 over PdAu Bimetallic Nanoparticles. *J. Am. Chem. Soc.* **2021**, *143*, 5445–5464. [CrossRef] [PubMed]
13. Lewis, R.J.; Ntainjua, E.N.; Morgan, D.J.; Davies, T.E.; Carley, A.F.; Freakley, S.J.; Hutchings, G.J. Improving the performance of Pd based catalysts for the direct synthesis of hydrogen peroxide via acid incorporation during catalyst synthesis. *Catal. Commun.* **2021**, *161*, 106358. [CrossRef]
14. Flaherty, D.W. Direct Synthesis of H_2O_2 from H_2 and O_2 on Pd Catalysts: Current Understanding, Outstanding Questions, and Research Needs. *ACS Catal.* **2018**, *8*, 1520–1527. [CrossRef]
15. Selinsek, M.; Deschner, B.J.; Doronkin, D.E.; Sheppard, T.L.; Grunwaldt, J.; Dittmeyer, R. Revealing the Structure and Mechanism of Palladium during Direct Synthesis of Hydrogen Peroxide in Continuous Flow Using Operando Spectroscopy. *ACS Catal.* **2018**, *8*, 2546–2557. [CrossRef]
16. Henkel, H.; Weber, W. (Henkel AG and Co KGaA), Manufacture of Hydrogen Peroxid. U.S. Patent 1,108,752, 25 August 1914.
17. Priyadarshini, P.; Ricciardulli, T.; Adams, J.S.; Yun, Y.S.; Flaherty, D.W. Effects of bromide adsorption on the direct synthesis of H_2O_2 on Pd nanoparticles: Formation rates, selectivities, and apparent barriers at steady-state. *J. Catal.* **2021**, *399*, 24–40. [CrossRef]
18. Pospelova, T.A.; Kobozev, N. *Russ. J. Phys. Chem.* **1961**, *35*, 1192–1197.
19. Wilson, N.M.; Priyadarshini, P.; Kunz, S.; Flaherty, D.W. Direct synthesis of H_2O_2 on Pd and Au_xPd_1 clusters: Understanding the effects of alloying Pd with Au. *J. Catal.* **2018**, *357*, 163–175. [CrossRef]
20. Richards, T.; Lewis, R.J.; Morgan, D.J.; Hutchings, G.J. The Direct Synthesis of Hydrogen Peroxide Over Supported Pd-Based Catalysts: An Investigation into the Role of the Support and Secondary Metal Modifiers. *Catal. Lett.* **2022**, 1–9. [CrossRef]
21. Brehm, J.; Lewis, R.J.; Morgan, D.J.; Davies, T.E.; Hutchings, G.J. The Direct Synthesis of Hydrogen Peroxide over AuPd Nanoparticles: An Investigation into Metal Loading. *Catal. Lett.* **2022**, *152*, 254–262. [CrossRef]
22. Freakley, S.J.; He, Q.; Harrhy, J.H.; Lu, L.; Crole, D.A.; Morgan, D.J.; Ntainjua, E.N.; Edwards, J.K.; Carley, A.F.; Borisevich, A.Y.; et al. Palladium-tin catalysts for the direct synthesis of H_2O_2 with high selectivity. *Science* **2016**, *351*, 965–968. [CrossRef]
23. Wang, S.; Lewis, R.J.; Doronkin, D.E.; Morgan, D.J.; Grunwaldt, J.; Hutchings, G.J.; Behrens, S. The direct synthesis of hydrogen peroxide from H_2 and O_2 using Pd-Ga and Pd-In catalysts. *Catal. Sci. Technol.* **2020**, *10*, 1925–1935. [CrossRef]
24. Crole, D.A.; Underhill, R.; Edwards, J.K.; Shaw, G.; Freakley, S.J.; Hutchings, G.J.; Lewis, R.J. The direct synthesis of hydrogen peroxide from H_2 and O_2 using $PdNi/TiO_2$ catalysts. *Philos. Trans. R. Soc. A* **2020**, *378*, 20200062. [CrossRef] [PubMed]
25. Crombie, C.M.; Lewis, R.J.; Taylor, R.L.; Morgan, D.J.; Davies, T.E.; Folli, A.; Murphy, D.M.; Edwards, J.K.; Qi, J.; Jiang, H.; et al. Enhanced Selective Oxidation of Benzyl Alcohol via In Situ H_2O_2 Production over Supported Pd-Based Catalysts. *ACS Catal.* **2021**, *11*, 2701–2714. [CrossRef]
26. Cao, K.; Yang, H.; Bai, S.; Xu, Y.; Yang, C.; Wu, Y.; Xie, M.; Cheng, T.; Shao, Q.; Huang, X. Efficient Direct H_2O_2 Synthesis Enabled by PdPb Nanorings via Inhibiting the O–O Bond Cleavage in O_2 and H_2O_2. *ACS Catal.* **2021**, *11*, 1106–1118. [CrossRef]
27. Tian, P.; Xuan, F.; Ding, D.; Sun, Y.; Xu, X.; Li, W.; Si, R.; Xu, J.; Han, Y. Revealing the role of tellurium in palladium-tellurium catalysts for the direct synthesis of hydrogen peroxide. *J. Catal.* **2020**, *385*, 21–29. [CrossRef]
28. Bernardotto, G.; Menegazzo, F.; Pinna, F.; Signoretto, M.; Cruciani, G.; Strukul, G. New Pd–Pt and Pd–Au catalysts for an efficient synthesis of H_2O_2 from H_2 and O_2 under very mild conditions. *Appl. Catal. A* **2009**, *358*, 129–135. [CrossRef]
29. Quon, S.; Jo, D.Y.; Han, G.; Han, S.S.; Seo, M.; Lee, K. Role of Pt atoms on Pd(1 1 1) surface in the direct synthesis of hydrogen peroxide: Nano-catalytic experiments and DFT calculations. *J. Catal.* **2018**, *368*, 237–247. [CrossRef]
30. Deguchi, T.; Yamano, H.; Takenouchi, S.; Iwamoto, M. Enhancement of catalytic activity of Pd-PVP colloid for direct H_2O_2 synthesis from H_2 and O_2 in water with addition of 0.5 atom% Pt or Ir. *Catal. Sci. Technol.* **2018**, *8*, 1002–1015. [CrossRef]

31. Kim, M.; Han, G.; Xiao, X.; Song, J.; Hong, J.; Jung, E.; Kim, H.; Ahn, J.; Han, S.S.; Lee, K.; et al. Anisotropic growth of Pt on Pd nanocube promotes direct synthesis of hydrogen peroxide. *Appl. Surf. Sci.* **2021**, *562*, 150031. [CrossRef]
32. Xu, J.; Ouyang, L.; Da, G.; Song, Q.; Yang, X.; Han, Y. Pt promotional effects on Pd–Pt alloy catalysts for hydrogen peroxide synthesis directly from hydrogen and oxygen. *J. Catal.* **2012**, *285*, 74–82. [CrossRef]
33. Chen, Q.; Beckman, E.J. Direct synthesis of H_2O_2 from O_2 and H_2 over precious metal loaded TS-1 in CO_2. *Green Chem.* **2007**, *9*, 802–808. [CrossRef]
34. Gong, X.; Lewis, R.J.; Zhou, S.; Morgan, D.J.; Davies, T.E.; Liu, X.; Kiely, C.J.; Zong, B.; Hutchings, G.J. Enhanced catalyst selectivity in the direct synthesis of H_2O_2 through Pt incorporation into TiO_2 supported AuPd catalysts. *Catal. Sci. Technol.* **2020**, *10*, 4635–4644. [CrossRef]
35. Lewis, R.J.; Ueura, K.; Fukuta, Y.; Davies, T.E.; Morgan, D.J.; Paris, C.B.; Singleton, J.; Edwards, J.K.; Freakley, S.J.; Yamamoto, Y.; et al. Cyclohexanone ammoximation via in situ H_2O_2 production using TS-1 supported catalysts. *Green Chem.* **2022**; Advance Article. [CrossRef]
36. Lewis, R.J.; Ueura, K.; Fukuta, Y.; Freakley, S.J.; Kang, L.; Wang, R.; He, Q.; Edwards, J.K.; Morgan, D.J.; Yamamoto, Y.; et al. The Direct Synthesis of H_2O_2 Using TS-1 Supported Catalysts. *ChemCatChem* **2019**, *11*, 1673–1680. [CrossRef]
37. Nguyen, H.V.; Kim, K.Y.; Nam, H.; Lee, S.Y.; Yu, T.; Seo, T.S. Centrifugal microfluidic device for the high-throughput synthesis of Pd@AuPt core–shell nanoparticles to evaluate the performance of hydrogen peroxide generation. *Lab Chip* **2020**, *20*, 3293–3301. [CrossRef]
38. Ham, H.C.; Hwang, G.S.; Han, J.; Nam, S.W.; Lim, T.H. On the Role of Pd Ensembles in Selective H_2O_2 Formation on PdAu Alloys. *J. Phys. Chem. C* **2009**, *113*, 12943–12945. [CrossRef]
39. Li, J.; Ishihara, T.; Yoshizawa, K. Theoretical Revisit of the Direct Synthesis of H_2O_2 on Pd and Au@Pd Surfaces: A Comprehensive Mechanistic Study. *J. Phys. Chem. C* **2011**, *115*, 25359–25367. [CrossRef]
40. Richards, T.; Harrhy, J.H.; Lewis, R.J.; Howe, A.G.R.; Suldecki, G.M.; Folli, A.; Morgan, D.J.; Davies, T.E.; Loveridge, E.J.; Crole, D.A.; et al. A residue-free approach to water disinfection using catalytic in situ generation of reactive oxygen species. *Nat. Catal.* **2021**, *4*, 575–585. [CrossRef]
41. Lewis, R.J.; Bara-Estaun, A.; Agarwal, N.; Freakley, S.J.; Morgan, D.J.; Hutchings, G.J. The Direct Synthesis of H_2O_2 and Selective Oxidation of Methane to Methanol Using HZSM-5 Supported AuPd Catalysts. *Catal. Lett.* **2019**, *149*, 3066–3075. [CrossRef]
42. Kang, J.; Puthiaraj, P.; Ahn, W.; Park, E.D. Direct synthesis of oxygenates via partial oxidation of methane in the presence of O_2 and H_2 over a combination of Fe-ZSM-5 and Pd supported on an acid-functionalized porous polymer. *Appl. Catal. A* **2020**, *602*, 117711. [CrossRef]
43. Lyu, J.; Wei, J.; Niu, L.; Lu, C.; Hu, Y.; Xiang, Y.; Zhang, G.; Zhang, Q.; Ding, C.; Li, X. Highly efficient hydrogen peroxide direct synthesis over a hierarchical TS-1 encapsulated subnano Pd/PdO hybrid. *RSC Adv.* **2019**, *9*, 13398–13402. [CrossRef] [PubMed]
44. Barnes, A.; Lewis, R.J.; Morgan, D.J.; Davies, T.E.; Hutchings, G.J. Enhancing catalytic performance of AuPd catalysts towards the direct synthesis of H_2O_2 through incorporation of base metals. *Catal. Sci. Technol.* **2022**, *12*, 1986–1995. [CrossRef]
45. Santos, A.; Lewis, R.J.; Malta, G.; Howe, A.G.R.; Morgan, D.J.; Hampton, E.; Gaskin, P.; Hutchings, G.J. Direct Synthesis of Hydrogen Peroxide over Au–Pd Supported Nanoparticles under Ambient Conditions. *Int. Eng. Chem. Res.* **2019**, *58*, 12623–12631. [CrossRef]
46. Edwards, J.K.; Thomas, A.; Carley, A.F.; Herzing, A.A.; Kiely, C.J.; Hutchings, G.J. Au–Pd supported nanocrystals as catalysts for the direct synthesis of hydrogen peroxide from H_2 and O_2. *Green Chem.* **2008**, *10*, 388–394. [CrossRef]
47. Xu, F.; Zhao, L.; Zhao, F.; Deng, L.; Hu, L.; Zeng, B. Electrodeposition of AuPdCu Alloy Nanoparticles on a Multi-Walled Carbon Nanotube Coated Glassy Carbon Electrode for the Electrocatalytic Oxidation and Determination of Hydrazine. *Int. J. Electrochem. Sci.* **2014**, *9*, 2832–2847.
48. Ab Rahim, M.H.; Armstrong, R.D.; Hammond, C.; Dimitratos, N.; Freakley, S.J.; Forde, M.M.; Morgan, D.J.; Lalev, G.; Jenkins, R.L.; Lopez-Sanchez, J.A.; et al. Low temperature selective oxidation of methane to methanol using titania supported gold palladium copper catalysts. *Catal. Sci. Technol.* **2016**, *6*, 3410–3418. [CrossRef]
49. Brehm, J.; Lewis, R.J.; Richards, T.; Qin, T.; Morgan, D.J.; Davies, T.E.; Chen, L.; Liu, X.; Hutchings, G.J. Enhancing the Chemo-Enzymatic One-Pot Oxidation of Cyclohexane via In Situ H_2O_2 Production over Supported Pd-Based Catalysts. *ACS Catal.* **2022**, *12*, 11776–11789. [CrossRef]
50. Joshi, A.M.; Delgass, W.N.; Thomson, K.T. Investigation of Gold−Silver, Gold−Copper, and Gold−Palladium Dimers and Trimers for Hydrogen Peroxide Formation from H_2 and O_2. *J. Phys. Chem. C* **2007**, *111*, 7384–7395. [CrossRef]
51. Wilson, A.R.; Sun, K.; Chi, M.; White, R.M.; LeBeau, J.M.; Lamb, H.H.; Wiley, B.J. From Core–Shell to Alloys: The Preparation and Characterization of Solution-Synthesized AuPd Nanoparticle Catalysts. *J. Phys. Chem. C* **2013**, *117*, 17557–17566. [CrossRef]
52. Zhu, B.; Thrimurthulu, G.; Delannoy, L.; Louis, C.; Mottet, C.; Creuze, J.; Legrand, B.; Guesmi, H. Evidence of Pd segregation and stabilization at edges of AuPd nano-clusters in the presence of CO: A combined DFT and DRIFTS study. *J. Catal.* **2013**, *308*, 272–281. [CrossRef]
53. Bollinger, M.A.; Vannice, M.A. A kinetic and DRIFTS study of low-temperature carbon monoxide oxidation over Au—TiO_2 catalysts. *Appl. Catal. B* **1996**, *8*, 417–443. [CrossRef]
54. Marx, S.; Krumeich, F.; Baiker, A. Surface Properties of Supported, Colloid-Derived Gold/Palladium Mono- and Bimetallic Nanoparticles. *J. Phys. Chem. C* **2011**, *115*, 8195. [CrossRef]

55. Carter, J.H.; Althahban, S.; Nowicka, E.; Freakley, S.J.; Morgan, D.J.; Shah, P.M.; Golunski, S.; Kiely, C.J.; Hutchings, G.J. Synergy and Anti-Synergy between Palladium and Gold in Nanoparticles Dispersed on a Reducible Support. *ACS Catal.* **2016**, *6*, 6623–6633. [CrossRef] [PubMed]
56. Tian, P.; Ding, D.; Sun, Y.; Xuan, F.; Xu, X.; Xu, J.; Han, Y. Theoretical study of size effects on the direct synthesis of hydrogen peroxide over palladium catalysts. *J. Catal.* **2019**, *369*, 95. [CrossRef]
57. Dissanayake, D.P.; Lunsford, J.H. Evidence for the Role of Colloidal Palladium in the Catalytic Formation of H_2O_2 from H_2 and O_2. *J. Catal.* **2002**, *206*, 173. [CrossRef]

MDPI AG
Grosspeteranlage 5
4052 Basel
Switzerland
Tel.: +41 61 683 77 34

Catalysts Editorial Office
E-mail: catalysts@mdpi.com
www.mdpi.com/journal/catalysts

Disclaimer/Publisher's Note: The title and front matter of this reprint are at the discretion of the Guest Editor. The publisher is not responsible for their content or any associated concerns. The statements, opinions and data contained in all individual articles are solely those of the individual Editor and contributors and not of MDPI. MDPI disclaims responsibility for any injury to people or property resulting from any ideas, methods, instructions or products referred to in the content.

www.ingramcontent.com/pod-product-compliance
Lightning Source LLC
LaVergne TN
LVHW072339090526
838202LV00019B/2445